A·N·N·U·A·L E·D·I·T·I·O·N·S

Environment

05/06

Twenty-Fourth Edition

EDITOR

John L. Allen
University of Wyoming

John L. Allen is professor and chair of geography at the University of Wyoming. He received his bachelor's degree in 1963 and his M.A. in 1964 from the University of Wyoming, and in 1969 he received his Ph.D. from Clark University. His special area of interest is the impact of contemporary human societies on environmental systems.

McGraw-Hill/Dushkin

2460 Kerper Blvd., Dubuque, IA 52001

Visit us on the Internet
http://www.dushkin.com

Credits

1. **The Global Environment: An Emerging World View**
   Unit photo—© The McGraw-Hill Companies, Inc./Christopher Kerrigan
2. **Population, Policy, and Economy**
   Unit photo—Photograph courtesy of USAID
3. **Energy: Present and Future Problems**
   Unit photo—© Getty Images/Sami Sarkis
4. **Biosphere: Endangered Species**
   Unit photo—© Getty Images/Alan and Sandy Carey
5. **Resources: Land and Water**
   Unit photo—© Getty Images/PhotoLink/Kent Knudson
6. **The Hazards of Growth: Pollution and Climate Change**
   Unit photo—© Getty Images/PhotoLink/D. Falconer

Copyright

Cataloging in Publication Data
Main entry under title: Annual Editions: Environment 2005/2006.
1. Environment—Periodicals. I. Allen, John L., *comp*. II. Title: Environment.
ISBN 0–07–352831–5   658'.05   ISSN 0272–9008

Twenty-Fourth Edition

Cover image © geneticist Robert Lewellen and technician Jose Orozco/Photo by Scott Bauer, USDA Agricultural Research Service
Printed in the United States of America   1234567890QPDQPD98765   Printed on Recycled Paper

Editors/Advisory Board

Members of the Advisory Board are instrumental in the final selection of articles for each edition of ANNUAL EDITIONS. Their review of articles for content, level, currentness, and appropriateness provides critical direction to the editor and staff. We think that you will find their careful consideration well reflected in this volume.

Preface

In publishing ANNUAL EDITIONS we recognize the enormous role played by the magazines, newspapers, and journals of the public press in providing current, first-rate educational information in a broad spectrum of interest areas. Many of these articles are appropriate for students, researchers, and professionals seeking accurate, current material to help bridge the gap between principles and theories and the real world. These articles, however, become more useful for study when those of lasting value are carefully collected, organized, indexed, and reproduced in a low-cost format, which provides easy and permanent access when the material is needed. That is the role played by ANNUAL EDITIONS.

At the beginning of our new millennium, environmental dilemmas long foreseen by natural and social scientists began to emerge in a number of guises: regional imbalances in numbers of people and the food required to feed them, international environmental crime, energy scarcity, acid rain, build-up of toxic and hazardous wastes, ozone depletion, water shortages, massive soil erosion, global atmospheric pollution, climate change, forest dieback and tropical deforestation, and the highest rates of plant and animal extinction the world has known in 65 million years.

These and other environmental problems continue to worsen in spite of an increasing amount of national and international attention to the issues surrounding them and increased environmental awareness and legislation at both global and national levels. The problems have resulted from centuries of exploitation and unwise use of resources, accelerated recently by the shortsighted public policies that have favored the short-term, expedient approach to problem-solving over longer-term economic and ecological good sense. In Africa, for example, the drive to produce enough food to support a growing population has caused the use of increasingly fragile and marginal resources, resulting in the dryland deterioration that brings famine to that troubled continent. Similar social and economic problems have contributed to massive deforestation in middle and South America and in Southeast Asia.

Part of the problem is that efforts to deal with environmental issues have been intermittent. During the decade of the 1980s, economic problems generated by resource scarcity caused the relaxation of environmental quality standards and contributed to the refusal of many of the world's governments and international organizations to develop environmentally sound protective measures, which were viewed as too costly. More recently, in the late 20th and early 21st century, as environmental protection policies were adopted, they were often cosmetic, designed for good press and TV sound bites, and—even worse—seemingly designed to benefit large corporations rather than to protect environmental systems. Even with these policies based more in public relations than in environmental ones, governments often lacked either the will or the means to implement them properly. The absence of effective environmental policy has been particularly apparent in those countries that are striving to become economically developed. But even in the more highly developed nations, economic concerns tend to favor a loosening of environmental controls. In the United States, for example, the interests of maintaining jobs for the timber industry imperil many of the last areas of old-growth forests, and the desire to maintain agricultural productivity at all costs causes the continued use of destructive and toxic chemicals on the nation's farmlands. In addition, concerns over energy availability have created the need for foreign policy and military action to protect the developed nations' access to cheap oil and have prompted increasing reliance on technological quick fixes, as well as the development of environmentally sensitive areas to new energy resource exploration and exploitation.

Despite the recent tendency of the U.S. government to turn its back on environmental issues and refuse to participate in international environmental accords, particularly those related to global warming, there is some reason to hope that a new environmental consciousness is awakening with the new global economic system. Unfortunately, increasing globalization of the economy has meant globalization of other things as well, such as internal conflict and disease transmission. The emergence of terrorism as an instrument of national or quasi-national policy—particularly where terrorism may employ environmental contamination as a weapon—has the potential to produce future environmental problems that are almost too frightening to think about.

In *Annual Editions: Environment 05/06* every effort has been made to choose articles that encourage an understanding of the nature of the environmental problems that beset us and how, with wisdom and knowledge and the proper perspective, they can be solved or at least mitigated. Accordingly, the selections in this book have been chosen more for their intellectual content than for their emotional tone. They have been arranged into an order of topics—the global environment; population, policy, and economy; energy; the biosphere; land and water resources; and pollution—that lends itself to a progressive understanding of the causes and effects of human modifications of Earth's environmental systems. We will not be protected against the ecological consequences of human actions by remaining ignorant of them.

Readers can have input into the next edition of *Annual Editions: Environment* by completing and returning the post-paid *article rating form* at the back of the book.

John L. Allen
Editor

Contents

UNIT 1
The Global Environment: An Emerging World View

The concepts in bold italics are developed in the article. For further expansion, please refer to the Topic Guide and the Index.

UNIT 2
Population, Policy, and Economy

The concepts in bold italics are developed in the article. For further expansion, please refer to the Topic Guide and the Index.

UNIT 3
Energy: Present and Future Problems

UNIT 4
Biosphere: Endangered Species

The concepts in bold italics are developed in the article. For further expansion, please refer to the Topic Guide and the Index.

UNIT 5
Resources: Land and Water

The concepts in bold italics are developed in the article. For further expansion, please refer to the Topic Guide and the Index.

UNIT 6
The Hazards of Growth: Pollution and Climate Change

The concepts in bold italics are developed in the article. For further expansion, please refer to the Topic Guide and the Index.

The concepts in bold italics are developed in the article. For further expansion, please refer to the Topic Guide and the Index.

Topic Guide

This topic guide suggests how the selections in this book relate to the subjects covered in your course. You may want to use the topics listed on these pages to search the Web more easily.

On the following pages a number of Web sites have been gathered specifically for this book. They are arranged to reflect the units of this *Annual Edition.* You can link to these sites by going to the DUSHKIN ONLINE support site at *http://www.dushkin.com/online/.*

ALL THE ARTICLES THAT RELATE TO EACH TOPIC ARE LISTED BELOW THE BOLD-FACED TERM.

World Wide Web Sites

The following World Wide Web sites have been carefully researched and selected to support the articles found in this reader. The easiest way to access these selected sites is to go to our DUSHKIN ONLINE support site at *http://www.dushkin.com/online/*.

AE: Environment 05/06

The following sites were available at the time of publication. Visit our Web site—we update DUSHKIN ONLINE regularly to reflect any changes.

General Sources

Britannica's Internet Guide
http://www.britannica.com

This site presents extensive links to material on world geography and culture, encompassing material on wildlife, human lifestyles, and the environment.

EnviroLink
http://www.envirolink.org/

One of the world's largest environmental information clearinghouses, EnviroLink is a grassroots nonprofit organization that unites organizations and volunteers around the world and provides up-to-date information and resources.

Library of Congress
http://www.loc.gov

Examine this extensive Web site to learn about resource tools, library services/resources, exhibitions, and databases in many different subfields of environmental studies.

The New York Times
http://www.nytimes.com

Browsing through the archives of the *New York Times* will provide a wide array of articles and information related to the different subfields of the environment.

SocioSite: Sociological Subject Areas
http://www.pscw.uva.nl/sociosite/TOPICS/

This huge sociological site from the University of Amsterdam provides many discussions and references of interest to students of the environment, such as the links to information on ecology and consumerism.

U.S. Geological Survey
http://www.usgs.gov

This site and its many links are replete with information and resources in environmental studies, from explanations of El Niño to discussion of concerns about water resources.

UNIT 1: The Global Environment: An Emerging World View

Alternative Energy Institute (AEI)
http://www.altenergy.org

The AEI will continue to monitor the transition from today's energy forms to the future in a "surprising journey of twists and turns." This site is the beginning of an incredible journey.

Earth Science Enterprise
http://www.earth.nasa.gov

Information about NASA's Mission to Planet Earth program and its Science of the Earth System can be found here. Surf to learn about satellites, El Niño, and even "strategic visions" of interest to environmentalists.

IISDnet
http://iisd.ca

The International Institute for Sustainable Development, a Canadian organization, presents information through gateways entitled Business, Climate Change, Measurement and Assessment, and Natural Resources. IISD Linkages is its multimedia resource for environment and development policy makers.

National Geographic Society
http://www.nationalgeographic.com

Links to *National Geographic*'s huge archive are provided here. There is a great deal of material related to the atmosphere, the oceans, and other environmental topics.

Research and Reference (Library of Congress)
http://lcweb.loc.gov/rr/

This research and reference site of the Library of Congress will lead to invaluable information on different countries. It provides links to numerous publications, bibliographies, and guides in area studies that can be of great help to environmentalists.

Santa Fe Institute
http://acoma.santafe.edu

This home page of the Santa Fe Institute—a nonprofit, multidisciplinary research and education center—will lead to many interesting links related to its primary goal: to create a new kind of scientific research community, pursuing emerging science.

Solstice: Documents and Databases
http://solstice.crest.org/index.html

In this online source for sustainable energy information, the Center for Renewable Energy and Sustainable Technology (CREST) offers documents and databases on renewable energy, energy efficiency, and sustainable living. The site also offers related Web sites, case studies, and policy issues.

United Nations
http://www.unsystem.org

Visit this official Web site Locator for the United Nations System of Organizations to get a sense of the scope of international environmental inquiry today. Various UN organizations concern themselves with everything from maritime law to habitat protection to agriculture.

United Nations Environment Programme (UNEP)
http://www.unep.ch

Consult this home page of UNEP for links to critical topics of concern to environmentalists, including desertification, migratory species, and the impact of trade on the environment. The site will direct you to useful databases and global resource information.

UNIT 2: Population, Policy, and Economy

The Hunger Project
http://www.thp.org

Browse through this nonprofit organization's site to explore the ways in which it attempts to achieve its goal: the sustainable end to global hunger through leadership at all levels of society. The Hunger Project contends that the persistence of hunger is at the heart of the major security issues that are threatening our planet.

www.dushkin.com/online/

Poverty Mapping
http://www.povertymap.net

Poverty maps can quickly provide information on the spatial distribution of poverty. This site provides maps, graphics, data, publications, news, and links that provide the public with poverty mapping from the global to the subnational level.

World Health Organization
http://www.who.int

The home page of the World Health Organization provides links to a wealth of statistical and analytical information about health and the environment in the developing world.

World Population and Demographic Data
http://geography.about.com/cs/worldpopulation/

On this site, information about world population and additional demographic data for all the countries of the world are provided.

WWW Virtual Library: Demography & Population Studies
http://demography.anu.edu.au/VirtualLibrary/

This is a definitive guide to demography and population studies. A multitude of important links to information about global poverty and hunger can be found here.

UNIT 3: Energy: Present and Future Problems

Alliance for Global Sustainability (AGS)
http://globalsustainability.org/

The AGS is a cooperative venture seeking solutions to today's urgent and complex environmental problems. Research teams from four universities study large-scale, multidisciplinary environmental problems that are faced by the world's ecosystems, economies, and societies.

Alternative Energy Institute, Inc.
http://www.altenergy.org

On this site created by a nonprofit organization, discover how the use of conventional fuels affects the environment. Also learn about research work on new forms of energy.

Energy and the Environment: Resources for a Networked World
http://zebu.uoregon.edu/energy.html

An extensive array of materials having to do with energy sources—both renewable and nonrenewable—as well as other topics of interest to students of the environment is found on this site.

Institute for Global Communication/EcoNet
http://www.igc.org/

This environmentally friendly site provides links to dozens of governmental, organizational, and commercial sites having to do with energy sources. Resources address energy efficiency, renewable generating sources, global warming, and more.

Nuclear Power Introduction
http://library.thinkquest.org/17658/pdfs/nucintro.pdf

Information regarding alternative energy forms can be accessed here. There is a brief introduction to nuclear power and a link to maps that show where nuclear power plants exist.

U.S. Department of Energy
http://www.energy.gov

Scrolling through the links provided by this Department of Energy home page will lead to information about fossil fuels and a variety of sustainable/renewable energy sources.

UNIT 4: Biosphere: Endangered Species

Endangered Species
http://www.endangeredspecie.com/

This site provides a wealth of information on endangered species anywhere in the world. Links providing data on the causes, interesting facts, law issues, case studies, and other issues on endangered species are available.

Friends of the Earth
http://www.foe.co.uk/index.html

Friends of the Earth, a nonprofit organization based in the United Kingdom, pursues a number of campaigns to protect the Earth and its living creatures. This site has links to many important environmental sites, covering such broad topics as ozone depletion, soil erosion, and biodiversity.

Smithsonian Institution Web Site
http://www.si.edu

Looking through this site, which will provide access to many of the enormous resources of the Smithsonian, offers a sense of the biological diversity that is threatened by humans' unsound environmental policies and practices.

World Wildlife Federation (WWF)
http://www.wwf.org

This home page of the WWF leads to an extensive array of information links about endangered species, wildlife management and preservation, and more. It provides many suggestions for how to take an active part in protecting the biosphere.

UNIT 5: Resources: Land and Water

Global Climate Change
http://www.puc.state.oh.us/consumer/gcc/index.html

The goal of this PUCO (Public Utilities Commission of Ohio) site is to serve as a clearinghouse of information related to global climate change. Its extensive links provide an explanation of the science and chronology of global climate change, acronyms, definitions, and more.

National Oceanic and Atmospheric Administration (NOAA)
http://www.noaa.gov

Through this home page of NOAA, you can find information about coastal issues, fisheries, climate, and more.

National Operational Hydrologic Remote Sensing Center (NOHRSC)
http://www.nohrsc.nws.gov

Flood images are available at this site of the NOHRSC, which works with the U.S. National Weather Service to track weather-related information.

Virtual Seminar in Global Political Economy/Global Cities & Social Movements
http://csf.colorado.edu/gpe/gpe95b/resources.html

Links to subjects of interest in regional environmental studies, covering topics such as sustainable cities, megacities, and urban planning are available here. Many international nongovernmental organizations are included.

Terrestrial Sciences
http://www.cgd.ucar.edu/tss/

The Terrestrial Sciences Section (TSS) is part of the Climate and Global Dynamics (CGD) Division at the National Center for Atmospheric Research (NCAR) in Boulder, Colorado. Scientists in the section study land-atmosphere interactions, in particular surface forcing of the atmosphere, through model development, application, and observational analyses. Here, you'll find a link to

VEMAP, The Vegetation/Ecosystem Modeling and Analysis Project.

UNIT 6: The Hazards of Growth: Pollution and Climate Change

Persistent Organic Pollutants (POP)
http://www.chem.unep.ch/pops/

Visit this site to learn more about persistent organic pollutants (POPs) and the issues and concerns surrounding them.

School of Labor and Industrial Relations (SLIR): Hot Links
http://www.lir.msu.edu/hotlinks/

Michigan State University's SLIR page connects to industrial relations sites throughout the world. It has links to U.S. government statistics, newspapers and libraries, international intergovernmental organizations, and more.

Space Research Institute
http://arc.iki.rssi.ru/eng/index.htm

For a change of pace, browse through this home page of Russia's Space Research Institute for information on its Environment Monitoring Information Systems, the IKI Satellite Situation Center, and its Data Archive.

Worldwatch Institute
http://www.worldwatch.org

The Worldwatch Institute, dedicated to fostering the evolution of an environmentally sustainable society, presents this site with access to *World Watch Magazine* and *State of the World 2000*. Click on In the News and Press Releases for discussions of current problems.

We highly recommend that you review our Web site for expanded information and our other product lines. We are continually updating and adding links to our Web site in order to offer you the most usable and useful information that will support and expand the value of your Annual Editions. You can reach us at: *http://www.dushkin.com/annualeditions/*.

UNIT 1

The Global Environment: An Emerging World View

Unit Selections

1. **How Many Planets? A Survey of the Global Environment**, The Economist
2. **Five Meta-Trends Changing the World**, David Pearce Snyder
3. **Crimes of (a) Global Nature**, Lisa Mastny and Hilary French
4. **Advocating For the Environment: Local Dimensions of Transnational Networks**, Maria Guadalupe Moog Rodrigues
5. **Rescuing a Planet Under Stress**, Lester R. Brown
6. **Globalizing Greenwash**, Pamela Foster

Key Points to Consider

- What are the connections between the attempts to develop sustainable systems and the quantity and quality of environmental data? Are there also relationships between data and the role of technology and economic systems in shaping the environmental future?

- What are some of the key "meta-trends" produced by increasing globalization of economic and other human systems? How can human societies and cultures adapt to such trends in order to prevent significant environmental disruption?

- How well do international agreements work in controlling "environmental crimes" such as the taking of endangered species, hazardous waste dumping, or emissions of harmful pollutants? Are there ways in which international environmental accords could be made more enforceable?

- How can you link the scales of environmental impact from the global to the local and still make sense of how local actions fit into global change and vice versa?

- Explain the analogy between human impact on environmental systems and an "economic bubble" in which demand so exceeds supply that the economic system collapses.

 **Links: www.dushkin.com/online/**
These sites are annotated in the World Wide Web pages.

Alternative Energy Institute (AEI)
http://www.altenergy.org

Earth Science Enterprise
http://www.earth.nasa.gov

IISDnet
http://iisd.ca

National Geographic Society
http://www.nationalgeographic.com

Research and Reference (Library of Congress)
http://lcweb.loc.gov/rr/

Santa Fe Institute
http://acoma.santafe.edu

Solstice: Documents and Databases
http://solstice.crest.org/index.html

United Nations
http://www.unsystem.org

United Nations Environment Programme (UNEP)
http://www.unep.ch

More than three decades after the celebration of the first Earth Day in 1970, public apprehension over the environmental future of the planet has reached levels unprecedented even during the late 1960's and early 1970's "Age of Aquarius." No longer are those concerned about the environment dismissed as "eco-freaks" and "tree-huggers." Most serious scientists have joined the rising clamor for environmental protection, as have the more traditional environmentally conscious public-interest groups. There are a number of reasons for this increased environmental awareness. Some of these reasons arise from environmental events; it is, for example, becoming increasingly difficult to deny the effects of global warming and atmospheric scientists are nearly unanimous in their attribution of human agencies as at least a partial cause for increasing global temperatures. But more reasons for environmental awareness arise simply from the process of globalization: the increasing unity of the world's economic, social, and information systems. Hailed by many as the salvation of the future, globalization has done little to make the world a better or safer place. Diseases once confined to specific regions now have increased capacity for widespread dissemination. Increasing human mobility has allowed human-caused disruptions to political, cultural, and economic systems to spread, and acts of terrorism now take place in locations once thought safe from such manifestations of hatred and despair. On the more positive side, the expansion of global information sys-

tems has fostered a maturation of concepts about the global nature of environmental processes.

Much of what has been learned through this increased information flow, particularly by American observers, has been of the environmentally ravaged world behind the old Iron Curtain—a chilling forecast of what other industrialized regions as well as the developing countries can become in the near future unless strict international environmental measures are put in place. For perhaps the first time ever, countries are beginning to recognize that environmental problems have no boundaries and that international cooperation is the only way to solve them.

The subtitle of this first unit, "An Emerging World View," is an optimistic assessment of the future: a future in which less money is spent on defense and more on environmental protection and cleanup—a new world order in which political influence might be based more on leadership in environmental and economic issues than on military might. It is probably far too early to make such optimistic predictions, to conclude that the world's nations—developed and underdeveloped—will begin to recognize that Earth's environment is a single unit. Thus far those nations have shown no tendency to recognize that humankind is a single unit and that what harms one harms all. The recent emergence of wide-scale terrorism as an instrument of political and social policy is evidence of such a failure of recognition. Nevertheless, there is a growing international realization—aided by the information superhighway—that we are all, as environmental activists have been saying for decades, inhabitants of Spaceship Earth and will survive or succumb together.

The articles selected for this unit have been chosen to illustrate the increasingly global perspective on environmental problems and the degree to which their solutions must be linked to political, economic, and social problems and solutions. In the lead piece of the unit, "How Many Planets?" the editors of *The Economist* attempt an analysis of what they admit is a very slippery subject by beginning with the observation that "it comes as a shock to discover how little information there is on the environment." They note the lip service paid everywhere to the concept of sustainability and acknowledge that economic growth and environmental health are not mutually inconsistent but that a great deal more work is necessary to make them compatible. They also conclude that governments, corporations, and individuals are more prepared now to think about how to use the planet than they were even 10 years ago.

Issues surrounding globalization form the subject of the next selection in the unit. In "Five Meta-Trends Changing the World," David Snyder, lifestyles editor of *The Futurist*, recognizes the "meta-trends" or multidimensional and evolutionary trends in the human-environment systems that are occurring as a result of a series of simultaneous demographic, economic, and technological trends. Snyder identifies these "meta-trends" as: (1) *Cultural modernization,* referring to the increasing "Westernization" of cultures in technological, economic, and demographic terms; (2) *Economic globalization,* meaning increasingly global competition for workers, resources, and markets; (3) Universal connectivity or what the editor of *The Economist* has called "the death of distance," a recognition that the cell phone, the Internet, and other connective technologies have truly made the world smaller by linking people together on a virtually-instantaneous basis; (4) *Transactional transparency,* in which corporate integrity and openness will be forced to grow as watchdog groups and citizens demand a new transparency of business that will allow closer public scrutiny of business practices that impact the environment; and (5) *Social adaptation,* or responses to new medical and other technologies that will allow people to remain productive in the workplace for longer periods of time and may even produce a return to the multigenerational family of children, parents, and grandchildren that formed society's safety net prior to the onset of industrialization. Each of these trends has profound implications for the ways in which human societies around the world relate to the environmental systems they inhabit.

Some of the impediments in making this new society one in which we would want to live are discussed in the third article in this unit. Lisa Mastny and Hilary French of the World Watch Institute describe the difficulties of developing and enforcing international environmental treaties and other agreements. In "Crimes of (a) Global Nature," Mastny and French focus on three types of international environmental agreements: those related to the taking and sale of endangered terrestrial and aquatic species; those governing the disposal of toxic and hazardous waste materials; and those mandating restrictions on the manufacture and use of certain chemicals that damage portions of the atmosphere. It is easy enough for countries to agree to international environmental accords. It is much more difficult for officials to en-

force those same accords and some of the authors' statistics are staggering—for example, illegally cut wood accounted for 65 percent of the world's supply in 2000. The future of the environment depends not just upon international agreement but upon enforcement.

The final selections in the opening section deal with those issues of human intellectual and political response to the global changes in society, economy, and environmental relations that are part of today's world. In "Advocating For The Environment," political scientist Maria Guadalupe Moog Rodrigues describes the increasing local focus of transnational environmental advocacy groups. While the concept of "think globally, act locally" has long been a watchphrase of the environmental movement, it is becoming increasingly apparent that small groups of dedicated persons directed toward specific local goals are beginning to make an impact on environmental policies. In places as different as India and Ecuador, local groups resistant to the construction of dams or the pollution resulting from petroleum extraction have forced both governments and industries to alter their approaches to locally-based projects. This priority of locally-based strategies consistent with the international environmental movement has, according to Rodrigues, provided strong links toward the transition to environmentally sustainable development. And in "Rescuing a Planet Under Stress" Lester Brown of the Earth Policy Institute provides an overview of "mega-threats" to continued environmental, social, and economic stability. Issues such as climate change, the worldwide HIV epidemic, soil erosion and desertification, and increasing demands for food, water, and fuel are foremost, according to Brown, who equates the increasing environmental pressures to the demands of a "bubble economy" in which demands exceed supply. The way to resolve the problems inherent in this economy is to deflate demand before it bursts the bubble of supply or environmental systems. This must be done at "wartime speed", says Brown, and international agreements must be reached to stabilize population, climate, water resources, soils, and other components of the human-environment system. If this is not done, then massive environmental and economic decline will be inevitable. The last article in the unit, "Globalizing Greenwash" by Pamela Foster of the Halifax Initiative (a Canadian coalition concerned about the international financial system and its impact on environments), furthers the economic concerns raised by Brown by focusing on the World Bank, one of the primary international agencies that has been responsible for the increasing economic growth that has produced such environmental stress. Foster notes that, despite criticisms of its environmental policies and promises to "green" its act, the World Bank's response to environmental problems has been anything but satisfactory.

All the articles in this opening section deal with environmental problems that were once confined to specific locales—as the world has grown increasingly smaller through advanced transportation, communication, and other technologies, and as a truly global economy has developed—but have now become problems on a global scale. While the potential for world-wide collapses of environmental systems is still a threat, the development of a global community with increasing awareness of the fragility of both economic and environmental systems, provides a promise for the future.

How many planets?

A survey of the global environment

The great race

Growth need not be the enemy of greenery. But much more effort is required to make the two compatible, says Vijay Vaitheeswaran

SUSTAINABLE development is a dangerously slippery concept. Who could possibly be against something that invokes such alluring images of untouched wildernesses and happy creatures? The difficulty comes in trying to reconcile the "development" with the "sustainable" bit: look more closely, and you will notice that there are no people in the picture.

That seems unlikely to stop a contingent of some of 60,000 world leaders, businessmen, activists, bureaucrats and journalists from travelling to South Africa next month for the UN-sponsored World Summit on Sustainable Development in Johannesburg. Whether the summit achieves anything remains to be seen, but at least it is asking the right questions. This survey will argue that sustainable development cuts to the heart of mankind's relationship with nature—or, as Paul Portney of Resources for the Future, an American think-tank, puts it, "the great race between development and degradation". It will also explain why there is reason for hope about the planet's future.

The best way known to help the poor today—economic growth—has to be handled with care, or it can leave a degraded or even devastated natural environment for the future. That explains why ecologists and economists have long held diametrically opposed views on development. The difficult part is to work out what we owe future generations, and how to reconcile that moral obligation with what we owe the poorest among us today.

It is worth recalling some of the arguments fielded in the run-up to the big Earth Summit in Rio de Janeiro a decade ago. A publication from UNESCO, a United Nations agency, offered the following vision of the future: "Every generation should leave water, air and soil resources as pure and unpolluted as when it came on earth. Each generation should leave undiminished all the species of animals it found existing on earth." Man, that suggests, is but a strand in the web of life, and the natural order is fixed and supreme. Put earth first, it seems to say.

Robert Solow, an economist at the Massachusetts Institute of Technology, replied at the time that this was "fundamentally the wrong way to go", arguing that the obligation to the future is "not to leave the world as we found it in detail, but rather to leave the option or the capacity to be as well off as we are." Implicit in that argument is the seemingly hard-hearted notion of "fungi-bility": that natural resources, whether petroleum or giant pandas, are substitutable.

Rio's fatal flaw

Champions of development and defenders of the environment have been locked in battle ever since a UN summit in Stockholm launched the sustainable-development debate three decades ago. Over the years, this debate often pitted indignant politicians and social activists from the poor world against equally indignant politicians and greens from the rich world. But by the time the Rio summit came along, it seemed they had reached a truce. With the help of a committee of grandees led by Gro Harlem Brundtland, a former Norwegian prime minister, the interested parties struck a deal in 1987: development and the environment, they declared, were inextricably linked. That compromise generated a good deal of euphoria. Green groups grew concerned over poverty, and development charities waxed lyrical about greenery. Even the World Bank joined in. Its World Development Report in 1992 gushed about "win-win" strategies, such as ending environmentally harmful subsidies, that would help both the economy and the environment.

By nearly universal agreement, those grand aspirations have fallen flat in the decade since that summit. Little headway has been made with environmental problems such as climate change and loss of biodiversity. Such progress as has been achieved has been largely due to three factors that this survey will explore in later sections: more decision-making at local level, technological innovation, and the rise of market forces in environmental matters.

The main explanation for the disappointment—and the chief lesson for those about to gather in South Africa—is that Rio overreached itself. Its participants were so anxious to reach a political consensus that they agreed to the Brundtland definition of sustainable development, which Daniel Esty of Yale University thinks has turned into "a buzz-word largely devoid of content". The biggest mistake, he reckons, is that it slides over the difficult trade-offs between environment and development in the real world. He is careful to note that there are plenty of cases where those goals are linked—but also many where they are not: "Environmental and economic policy goals are distinct, and the actions needed to achieve them are not the same."

No such thing as win-win

To insist that the two are "impossible to separate", as the Brundtland commission claimed, is nonsense. Even the World Bank now accepts that its much-trumpeted 1992 report was much too optimistic. Kristalina Georgieva, the Bank's director for the environment, echoes comments from various colleagues when she says: "I've never seen a real win-win in my life. There's always somebody, usually an elite group grabbing rents, that loses. And we've learned in the past decade that those losers fight hard to make sure that technically elegant win-win policies do not get very fat."

So would it be better to ditch the concept of sustainable development altogether? Probably not. Even people with their feet firmly planted on the ground think one aspect of it is worth salvaging: the emphasis on the future.

Nobody would accuse John Graham of jumping on green bandwagons. As an official in President George Bush's Office of Management and Budget, and previously as head of Harvard University's Centre for Risk Analysis, he has built a reputation for evidence-based policymaking. Yet he insists sustainable development is a worthwhile concept: "It's good therapy for the tunnel vision common in government ministries, as it forces integrated policymaking. In practical terms, it means that you have to take economic cost-benefit trade-offs into account in environmental laws, and keep environmental trade-offs in mind with economic development."

Jose Maria Figueres, a former president of Costa Rica, takes a similar view. "As a politician, I saw at first hand how often policies were dictated by short-term considerations such as elections or partisan pressure. Sustainability is a useful template to align short-term policies with medium- to long-term goals."

It is not only politicians who see value in saving the sensible aspects of sustainable development. Achim Steiner, head of the International Union for the Conservation of Nature, the world's biggest conservation group, puts it this way: "Let's be honest: greens and businesses do not have the same objective, but they can find common ground. We look for pragmatic ways to save species. From our own work on the ground on poverty, our members—be they bird watchers or passionate ecologists—have learned that 'sustainable use' is a better way to conserve."

Sir Robert Wilson, boss of Rio Tinto, a mining giant, agrees. He and other business leaders say it forces hard choices about the future out into the open: "I like this concept because it frames the trade-offs inherent in a business like ours. It means that single-issue activism is simply not as viable."

Kenneth Arrow and Larry Goulder, two economists at Stanford University, suggest that the old ideological enemies are converging: "Many economists now accept the idea that natural capital has to be valued, and that we need to account for ecosystem services. Many ecologists now accept that prohibiting everything in the name of protecting nature is not useful, and so are being selective." They think the debate is narrowing to the more empirical question of how far it is possible to substitute natural capital with the man-made sort, and specific forms of natural capital for one another.

The job for Johannesburg

So what can the Johannesburg summit contribute? The prospects are limited. There are no big, set-piece political treaties to be signed as there were at Rio. America's acrimonious departure from the Kyoto Protocol, a UN treaty on climate change, has left a bitter taste in many mouths. And the final pre-summit gathering, held in early June in Indonesia, broke up in disarray. Still, the gathered worthies could usefully concentrate on a handful of areas where international co-operation can help deal with environmental problems. Those include improving access for the poor to cleaner energy and to safe drinking water, two areas where concerns about human health and the environmental overlap. If rich countries want to make progress, they must agree on firm targets and offer the money needed to meet them. Only if they do so will poor countries be willing to cooperate on problems such as global warming that rich countries care about.

That seems like a modest goal, but it just might get the world thinking seriously about sustainability once again. If the Johannesburg summit helps rebuild a bit of faith in international environmental cooperation, then it will have been worthwhile. Minimising the harm that future economic growth does to the environment will require the rich world to work hand in glove with the poor world—which seems nearly unimaginable in today's atmosphere poisoned by the shortcomings of Rio and Kyoto.

To understand why this matters, recall that great race between development and degradation. Mankind has stayed comfortably ahead in that race so far, but can it go on doing so? The sheer magnitude of the economic growth that is hoped for in the coming decades (see chart 1) makes it seem inevitable that the clashes between mankind and nature will grow worse. Some are now asking whether all this economic growth is really necessary or useful in the first place, citing past advocates of the simple life.

"God forbid that India should ever take to industrialism after the manner of the West… It took Britain half the resources of the planet to achieve this prosperity. How many planets will a country like India require?", Mahatma Gandhi asked half a century ago. That question encapsulated the bundle of worries that haunts the sustainable-development debate to this day. Today, the vast majority of Gandhi's countrymen are still living the simple life—full of simple misery, malnourishment and material want. Grinding poverty, it turns out, is pretty sustainable.

If Gandhi were alive today, he might look at China next door and find that the country, once as poor as India, has been transformed beyond recognition by two decades of roaring economic growth. Vast numbers of people have been lifted out of poverty and into middle-class comfort. That could prompt him to reframe his question: how many planets will it take to satisfy China's needs if it ever achieves profligate America's affluence? One green group reckons the answer is three. The next section looks at the environmental data that might underpin such claims. It makes for alarming reading—though not for the reason that first springs to mind.

Flying blind

It comes as a shock to discover how little information there is on the environment

W**HAT** is the true state of the planet? It depends from which side you are peering at it. "Things are really looking up," comes the cry from one corner (usually overflowing with economists and technologists), pointing to a set of rosy statistics. "Disaster is nigh," shouts the other corner (usually full of ecologists and environmental lobbyists), holding up a rival set of troubling indicators.

According to the optimists, the 20th century marked a period of unprecedented economic growth that lifted masses of people out of abject poverty. It also brought technological innovations such as vaccines and other advances in public health that tackled many preventable diseases. The result has been a breath-taking enhancement of human welfare and longer, better lives for people everywhere on earth (see chart 2).

At this point, the pessimists interject: "Ah, but at what ecological cost?" They note that the economic growth which made all these gains possible sprang from the rapid spread of industrialisation and its resource-guzzling cousins, urbanisation, motorisation and electrification. The earth provided the necessary raw materials, ranging from coal to pulp to iron. Its ecosystems—rivers, seas, the atmosphere—also absorbed much of the noxious fallout from that process. The sheer magnitude of ecological change resulting directly from the past century's economic activity is remarkable (see table 3).

To answer that Gandhian question about how many planets it would take if everybody lived like the West, we need to know how much—or how little—damage the West's transformation from poverty to plenty has done to the planet to date. Economists point to the remarkable improvement in local air and water pollution in the rich world in recent decades. "It's Getting Better All the Time", a cheerful tract co-written by the late Julian Simon, insists that: "One of the greatest trends of the past 100 years has been the astonishing rate of progress in reducing almost every form of pollution." The conclusion seems unavoidable: "Relax! If we keep growing as usual, we'll inevitably grow greener."

The ecologically minded crowd takes a different view. "GEO3", a new report from the United Nations Environment Programme, looks back at the past few decades and sees much reason for concern. Its thoughtful boss, Klaus Töpfer (a former German environment minister), insists that his report is not "a document of doom and gloom". Yet, in summing it up, UNEP decries "the declining environmental quality of planet earth", and wags a finger at economic prosperity: "Currently, one-fifth of the world's population enjoys high, some would say excessive, levels of affluence." The conclusion seems unavoidable: "Panic! If we keep growing as usual, we'll inevitably choke the planet to death."

"People and Ecosystems", a collaboration between the World Resources Institute, the World Bank and the United Nations, tried to gauge the condition of ecosystems by examining the goods and services they produce—food, fibre, clean water, carbon storage and so on—and their capacity to continue producing them. The authors explain why ecosystems matter: half of all jobs worldwide are in agriculture, forestry and fishing, and the output from those three commodity businesses still dominates the economies of a quarter of the world's countries.

The report reached two chief conclusions after surveying the best available environmental data. First, a number of ecosystems are "fraying" under the impact of human activity. Second, ecosystems in future will be less able than in the past to deliver the goods and services human life depends upon, which points to unsustainability. But it took care to say: "It's hard, of course, to know what will be truly sustainable." The reason this collection of leading experts could not reach a firm conclusion was that, remarkably, much of the information they needed was incomplete or missing altogether: "Our knowledge of ecosystems has increased dramatically, but it simply has not kept pace with our ability to alter them."

Another group of experts, this time organised by the World Economic Forum, found itself similarly frustrated. The leader of that project, Daniel Esty of Yale, exclaims, throwing his arms in the air: "Why hasn't anyone done careful environmental measurement before? Businessmen always say, 'what matters gets measured.' Social scientists started quantitative measurement 30 years ago, and even political science turned to hard numbers 15 years ago. Yet look at environmental policy, and the data are lousy."

Gaping holes

At long last, efforts are under way to improve environmental data collection. The most ambitious of these is the Millennium Ecosystem Assessment, a joint effort among leading development agencies and environmental groups. This four-year effort is billed as an attempt to establish systematic data sets on all environmental matters across the world. But one of the researchers involved grouses that it "has very, very little new money to collect or analyse new data". It seems astonishing that governments have been making sweeping decisions on environmental policy for decades without such a baseline in the first place.

One positive sign is the growing interest of the private sector in collecting environmental data. It seems plain that leaving the task to the public sector has not worked. Information on the environment comes far lower on the bureaucratic pecking order than data on education or social affairs, which tend to be overseen by ministries with bigger budgets and more political clout. A number of countries, ranging from New Zealand to Austria, are now looking to the private sector to help collect and manage data in areas such as climate. Development banks are also considering using private contractors to monitor urban air quality, in part to get around the corruption and apathy in some city governments.

"I see a revolution in environmental data collection coming because of computing power, satellite mapping, remote sensing and other such information technologies," says Mr Esty. The arrival of hard data in this notoriously fuzzy area could cut down on environmental disputes by reducing uncertainty. One example is the long-running squabble between America's mid-western states, which rely heavily on coal, and the north-eastern states, which suffer from acid rain. Technology helped disprove claims by the mid-western states that New York's problems all resulted from home-grown pollution.

The arrival of good data would have other benefits as well, such as helping markets to work more robustly: witness America's pioneering scheme to trade emissions of sulphur dioxide, made possible by fancy equipment capable of monitoring emissions in real time. Mr Esty raises an even more intriguing possibility: "Like in the American West a hundred years ago, when barbed wire helped establish rights and prevent overgrazing, information technology can help establish 'virtual barbed wire' that secures property rights and so prevents overexploitation of the commons." He points to fishing in the waters between Australia and New Zealand, where tracking and monitoring devices have reduced over-exploitation.

Best of all, there are signs that the use of such fancy technology will not be confined to rich countries. Calestous Juma of Harvard University shares Mr Esty's excitement about the possibility of such a technology-driven revolution even in Africa: "In the past, the only environmental 'database' we had in Africa was our grandmothers. Now, with global information systems and such, the potential is enormous." Conservationists in Namibia, for example, already use satellite tracking to keep count of their elephants. Farmers in Mali receive satellite updates about impending storms on hand-wound radios. Mr Juma thinks the day is not far off when such technology, combined with ground-based monitoring, will help Africans measure trends in deforestation, soil erosion and climate change, and assess the effects on their local environment.

Make a start

That is at once a sweeping vision and a modest one. Sweeping, because it will require heavy investment in both sophisticated hardware and nuts-and-bolts information infrastructure on the ground to make sense of all these new data. As the poor world clearly cannot afford to pay for all this, the rich world must help—partly for altruistic reasons, partly with the selfish aim of discovering

in good time whether any global environmental calamities are in the making. A number of multilateral agencies now say they are willing to invest in this area as a "neglected global public good"—neglected especially by those agencies themselves. Even President Bush's administration has recently indicated that it will give environmental satellite data free to poor countries.

But that vision is also quite a modest one. Assuming that this data "revolution" does take place, all it will deliver is a reliable assessment of the health of the planet today. We will still not be able to answer the broader question of whether current trends are sustainable or not.

To do that, we need to look more closely at two very different sorts of environmental problems: global crises and local troubles. The global sort is hard to pin down, but can involve irreversible changes. The local kind is common and can have a big effect on the qualify of life, but is usually reversible. Data on both are predictably inadequate. We turn first to the most elusive environmental problem of all, global warming.

Blowing hot and cold

Climate change may be slow and uncertain, but that is no excuse for inaction

WHAT would Winston Churchill have done about climate change? Imagine that Britain's visionary wartime leader had been presented with a potential time bomb capable of wreaking global havoc, although not certain to do so. Warding it off would require concerted global action and economic sacrifice on the home front. Would he have done nothing?

Not if you put it that way. After all, Churchill did not dismiss the Nazi threat for lack of conclusive evidence of Hitler's evil intentions. But the answer might be less straightforward if the following provisos had been added: evidence of this problem would remain cloudy for decades; the worst effects might not be felt for a century; but the costs of tackling the problem would start biting immediately. That, in a nutshell, is the dilemma of climate change. It is asking a great deal of politicians to take action on behalf of voters who have not even been born yet.

One reason why uncertainty over climate looks to be with us for a long time is that the oceans, which absorb carbon from the atmosphere, act as a time-delay mechanism. Their massive thermal inertia means that the climate system responds only very slowly to changes in the composition of the atmosphere. Another complication arises from the relationship between carbon dioxide (CO_2), the principal greenhouse gas (GHG), and sulphur dioxide (SO_2), a common pollutant. Efforts to reduce manmade emissions of GHGs by cutting down on fossil-fuel use will reduce emissions of both gases. The reduction in

CO_2 will cut warming, but the concurrent SO_2 cut may mask that effect by contributing to the warming.

There are so many such fuzzy factors—ranging from aerosol particles to clouds to cosmic radiation—that we are likely to see disruptions to familiar climate patterns for many years without knowing why they are happening or what to do about them. Tom Wigley, a leading climate scientist and member of the UN's Intergovernmental Panel on Climate Change (IPCC), goes further. He argues in an excellent book published by the Aspen Institute, "US Policies on Climate Change: What Next?", that whatever policy changes governments pursue, scientific uncertainties will "make it difficult to detect the effects of such changes, probably for many decades."

As evidence, he points to the negligible short- to medium-term difference in temperature resulting from an array of emissions "pathways" on which the world could choose to embark if it decided to tackle climate change (see chart 4). He plots various strategies for reducing GHGs (including the Kyoto one) that will lead in the next century to the stabilisation of atmospheric concentrations of CO_2 at 550 parts per million (ppm). That is roughly double the level which prevailed in pre-industrial times, and is often mooted by climate scientists as a reasonable target. But even by 2040, the temperature differences between the various options will still be tiny—and certainly within the magnitude of natural climatic variance. In short, in another four decades we will probably still not know if we have over- or undershot.

Ignorance is not bliss

However, that does not mean we know nothing. We do know, for a start, that the "greenhouse effect" is real: without the heat-trapping effect of water vapour, CO_2, methane and other naturally occurring GHGs, our planet would be a lifeless 30°C or so colder. Some of these GHG emissions are captured and stored by "sinks", such as the oceans, forests and agricultural land, as part of nature's carbon cycle.

We also know that since the industrial revolution began, mankind's actions have contributed significantly to that greenhouse effect. Atmospheric concentrations of GHGs have risen from around 280ppm two centuries ago to around 370ppm today, thanks chiefly to mankind's use of fossil fuels and, to a lesser degree, to deforestation and other land-use changes. Both surface temperatures and sea levels have been rising for some time.

There are good reasons to think temperatures will continue rising. The IPCC has estimated a likely range for that increase of 1.4°C–5.8°C over the next century, although the lower end of that range is more likely. Since what matters is not just the absolute temperature level but the rate of change as well, it makes sense to try to slow down the increase.

The worry is that a rapid rise in temperatures would lead to climate changes that could be devastating for many (though not all) parts of the world. Central America, most of Africa, much of south Asia and northern China could all be hit by droughts, storms and floods and otherwise made miserable. Because they are poor and have the misfortune to live near the tropics, those most likely to be affected will be least able to adapt.

The colder parts of the world may benefit from warming, but they too face perils. One is the conceivable collapse of the Atlantic "conveyor belt", a system of currents that gives much of Europe its relatively mild climate; if temperatures climb too high, say scientists, the system may undergo radical changes that damage both Europe and America. That points to the biggest fear: warming may trigger irreversible changes that transform the earth into a largely uninhabitable environment.

Given that possibility, extremely remote though it is, it is no comfort to know that any attempts to stabilise atmospheric concentrations of GHGs at a particular level will take a very long time. Because of the oceans' thermal inertia, explains Mr Wigley, even once atmospheric concentrations of GHGs are stabilised, it will take decades or centuries for the climate to follow suit. And even then the sea level will continue to rise, perhaps for millennia.

This is a vast challenge, and it is worth bearing in mind that mankind's contribution to warming is the only factor that can be controlled. So the sooner we start drawing up a long-term strategy for climate change, the better.

What should such a grand plan look like? First and foremost, it must be global. Since CO_2 lingers in the atmosphere for a century or more, any plan must also extend across several generations.

The plan must recognise, too, that climate change is nothing new: the climate has fluctuated through history, and mankind has adapted to those changes—and must continue doing so. In the rich world, some of the more obvious measures will include building bigger dykes and flood defences. But since the most vulnerable people are those in poor countries, they too have to be helped to adapt to rising seas and unpredictable storms. Infrastructure improvements will be useful, but the best investment will probably be to help the developing world get wealthier.

It is essential to be clear about the plan's long-term objective. A growing chorus of scientists now argues that we need to keep temperatures from rising by much more than 2–3°C in all. That will require the stabilisation of atmospheric concentrations of GHGs. James Edmonds of the University of Maryland points out that because of the long life of CO_2, stabilisation of CO_2 concentrations is not at all the same thing as stabilisation of CO_2 emissions. That, says Mr Edmonds, points to an unavoidable conclusion: "In the very long term, global net CO_2 emissions must eventually peak and gradually decline toward zero, regardless of whether we go for a target of 350ppm or 1,000ppm."

A low-carbon world

That is why the long-term objective for climate policy must be a transition to a low-carbon energy system. Such a transition can be very gradual and need not necessarily lead to a world powered only by bicycles and windmills, for two reasons that are often overlooked.

One involves the precise form in which the carbon in the ground is distributed. According to Michael Grubb of the Carbon Trust, a British quasi-governmental body, the long-term problem is coal. In theory, we can burn all of the conventional oil and natural gas in the ground and still meet the most ambitious goals for tackling climate change. If we do that, we must ensure that the far greater amounts of carbon trapped as coal (and unconventional resources like tar sands) never enter the atmosphere.

The snag is that poor countries are likely to continue burning cheap domestic reserves of coal for decades. That suggests the rich world should speed the development and diffusion of "low carbon" technologies using the energy content of coal without releasing its carbon into the atmosphere. This could be far off, so it still makes sense to keep a watchful eye on the soaring carbon emissions from oil and gas.

The other reason, as Mr Edmonds took care to point out, is that it is net emissions of CO_2 that need to peak and decline. That leaves scope for the continued use of fossil fuels as the main source of modern energy if only some magical way can be found to capture and dispose of the associated CO_2. Happily, scientists already have some magic in the works.

One option is the biological "sequestration" of carbon in forests and agricultural land. Another promising idea is capturing and storing CO_2—underground, as a solid or even at the bottom of the ocean. Planting "energy crops" such as switch-grass and using them in conjunction with sequestration techniques could even result in negative net CO_2 emissions, because such plants use carbon from the atmosphere. If sequestration is combined with techniques for stripping the hydrogen out of this hydrocarbon, then coal could even offer a way to sustainable hydrogen energy.

But is anyone going to pay attention to these long-term principles? After all, over the past couple of years all participants in the Kyoto debate have excelled at producing short-sighted, selfish and disingenuous arguments. And the political rift continues: the EU and Japan pushed ahead with ratification of the Kyoto treaty a month ago, whereas President Bush reaffirmed his opposition.

However, go back a decade and you will find precisely those principles enshrined in a treaty approved by the elder George Bush and since reaffirmed by his son: the UN Framework Convention on Climate Change (FCCC). This treaty was perhaps the most important outcome of the Rio summit, and it remains the basis for the international climate-policy regime, including Kyoto.

The treaty is global in nature and long-term in perspective. It commits signatories to pursuing "the stabilisation of GHG concentrations in the atmosphere at a level that would prevent dangerous interference with the climate system." Note that the agreement covers GHG concentrations, not merely emissions. In effect, this commits even gas-guzzling America to the goal of declining emissions.

Better than Kyoto

Crucially, the FCCC treaty not only lays down the ends but also specifies the means: any strategy to achieve stabilisation of GHG concentrations, it insists, "must not be disruptive of the global economy". That was the stumbling block for the Kyoto treaty, which is built upon the FCCC agreement: its targets and timetables proved unrealistic.

Any revised Kyoto treaty or follow-up accord (which must include the United States and the big developing countries) should rest on the three basic pillars. First, governments everywhere (but especially in Europe) must understand that a reduction in emissions has to start modestly. That is because the capital stock involved in the global energy system is vast and long-lived, so a dash to scrap fossil-fuel production would be hugely expensive. However, as Mr Grubb points out, that pragmatism must be flanked by policies that encourage a switch to low-carbon technologies when replacing existing plants.

Second, governments everywhere (but especially in America) must send a powerful signal that carbon is going out of fashion. The best way to do this is to levy a carbon tax. However, whether it is done through taxes, mandated restrictions on GHG emissions or market mechanisms is less important than that the signal is sent clearly, forcefully and unambiguously. This is where President Bush's mixed signals have done a lot of harm: America's industry, unlike Europe's, has little incentive to invest in low-carbon technology. The irony is that even some coal-fired utilities in America are now clamouring for CO_2 regulation so that they can invest in new plants with confidence.

The third pillar is to promote science and technology. That means encouraging basic climate and energy research, and giving incentives for spreading the results. Rich countries and aid agencies must also find ways to help the poor world adapt to climate change. This is especially important if the world starts off with small cuts in emissions, leaving deeper cuts for later. That, observes Mr Wigley, means that by mid-century "very large investments would have to have been made—and yet the 'return' on these investments would not be visible. Continued investment is going to require more faith in climate science than currently appears to be the case."

Even a visionary like Churchill might have lost heart in the face of all this uncertainty. Nevertheless, there is a glimmer of hope that today's peacetime politicians may rise to the occasion.

Miracles sometimes happen

Two decades ago, the world faced a similar dilemma: evidence of a hole in the ozone layer. Some inconclusive signs suggested that it was man-made, caused by the use of chlorofluorocarbons (CFCs). There was the distant threat of disaster, and the knowledge about a concerted global response was required. Industry was reluctant at first, yet with leadership from Britain and America the Montreal Protocol was signed in 1987. That deal has proved surprisingly successful. The manufacture of CFCs is nearly phased out, and there are already signs that the ozone layer is on the way to recovery.

This story holds several lessons for the admittedly far more complex climate problem. First, it is the rich world which has caused the problem and which must lead the way in solving it. Second, the poor world must agree to help, but is right to insist on being given time—as well as money and technology—to help it adjust. Third, industry holds the key: in the ozone-depletion story, it was only after DuPont and ICI broke ranks with the rest of the CFC manufacturers that a deal became possible. On the climate issue, BP and Shell have similarly broken ranks with Big Oil, but the American energy industry—especially the coal sector—remains hostile.

9

The final lesson is the most important: that the uncertainty surrounding a threat such as climate change is no excuse for inaction. New scientific evidence shows that the threat from ozone depletion had been much deadlier than was thought at the time when the world decided to act. Churchill would surely have approved.

Local difficulties

Greenery is for the poor too, particularly on their own doorstep

WHY should we care about the environment? Ask a European, and he will probably point to global warming. Ask the two little boys playing outside a newsstand in Da Shilan, a shabby neighbourhood in the heart of Beijing, and they will tell you about the city's notoriously foul air: "It's bad—like a virus!"

Given all the media coverage in the rich world, people there might believe that global scares are the chief environmental problems facing humanity today. They would be wrong. Partha Dasgupta, an economics professor at Cambridge University, thinks the current interest in global, future-oriented problems has "drawn attention away from the economic misery and ecological degradation endemic in large parts of the world today. Disaster is not something for which the poorest have to wait; it is a frequent occurrence."

Every year in developing countries, a million people die from urban air pollution and twice that number from exposure to stove smoke inside their homes. Another 3m unfortunates die prematurely every year from water-related diseases. All told, premature deaths and illnesses arising from environmental factors account for about a fifth of all diseases in poor countries, bigger than any other preventable factor, including malnutrition. The problem is so serious that Ian Johnson, the World Bank's vice-president for the environment, tells his colleagues, with a touch of irony, that he is really the bank's vice-president for health: "I say tackling the underlying environmental causes of health problems will do a lot more good than just more hospitals and drugs."

The link between environment and poverty is central to that great race for sustainability. It is a pity, then, that several powerful fallacies keep getting in the way of sensible debate. One popular myth is that trade and economic growth make poor countries' environmental problems worse. Growth, it is said, brings with it urbanisation, higher energy consumption and industrialisation—all factors that contribute to pollution and pose health risks.

In a static world, that would be true, because every new factory causes extra pollution. But in the real world, economic growth unleashes many dynamic forces that, in the longer run, more than offset that extra pollution. As chart 5 makes clear, traditional environmental risks (such as water-borne diseases) cause far more health problems in poor countries than modern environmental risks (such as industrial pollution).

Rigged rules

However, this is not to say that trade and economic growth will solve all environmental problems. Among the reasons for doubt are the "perverse" conditions under which world trade is carried on, argues Oxfam. The British charity thinks the rules of trade are "unfairly rigged against the poor", and cites in evidence the enormous subsidies lavished by rich countries on industries such as agriculture, as well as trade protection offered to manufacturing industries such as textiles. These measurements hurt the environment because they force the world's poorest countries to rely heavily on commodities—a particularly energy-intensive and ungreen sector.

Mr Dasgupta argues that this distortion of trade amounts to a massive subsidy of rich-world consumption paid by the world's poorest people. The most persuasive critique of all goes as follows: "Economic growth is not sufficient for turning environmental degradation around. If economic incentives facing producers and consumers do not change with higher incomes, pollution will continue to grow unabated with the growing scale of economic activity." Those words come not from some anti-globalist green group, but from the World Trade Organisation.

Another common view is that poor countries, being unable to afford greenery, should pollute now and clean up later. Certainly poor countries should not be made to adopt American or European environmental standards. But there is evidence to suggest that poor countries can and should try to tackle some environmental problems now, rather than wait till they have become richer.

This so-called "smart growth" strategy contradicts conventional wisdom. For many years, economists have observed that as agrarian societies industrialised, pollution increased at first, but as the societies grew wealthier it declined again. The trouble is that this applies only to some pollutants, such as sulphur dioxide, but not to others, such as carbon dioxide. Even more troublesome, those smooth curves going up, then down, turn out to be misleading. They are what you get when you plot data for poor and rich countries together at a given moment in time, but actual levels of various pollutants in any individual country plotted over time wiggle around a lot more. This suggests that the familiar bell-shaped curve reflects no immutable law, and that intelligent government policies might well help to reduce pollution levels even while countries are still relatively poor.

Developing countries are getting the message. From Mexico to the Philippines, they are now trying to curb the worst of the air and water pollution that typically accompanies industrialisation. China, for example, was persuaded by outside experts that it was losing so much potential economic output through health troubles caused by pollution (according to one World Bank study, somewhere between 3.5% and 7.7% of GDP) that tackling it was cheaper than ignoring it.

One powerful—and until recently ignored—weapon in the fight for a better environment is local people. Old-fashioned paternalists in the capitals of developing countries used to argue that poor villagers could not be relied on to look after natural resources. In fact, much academic research has shown that the poor are more often victims than perpetrators of resource depletion: it tends to be rich locals or outsiders who are responsible for the worst exploitation.

Local people usually have a better knowledge of local ecological conditions than experts in faraway capitals, as well as a direct interest in improving the quality of life in their village. A good example of this comes from the bone-dry state of Rajasthan in India, where local activism and indigenous know-how about rainwater "harvesting" provided the people with reliable water supplies—something the government had failed to do. In Bangladesh, villages with active community groups or concerned mullahs proved greener than less active neighbouring villages.

Community-based forestry initiatives from Bolivia to Nepal have shown that local people can be good custodians of nature. Several hundred million of the world's poorest people live in and around forests. Giving those villagers an incentive to preserve forests by allowing sustainable levels of harvesting, it turns out, is a far better way to save those forests than erecting tall fences around them.

To harness local energies effectively, it is particularly important to give local people secure property rights, argues Mr Dasgupta. In most parts of the developing world, control over resources at the village level is ill-defined. This often means that local elites usurp a disproportionate share of those resources, and that individuals have little incentive to maintain and upgrade forests or agricultural land. Authorities in Thailand tried to remedy this problem by distributing 5.5m land titles over a 20-year period. Agricultural output increased, access to credit improved and the value of the land shot up.

Name and shame

Another powerful tool for improving the local environment is the free flow of information. As local democracy flourishes, ordinary people are pressing for greater environmental disclosure by companies. In some countries, such as Indonesia, governments have adopted a "sunshine" policy that involves naming and shaming companies that do not meet environmental regulations. It seems to achieve results.

Bringing greenery to the grass roots is good, but on its own it will not avert perceived threats to global "public goods" such as the climate or biodiversity. Paul Portney of Resources for the Future explains: "Brazilian villagers may think very carefully and unselfishly about their future descendants, but there's no reason for them to care about and protect species or habitats that no future generation of Brazilians will care about."

That is why rich countries must do more than make pious noises about global threats to the environment. If they believe that scientific evidence suggests a credible threat, they must be willing to pay poor countries to protect such things as their tropical forests. Rather than thinking of this as charity, they should see it as payment for environmental services (say, for carbon storage) or as a form of insurance.

In the case of biodiversity, such payments could even be seen as a trade in luxury goods: rich countries would pay poor countries to look after creatures that only the rich care about. Indeed, private green groups are already buying up biodiversity "hot spots" to protect them. One such initiative, led by Conservation International and the International Union for the Conservation of Nature (IUCN), put the cost of buying and preserving 25 hot spots exceptionally rich in species diversity at less than $30 billion. Sceptics say it will cost more, as hot spots will need buffer zones of "sustainable harvesting" around them. Whatever the right figure, such creative approaches are more likely to achieve results than bullying the poor into conservation.

It is not that the poor do not have green concerns, but that those concerns are very different from those of the rich. In Beijing's Da Shilan, for instance, the air is full of soot from the many tiny coal boilers. Unlike most of the neighbouring districts, which have recently converted from coal to natural gas, this area has been considered too poor to make the transition. Yet ask Liu Shihua, a shopkeeper who has lived in the same spot for over 20 years, and he insists he would readily pay a bit more for the cleaner air that would come from using natural gas. So would his neighbours.

To discover the best reason why poor countries should not ignore pollution, ask those two little boys outside Mr

Liu's shop what colour the sky is. "Grey!" says one tyke, as if it were the most obvious thing in the world. "No, stupid, it's blue!" retorts the other. The children deserve blue skies and clean air. And now there is reason to think they will see them in their lifetime.

Working miracles

Can technology save the planet?

"NOTHING endures but change." That observation by Heraclitus often seems lost on modern environmental thinkers. Many invoke scary scenarios assuming that resources—both natural ones, like oil, and man-made ones, like knowledge—are fixed. Yet in real life man and nature are entwined in a dynamic dance of development, scarcity, degradation, innovation and substitution.

The nightmare about China turning into a resource-guzzling America raises two questions: will the world run out of resources? And even if it does not, could the growing affluence of developing nations lead to global environmental disaster?

The first fear is the easier to refute; indeed, history has done so time and again. Malthus, Ricardo and Mill all worried that scarcity of resources would snuff out growth. It did not. A few decades ago, the limits-to-growth camp raised worries that the world might soon run out of oil, and that it might not be able to feed the world's exploding population. Yet there are now more proven reserves of petroleum than three decades ago; there is more food produced than ever; and the past decade has seen history's greatest economic boom.

What made these miracles possible? Fears of oil scarcity prompted investment that led to better ways of producing oil, and to more efficient engines. In food production, technological advances have sharply reduced the amount of land required to feed a person in the past 50 years. Jesse Ausubel of Rockefeller University calculates that if in the next 60 to 70 years the world's average farmer reaches the yield of today's average (not best) American maize grower, then feeding 10 billion people will require just half of today's cropland. All farmers need to do is maintain the 2%-a-year productivity gain that has been the global norm since 1960.

"Scarcity and Growth", a book published by Resources for the Future, sums it up brilliantly: "Decades ago Vermont granite was only building and tombstone material; now it is a potential fuel, each ton of which has a usable energy content (uranium) equal to 150 tons of coal. The notion of an absolute limit to natural resource availability is untenable when the definition of resources changes drastically and unpredictably over time." Those words were written by Harold Barnett and Chandler Morse in 1963, long before the limits-to-growth bandwagon got rolling.

Giant footprint

Not so fast, argue greens. Even if we are not going to run out of resources, guzzling ever more resources could still do irreversible damage to fragile ecosystems.

WWF, an environmental group, regularly calculates mankind's "ecological footprint", which it defines as the "biologically productive land and water areas required to produce the resources consumed and assimilate the wastes generated by a given population using prevailing technology." The group reckons the planet has around 11.4 billion "biologically productive" hectares of land available to meet continuing human needs. As chart 6 overleaf shows, WWF thinks mankind has recently been using more than that. This is possible because a forest harvested at twice its regeneration rate, for example, appears in the footprint accounts at twice its area—an unsustainable practice which the group calls "ecological overshoot."

Any analysis of this sort must be viewed with scepticism. Everyone knows that environmental data are incomplete. What is more, the biggest factor by far is the land required to absorb CO_2 emissions of fossil fuels. If that problem could be managed some other way, then mankind's ecological footprint would look much more sustainable.

Even so, the WWF analysis makes an important point: if China's economy were transformed overnight into a clone of America's, an ecological nightmare could ensue. If a billion eager new consumers were suddenly to produce CO_2 emissions at American rates, they would be bound to accelerate global warming. And if the whole of the developing world were to adopt an American lifestyle tomorrow, local environmental crises such as desertification, aquifer depletion and topsoil loss could make humans miserable.

So is this cause for concern? Yes, but not for panic. The global ecological footprint is determined by three factors: population size, average consumption per person and technology. Fortunately, global population growth now appears to be moderating. Consumption per person in poor countries is rising as they become better off, but there are signs that the rich world is reducing the footprint of its consumption (as this survey's final section explains). The most powerful reason for hope—innovation—was foreshadowed by WWF's own definition. Today's "prevailing technologies" will, in time, be displaced by tomorrow's greener ones.

"The rest of the world will not live like America," insists Mr Ausubel. Of course poor people around the world covet the creature comforts that Americans enjoy, but they know full well that the economic growth needed to improve their lot will take time. Ask Wu Chengjian, an environmental official in booming Shanghai, what he thinks of the popular notion that his city might become as rich as today's Hong Kong by 2020: "Impossible—that's just not enough time." And that is Shanghai, not the impoverished countryside.

Leaps of faith

This extra time will allow poor countries to embrace new technologies that are more efficient and less environmentally damaging. That still does not guarantee a smaller ecological footprint for China in a few decades' time than for America now, but it greatly improves the chances. To see why, consider the history of "dematerialisation" and "decarbonisation" (see chart 7). Viewed across very long spans of time, productivity improvements allow economies to use ever fewer material inputs—and to emit ever fewer pollutants—per unit of economic output. Mr Ausubel concludes: "When China has today's American mobility, it will not have today's American cars," but the cleaner and more efficient cars of tomorrow.

The snag is that consumers in developing countries want to drive cars not tomorrow but today. The resulting emissions have led many to despair that technology (in the form of vehicles) is making matters worse, not better.

Can they really hope to "leapfrog" ahead to cleaner air? The evidence from Los Angeles—a pioneer in the fight against air pollution—suggests the answer is yes. "When I moved to Los Angeles in the 1960s, there was so much soot in the air that it felt like there was a man standing on your chest most of the time," says Ron Loveridge, the mayor of Riverside, a city to the east of LA that suffers the worst of the region's pollution. But, he says, "We have come an extraordinary distance in LA."

Four decades ago, the city had the worst air quality in America. The main problem was the city's infamous "smog" (an amalgam of "smoke" and "fog"). It took a while to figure out that this unhealthy ozone soup developed as a result of complex chemical reactions between nitrogen oxides and volatile organic compounds that need sunlight to trigger them off.

Arthur Winer, an atmospheric chemist at the University of California at Los Angeles, explains that tackling smog required tremendous perseverance and political will. Early regulatory efforts met stiff resistance from business interests, and began to falter when they failed to show dramatic results.

Clean-air advocates like Mr Loveridge began to despair: "We used to say that we needed a 'London fog' [a reference to an air-pollution episode in 1952 that may have killed 12,000 people in that city] here to force change." Even so, Californian officials forged ahead with an ambitious plan that combined regional regulation with stiff mandates for cleaner air. Despite uncertainties about the cause of the problem, the authorities introduced a sequence of controversial measures: unleaded and low-sulphur petrol, on-board diagnostics for cars to minimise emissions, three-way catalytic converters, vapour-recovery attachments for petrol nozzles and so on.

As a result, the city that two decades ago hardly ever met federal ozone standards has not had to issue a single alert in the past three years. Peak ozone levels are down by 50% since the 1960s. Though the population has shot up in recent years, and the vehicle-miles driven by car-crazy Angelenos have tripled, ozone levels have fallen by two-thirds. The city's air is much cleaner than it was two decades ago.

"California, in solving its air-quality problem, has solved it for the rest of the United States and the world—but it doesn't get credit for it," says Joe Norbeck of the University of California at Riverside. He is adamant that the poor world's cities can indeed leapfrog ahead by embracing some of the cleaner technologies developed specifically for the Californian market. He points to China's vehicle fleet as an example: "China's typical car has the emissions of a 1974 Ford Pinto, but the new Buicks sold there use 1990s emissions technology." The typical car sold today produces less than a tenth of the local pollution of a comparable model from the 1970s.

That suggests one lesson for poor cities such as Beijing that are keen to clean up: they can order polluters to meet high emissions standards. Indeed, from Beijing to Mexico city, regulators are now imposing rich-world rules, mandating new, cleaner technologies. In China's cities, where pollution from sooty coal fires in homes and industrial boilers had been a particular hazard, officials are keen to switch to natural-gas furnaces.

However, there are several reasons why such mandates—which worked wonders in LA—may be trickier to achieve in impoverished or politically weak cities. For a start, city officials must be willing to pay the political price of reforms that raise prices for voters. Besides, higher standards for new cars, useful though they are, cannot do the trick on their own. Often, clean technologies such as catalytic converters will require cleaner grades of petrol too. Introducing cleaner fuels, say experts, is an essential lesson from LA for poor countries. This will not come free either.

There is another reason why merely ordering cleaner new cars is inadequate: it does nothing about the vast stock of dirty old ones already on the streets. In most cities of the developing world, the oldest fifth of the vehicles on the road is likely to produce over half of the total pollution caused by all vehicles taken together. Policies that encourage a speedier turnover of the fleet therefore make more sense than "zero emissions" mandates.

Policy matters

In sum, there is hope that the poor can leapfrog at least some environmental problems, but they need more than just technology. Luisa and Mario Molina of the Massachusetts Institute of Technology, who have studied such questions closely, reckon that technology is less important than the institutional capacity, legal safeguards and financial resources to back it up: "The most important underlying factor is political will." And even a techno-optimist such as Mr Ausubel accepts that: "There is nothing automatic about technological innovation and adoption; in fact, at the micro level, it's bloody."

Clearly innovation is a powerful force, but government policy still matters. That suggests two rules for policymakers. First, don't do stupid things that inhibit innovation. Second, do sensible things that reward the development and adoption of technologies that enhance, rather than degrade, the environment.

The greatest threat to sustainability may well be the rejection of science. Consider Britain's hysterical reaction to genetically modified crops, and the European Commission's recent embrace of a woolly "precautionary principle". Precaution applied case-by-case is undoubtedly a good thing, but applying any such principle across the board could prove disastrous.

Explaining how not to stifle innovation that could help the environment is a lot easier than finding ways to encourage it. Technological change often goes hand-in-hand with greenery by saving resources, as the long history of dematerialisation shows—but not always. Sports utility vehicles, for instance, are technologically innovative, but hardly green. Yet if those SUVs were to come with hydrogen-powered fuel cells that emit little pollution, the picture would be transformed.

The best way to encourage such green innovations is to send powerful signals to the market that the environment matters. And there is no more powerful signal than price, as the next section explains.

The invisible green hand

Markets could be a potent force for greenery—if only greens could learn to love them

"MANDATE, regulate and litigate." That has been the environmentalists' rallying cry for ages. Nowhere in the green manifesto has there been much mention of the market. And, oddly, it was market-minded America that led the dirigiste trend. Three decades ago, Congress passed a sequence of laws, including the Clean Air Act, which set lofty goals and generally set rigid technological standards. Much of the world followed America's lead.

This top-down approach to greenery has long been a point of pride for groups such as the Natural Resources Defence Council (NRDC), one of America's most influential environmental outfits. And with some reason, for it has had its successes: the air and water in the developed world is undoubtedly cleaner than it was three decades ago, even though the rich world's economies have grown by leaps and bounds. This has convinced such groups stoutly to defend the green status quo.

But times may be changing. Gus Speth, now head of Yale University's environment school and formerly head of the World Resources Institute and the UNDP, as well as one of the founders of the NRDC, recently explained how he was converted to market economics: "Thirty years ago, the economists at Resources for the Future were pushing the idea of pollution taxes. We lawyers at NRDC thought they were nuts, and feared that they would derail command-and-control measures like the Clean Air Act, so we opposed them. Looking back, I'd have to say this was the single biggest failure in environmental management—not getting the prices right."

A remarkable mea culpa; but in truth, the command-and-control approach was never as successful as its advocates claimed. For example, although it has cleaned up the air and water in rich countries, it has notably failed in dealing with waste management, hazardous emissions and fisheries depletion. Also, the gains achieved have come at a needlessly high price. That is because technology mandates and bureaucratic edicts stifle innovation and ignore local realities, such as varying costs of abatement. They also fail to use cost-benefit analysis to judge trade-offs.

Command-and-control methods will also be ill-suited to the problems of the future, which are getting trickier. One reason is that the obvious issues—like dirty air and water—have been tackled already. Another is increasing technological complexity: future problems are more likely to involve subtle linkages—like those involved in ozone depletion and global warming—that will require sophisticated responses. The most important factor may be society's ever-rising expectations; as countries grow wealthier, their people start clamouring for an ever-cleaner environment. But because the cheap and simple things have been done, that is proving increasingly expensive. Hence the greens' new interest in the market.

Carrots, not just sticks

In recent years, market-based greenery has taken off in several ways. With emissions trading, officials decide on a pollution target and then allocate tradable credits to companies based on that target. Those that find it expensive to cut emissions can buy credits from those that find it cheaper, so the target is achieved at the minimum cost and disruption.

The greatest green success story of the past decade is probably America's innovative scheme to cut emissions of sulphur dioxide (SO_2). Dan Dudek of Environmental Defence, a most unusual green group, and his market-minded colleagues persuaded the elder George Bush to agree to an amendment to the sacred Clean Air Act that would introduce an emissions-trading system to achieve sharp cuts in SO_2. At the time, this was hugely controversial: America's power industry insisted the cuts were prohibitively costly, while nearly every other green group decried the measure as a sham. In the event, ED has been vindicated. America's scheme has surpassed its initial objectives, and at far lower cost than expected. So great is the interest worldwide in trading that ED is now advising groups ranging from hard-nosed oilmen at BP to bureaucrats in China and Russia.

Europe, meanwhile, is forging ahead with another sort of market-based instrument: pollution taxes. The idea is to levy charges on goods and services so that their price reflects their "externalities"—jargon for how much harm they do to the environment and human health. Sweden introduced a sulphur tax a decade ago, and found that the sulphur content of fuels dropped 50% below legal requirements.

Though "tax" still remains a dirty word in America, other parts of the world are beginning to embrace green tax reform by shifting taxes from employment to pollu-

tion. Robert Williams of Princeton University has looked at energy use (especially the terrible effects on health of particulate pollution) and concluded that such externalities are comparable in size to the direct economic costs of producing that energy.

Externalities are only half the battle in fixing market distortions. The other half involves scrapping environmentally harmful subsidies. These range from prices below market levels for electricity and water to shameless cash handouts for industries such as coal. The boffins at the OECD reckon that stripping away harmful subsidies, along with introducing taxes on carbon-based fuels and chemicals use, would result in dramatically lower emissions by 2020 than current policies would be able to achieve. If the revenues raised were then used to reduce other taxes, the cost of these virtuous policies would be less than 1% of the OECD's economic output in 2020.

Such subsidies are nothing short of perverse, in the words of Norman Myers of Oxford University. They do double damage, by distorting markets and by encouraging behaviour that harms the environment. Development banks say such subsidies add up to $700 billion a year, but Mr Myers reckons the true sum is closer to $2 trillion a year. Moreover, the numbers do not fully reflect the harm done. For example, EU countries subsidise their fishing fleets to the tune of $1 billion a year, but that has encouraged enough overfishing to drive many North Atlantic fishing grounds to near-collapse.

Fishing is an example of the "tragedy of the commons", which pops up frequently in the environmental debate. A resource such as the ocean is common to many, but an individual "free rider" can benefit from plundering that commons or dumping waste into it, knowing that the costs of his actions will probably be distributed among many neighbours. In the case of shared fishing grounds, the absence of individual ownership drives each fisherman to snatch as many fish as he can—to the detriment of all.

Of rights and wrongs

Assigning property rights can help, because providing secure rights (set at a sustainable level) aligns the interests of the individual with the wider good of preserving nature. This is what sceptical conservationists have observed in New Zealand and Iceland, where schemes for tradable quotas have helped revive fishing stocks. Similar rights-based approaches have led to revivals in stocks of African elephants in southern Africa, for example, where the authorities stress property rights and private conservation.

All this talk of property rights and markets makes many mainstream environmentalists nervous. Carl Pope, the boss of the Sierra Club, one of America's biggest green groups, does not reject market forces out of hand, but expresses deep scepticism about their scope. Pointing to the difficult problem of climate change, he asks: "Who has property rights over the commons?"

Even so, some greens have become converts. Achim Steiner of the IUCN reckons that the only way forward is rights-based conservation, allowing poor people "sustainable use" of their local environment. Paul Faeth of the World Resources Institute goes further. He says he is convinced that market forces could deliver that holy grail of environmentalism, sustainability—"but only if we get prices right."

The limits to markets

Economic liberals argue that the market itself is the greatest price-discovery mechanism known to man. Allow it to function freely and without government meddling, goes the argument, and prices are discovered and internalised automatically. Jerry Taylor of the Cato Institute, a libertarian think-tank, insists that "The world today is already sustainable—except those parts where western capitalism doesn't exist." He notes that countries that have relied on central planning, such as the Soviet Union, China and India, have invariably misallocated investment, stifled innovation and fouled their environment far more than the prosperous market economies of the world have done.

All true. Even so, markets are currently not very good at valuing environmental goods. Noble attempts are under way to help them do better. For example, the Katoomba Group, a collection of financial and energy companies that have linked up with environmental outfits, is trying to speed the development of markets for some of forestry's ignored "co-benefits" such as carbon storage and watershed management, thereby producing new revenue flows for forest owners. This approach shows promise: water consumers ranging from officials in New York City to private hydro-electric operators in Costa Rica are now paying people upstream to manage their forests and agricultural land better. Paying for greenery upstream turns out to be cheaper than cleaning up water downstream after it has been fouled.

Economists too are getting into the game of helping capitalism "get prices right." The World Bank's Ian Johnson argues that conventional economic measures such as gross domestic product are not measuring wealth creation properly because they ignore the effects of environmental degradation. He points to the positive contribution to China's GDP from the logging industry, arguing that such a calculation completely ignores the billions of dollars-worth of damage from devastating floods caused

by over-logging. He advocates a more comprehensive measure the Bank is working on, dubbed "genuine GDP", that tries (imperfectly, he accepts) to measure depletion of natural resources.

That could make a dramatic difference to how the welfare of the poor is assessed. Using conventional market measures, nearly the whole of the developing world save Africa has grown wealthier in the past couple of decades. But when the degradation of nature is properly accounted for, argues Mr Dasgupta at Cambridge, the countries of Africa and south Asia are actually much worse off today than they were a few decades ago—and even China, whose economic "miracle" has been much trumpeted, comes out barely ahead.

The explanation, he reckons, lies in a particularly perverse form of market distortion: "Countries that are exporting resource-based products (often among the poorest) may be subsidising the consumption of countries that are doing the importing (often among the richest)." As evidence, he points to the common practice in poor countries of encouraging resource extraction. Whether through licenses granted at below-market rates, heavily subsidised exports or corrupt officials tolerating illegal exploitation, he reckons the result is the same: "The cruel paradox we face may well be that contemporary economic development is unsustainable in poor countries because it is sustainable in rich countries."

One does not have to agree with Mr Dasgupta's conclusion to acknowledge that markets have their limits. That should not dissuade the world from attempting to get prices right—or at least to stop getting them so wrong. For grotesque subsidies, the direction of change should be obvious. In other areas, the market itself may not provide enough information to value nature adequately. This is true of threats to essential assets, such as nature's ability to absorb and "recycle" CO_2, that have no substitute at any price. That is when governments must step in, ensuring that an informed public debate takes place.

Robert Stavins of Harvard University argues that the thorny notion of sustainable development can be reduced to two simple ideas: efficiency and intergenerational equity. The first is about making the economic pie as large as possible; he reckons that economists are well equipped to handle it, and that market-based policies can be used to achieve it. On the second (the subject of the next section), he is convinced that markets must yield to public discourse and government policy: "Markets can be efficient, but nobody ever said they're fair. The question is, what do we owe the future?"

Insuring a brighter future

How to hedge against tomorrow's environmental risks

So WHAT do we owe the future? A precise definition for sustainable development is likely to remain elusive but, as this survey has argued, the hazy outline of a useful one is emerging from the experience of the past decade.

For a start, we cannot hope to turn back the clock and return nature to a pristine state. Nor must we freeze nature in the state it is today, for that gift to the future would impose an unacceptable burden on the poorest alive today. Besides, we cannot forecast the tastes, demands or concerns of future generations. Recall that the overwhelming pollution problem a century ago was horse manure clogging up city streets: a century hence, many of today's problems will surely seem equally irrelevant. We should therefore think of our debt to the future as including not just natural resources but also technology, institutions and especially the capacity to innovate. Robert Solow got it mostly right a decade ago: the most important thing to leave future generations, he said, is the capacity to live as well as we do today.

However, as the past decade has made clear, there is a limit to that argument. If we really care about the "sustainable" part of sustainable development, we must be much more watchful about environmental problems with critical thresholds. Most local problems are reversible and hence no cause for alarm. Not all, however: the depletion of aquifers and the loss of topsoil could trigger irreversible changes that would leave future generations worse off. And global or long-term threats, where victims are far removed in time and space, are easy to brush aside.

In areas such as biodiversity, where there is little evidence of a sustainability problem, a voluntary approach is best. Those in the rich world who wish to preserve pandas, or hunt for miracle drugs in the rainforest, should pay for their predilections. However, where there are strong scientific indications of unsustainability, we must act on behalf of the future—even at the price of today's development. That may be expensive, so it is prudent to try to minimise those risks in the first place.

A riskier world

Human ingenuity and a bit of luck have helped mankind stay a few steps ahead of the forces degrading the environment this past century, the first full one in which the planet has been exposed to industrialisation. In the century ahead, the great race between development and degradation could well become a closer call.

On one hand, the demands of development seem sure to grow at a cracking pace in the next few decades as the Chinas, Indias and Brazils of this world grow wealthy enough to start enjoying not only the necessities but also some of the luxuries of life. On the other hand, we seem to be entering a period of huge technological advances in emerging fields such as biotechnology that could greatly increase resource productivity and more than offset the effect of growth on the environment. The trouble is, nobody knows for sure.

Since uncertainty will define the coming era, it makes sense to invest in ways that reduce that risk at relatively low cost. Governments must think seriously about the future implications of today's policies. Their best bet is to encourage the three powerful forces for sustainability outlined in this survey: the empowerment of local people to manage local resources and adapt to environmental change; the encouragement of science and technology, especially innovations that reduce the ecological footprint of consumption; and the greening of markets to get prices right.

To advocate these interventions is not to call for a return to the hubris of yesteryear's central planners. These measures would merely give individuals the power to make greener choices if they care to. In practice, argues Chris Heady of the OECD, this may still not add up to sustainability "because we might still decide to be greedy, and leave less for our children."

Happily, there are signs of an emerging bottom-up push for greenery. Even such icons of western consumerism as Unilever and Procter & Gamble now sing the virtues of "sustainable consumption." Unilever has vowed that by 2005 it will be buying fish only from sustainable sources, and P&G is coming up with innovative products such as detergents that require less water, heat and packaging. It would be naive to label such actions as expressions of "corporate social responsibility": in the long run, firms will embrace greenery only if they see profit in it. And that, in turn, will depend on choices made by individuals.

Such interventions should really be thought of as a kind of insurance that tilts the odds of winning that great race just a little in humanity's favour. Indeed, even some of the world's most conservative insurance firms increasingly see things this way. As losses from weather-related disasters have risen of late (see chart 8), the industry is getting more involved in policy debates on long-term environmental issues such as climate change.

Bruno Porro, chief risk officer at Swiss Re, argues that: "The world is entering a future in which risks are more concentrated and more complex. That is why we are pressing for policies that reduce those risks through preparation, adaptation and mitigation. That will be cheaper than covering tomorrow's losses after disaster strikes."

Jeffrey Sachs of Columbia University agrees: "When you think about the scale of risk that the world faces, it is clear that we grossly underinvest in knowledge... we have enough income to live very comfortably in the developed world and to prevent dire need in the developing world. So we should have the confidence to invest in longer-term issues like the environment. Let's help insure the sustainability of this wonderful situation."

He is right. After all, we have only one planet, now and in the future. We need to think harder about how to use it wisely.

Acknowledgements
In addition to those cited in the text, the author would like to thank Robert Socolow, David Victor, Geoffrey Heal, and experts at Tsinghua University, Friends of the Earth, the European Commission, the World Business Council for Sustainable Development, the International Energy Agency, the OECD and the UN for sharing their ideas with him. A list of sources can be found on *The Economist's* website.

Five Meta-Trends Changing the World

Global, overarching forces such as modernization and widespread interconnectivity are converging to reshape our lives. But human adaptability—itself a "meta-trend"—will help keep our future from spinning out of control, assures THE FUTURIST's lifestyles editor.

By David Pearce Snyder

Last year, I received an e-mail from a long-time Australian client requesting a brief list of the "meta-trends" having the greatest impact on global human psychology. What the client wanted to know was, which global trends would most powerfully affect human consciousness and behavior around the world?

The Greek root *meta* denotes a transformational or transcendent phenomenon, not simply a big, pervasive one. A meta-trend implies multidimensional or catalytic change, as opposed to a linear or sequential change.

What follows are five meta-trends I believe are profoundly changing the world. They are evolutionary, system-wide developments arising from the simultaneous occurrence of a number of individual demographic, economic, and technological trends. Instead of each being individual freestanding global trends, they are composites of trends.

Trend 1—Cultural Modernization

Around the world over the past generation, the basic tenets of modern cultures—including equality, personal freedom, and self-fulfillment—have been eroding the domains of traditional cultures that value authority, filial obedience, and self-discipline. The children of traditional societies are growing up wearing Western clothes, eating Western food, listening to Western music, and (most importantly of all) thinking Western thoughts. Most Westerners—certainly most Americans—have been unaware of the personal intensities of this culture war because they are so far away from the "battle lines." Moreover, people in the West regard the basic institutions of modernization, including universal education, meritocracy, and civil law, as benchmarks of social progress, while the defenders of traditional cultures see them as threats to social order.

Demographers have identified several leading social indicators as key measures of the extent to which a nation's culture is modern. They cite the average level of education for men and for women, the percentage of the salaried workforce that is female, and the percentage of population that lives in urban areas. Other indicators include the percentage of the workforce that is salaried (as opposed to self-employed) and the percentage of GDP spent on institutionalized socioeconomic support services, including insurance, pensions, social security, civil law courts, worker's compensation, unemployment benefits, and welfare.

As each of these indicators rises in a society, the birthrate in that society goes down. The principal measurable consequence of cultural modernization is declining fertility. As the world's developing nations have become better educated, more urbanized, and more institutionalized during the past 20 years, their birthrates have fallen dramatically. In 1988, the United Nations forecast that the world's population would double to 12 billion by 2100. In 1992, their estimate dropped to 10 billion, and they currently expect global population to peak at 9.1 billion in 2100. After that, demographers expect the world's population will begin to slowly decline, as has already begun to happen in Europe and Japan.

Three signs that a culture is modern: its citizens' average level of education, the number of working women, and the percentage of the population that is urban. As these numbers increase, the birthrate in a society goes down, writes author David Pearce Snyder.

The effects of cultural modernization on fertility are so powerful that they are reflected clearly in local vital statistics. In India, urban birthrates are similar to those in the United States, while rural birthrates remain unmanageably high. Cultural modernization is the linchpin of human sustainability on the planet.

The forces of cultural modernization, accelerated by economic globalization and the rapidly spreading wireless telecommunications infostructure, are likely to marginalize the world's traditional cultures well before the century is over. And because the wellsprings of modernization—secular industrial economies—are so unassailably powerful, terrorism is the only means by which the defenders of traditional culture can fight to preserve their values and way of life. In the near-term future, most observers believe that ongoing cultural conflict is likely to produce at least a few further extreme acts of terrorism, security measures not withstanding. But the eventual intensity and duration of the overt, vi-

olent phases of the ongoing global culture war are largely matters of conjecture. So, too, are the expert pronouncements of the probable long-term impacts of September 11, 2001, and terrorism on American priorities and behavior.

After the 2001 attacks, social commentators speculated extensively that those events would change America. Pundits posited that we would become more motivated by things of intrinsic value—children, family, friends, nature, personal self-fulfillment—and that we would see a sharp increase in people pursuing *pro bono* causes and public-service careers. A number of media critics predicted that popular entertainment such as television, movies, and games would feature much less gratuitous violence after September 11. None of that has happened. Nor have Americans become more attentive to international news coverage. Media surveys show that the average American reads less international news now than before September 11. Event-inspired changes in behavior are generally transitory. Even if current conflicts produce further extreme acts of terrorist violence, these seem unlikely to alter the way we live or make daily decisions. Studies in Israel reveal that its citizens have become habituated to terrorist attacks. The daily routine of life remains the norm, and random acts of terrorism remain just that: random events for which no precautions or mind-set can prepare us or significantly reduce our risk.

In summary, cultural modernization will continue to assault the world's traditional cultures, provoking widespread political unrest, psychological stress, and social tension. In developed nations, where the great majority embrace the tenets of modernization and where the threats from cultural conflict are manifested in occasional random acts of violence, the ongoing confrontation between tradition and modernization seems likely to produce security mea-

sures that are inconvenient, but will do little to alter our basic personal decision making, values, or day-to-day life. Developed nations are unlikely to make any serious attempts to restrain the spread of cultural modernization or its driving force, economic globalization.

Trend 2—Economic Globalization

On paper, globalization poses the long-term potential to raise living standards and reduce the costs of goods and services for people everywhere. But the short-term marketplace consequences of free trade threaten many people and enterprises in both developed and developing nations with potentially insurmountable competition. For most people around the world, the threat from foreign competitors is regarded as much greater than the threat from foreign terrorists. Of course, risk and uncertainty in daily life is characteristically high in developing countries. In developed economies, however, where formal institutions sustain order and predictability, trade liberalization poses unfamiliar risks and uncertainties for many enterprises. It also appears to be affecting the collective psychology of both blue-collar and white-collar workers—especially males—who are increasingly unwilling to commit themselves to careers in fields that are likely to be subject to low-cost foreign competition.

Strikingly, surveys of young Americans show little sign of xenophobia in response to the millions of new immigrant workers with whom they are competing in the domestic job market. However, they feel hostile and helpless at the prospect of competing with Chinese factory workers and Indian programmers overseas. And, of course, economic history tells us that they are justifiably concerned. In those job markets that supply untariffed international industries, a "comparable global wage" for comparable types of work can be expected to emerge worldwide. This will raise workers' wages for freely traded goods and services in developing nations, while depressing wages for comparable work in mature industrial economies. To earn more than the comparable global wage, labor in developed nations will have to perform *incomparable* work, either in terms of their productivity or the superior characteristics of the goods and services that they produce. The assimilation of mature information technology throughout all production and education levels should make this possible, but developed economies have not yet begun to mass-produce a new generation of high-value-adding, middle-income jobs.

Meanwhile, in spite of the undeniable short-term economic discomfort that it causes, the trend toward continuing globalization has immense force behind it. Since World War II, imports have risen from 6% of world GDP to more than 22%, growing steadily throughout the Cold War, and even faster since 1990. The global dispersion of goods production and the uneven distribution of oil, gas, and critical minerals worldwide have combined to make international interdependence a fundamental economic reality, and corporate enterprises are building upon that reality. Delays in globalization, like the September 2003 World Trade Organization contretemps in Cancun, Mexico, will arise as remaining politically sensitive issues are resolved, including trade in farm products, professional and financial services, and the need for corporate social responsibility. While there will be enormous long-term economic benefits from globalization in both developed and developing nations, the short-term disruptions in local domestic employment will make free trade an ongoing political issue that will be manageable only so long as domestic economies continue to grow.

Trend 3—Universal Connectivity

While information technology (IT) continues to inundate us with miraculous capabilities, it has given

us, so far, only one new power that appears to have had a significant impact on our collective behavior: our improved ability to communicate with each other, anywhere, anytime. Behavioral researchers have found that cell phones have blurred or changed the boundaries between work and social life and between personal and public life. Cell phones have also increased users' propensity to "micromanage their lives, to be more spontaneous, and, therefore, to be late for everything," according to Leysia Palen, computer science professor at the University of Colorado at Boulder.

Cell phones have blurred the lines between the public and the private. Nearly everyone is available anywhere, anytime—and in a decade cyberspace will be a town square, writes Snyder.

Most recently, instant messaging—via both cell phones and online computers—has begun to have an even more powerful social impact than cell phones themselves. Instant messaging initially tells you whether the person you wish to call is "present" in cyberspace—that is, whether he or she is actually online at the moment. Those who are present can be messaged immediately, in much the same way as you might look out the window and call to a friend you see in the neighbor's yard. Instant messaging gives a physical reality to cyberspace. It adds a new dimension to life: A person can now be "near," "distant," or "in cyberspace." With video instant messaging—available now, and widely available in three years—the illusion will be complete. We will have achieved what Frances Cairncross, senior editor of *The Economist*, has called "the death of distance."

Universal connectivity will be accelerated by the integration of the

telephone, cell phone, and other wireless telecom media with the Internet. By 2010, all long-distance phone calls, plus a third of all local calls, will be made via the Internet, while 80% to 90% of all Internet access will be made from Web-enabled phones, PDAs, and wireless laptops. Most important of all, in less than a decade, one-third of the world's population—2 billion people—will have access to the Internet, largely via Web-enabled telephones. In a very real sense, the Internet will be the "Information Highway"—the infrastructure, or infostructure, for the computer age. The infostructure is already speeding the adoption of flexplace employment and reducing the volume of business travel, while making possible increased "distant collaboration," outsourcing, and offshoring.

Corporate integrity and openness will grow steadily under pressure from watchdog groups and ordinary citizens demanding business transparency. The leader of tomorrow must adapt to this new openness or risk business disaster.

As the first marketing medium with a truly global reach, the Internet will also be the crucible from which a global consumer culture will be forged, led by the first global youth peer culture. By 2010, we will truly be living in a global village, and cyberspace will be the town square.

Trend 4—Transactional Transparency

Long before the massive corporate malfeasance at Enron, Tyco, and WorldCom, there was a rising global movement toward greater transparency in all private and public enterprises. Originally aimed at kleptocratic regimes in Africa and the former Soviet states, the movement has now become universal, with the establishment of more stringent international accounting standards and more comprehensive rules for corporate oversight and record keeping, plus a new UN treaty on curbing public-sector corruption. Because secrecy breeds corruption and incompetence, there is a growing worldwide consensus to expose the principal transactions and decisions of *all* enterprises to public scrutiny.

But in a world where most management schools have dropped all ethics courses and business professors routinely preach that government regulation thwarts the efficiency of the marketplace, corporate and government leaders around the world are lobbying hard against transparency mandates for the private sector. Their argument: Transparency would "tie their hands," "reveal secrets to their competition," and "keep them from making a fair return for their stockholders."

Most corporate management is resolutely committed to the notion that secrecy is a necessary concomitant of leadership. But pervasive, ubiquitous computing and comprehensive electronic documentation will ultimately make all things transparent, and this may leave many leaders and decision makers feeling uncomfortably exposed, especially if they were not provided a moral compass prior to adolescence. Hill and Knowlton, an international public-relations firm, recently surveyed 257 CEOs in the United States, Europe, and Asia regarding the impact of the Sarbanes-Oxley Act's reforms on corporate accountability and governance. While more than 80% of respondents felt that the reforms would significantly improve corporate integrity, 80% said they also believed the reforms would not increase ethical behavior by corporate leaders.

While most consumer and public-interest watchdog groups are demanding even more stringent regulation of big business, some corporate reformers argue that regu-

lations are often counterproductive and always circumventable. They believe that only 100% transparency can assure both the integrity and competency of institutional actions. In the world's law courts—and in the court of public opinion—the case for transparency will increasingly be promoted by nongovernmental organizations (NGOs) who will take advantage of the global infostructure to document and publicize environmentally and socially abusive behaviors by both private and public enterprises. The ongoing battle between institutional and socioecological imperatives will become a central theme of Web newscasts, Netpress publications, and Weblogs that have already begun to supplant traditional media networks and newspaper chains among young adults worldwide. Many of these young people will sign up with NGOs to wage undercover war on perceived corporate criminals.

In a global marketplace where corporate reputation and brand integrity will be worth billions of dollars, businesses' response to this guerrilla scrutiny will be understandably hostile. In their recently released *Study of Corporate Citizenship*, Cone/Roper, a corporate consultant on social issues, found that a majority of consumers "are willing to use their individual power to punish those companies that do not share their values." Above all, our improving comprehension of humankind's innumerable interactions with the environment will make it increasingly clear that total transparency will be crucial to the security and sustainability of a modern global economy. But there will be skullduggery, bloodshed, and heroics before total transparency finally becomes international law—15 to 20 years from now.

Trend 5—Social Adaptation

The forces of cultural modernization—education, urbanization, and institutional order—are producing social change in the developed world

as well as in developing nations. During the twentieth century, it became increasingly apparent to the citizens of a growing number of modern industrial societies that neither the church nor the state was omnipotent and that their leaders were more or less ordinary people. This realization has led citizens of modern societies to assign less weight to the guidance of their institutions and their leaders and to become more self-regulating. U.S. voters increasingly describe themselves as independents, and the fastest-growing Christian congregations in America are nondenominational.

Since the dawn of recorded history, societies have adapted to their changing circumstances. Moreover, cultural modernization has freed the societies of mature industrial nations from many strictures of church and state, giving people much more freedom to be individually adaptive. And we can be reasonably certain that modern societies will be confronted with a variety of fundamental changes in circumstance during the next five, 10, or 15 years that will, in turn, provoke continuous widespread adaptive behavior, especially in America.

Reaching retirement age no longer always means playing golf and spoiling the grandchildren. Seniors in good health who enjoy working probably won't retire, slowing the prophesied workforce drain, according to author David Pearce Snyder

---

During the decade ahead, *infomation*—the automated collection, storage, and application of electronic data—will dramatically reduce paperwork. As outsourcing and offshoring eliminate millions of U.S. middle-income jobs, couples are likely to work two lower-pay/lower-skill jobs to replace lost income. If our employers ask us to work from home to reduce the company's office rental costs, we will do so, especially if the arrangement permits us to avoid two hours of daily commuting or to care for our offspring or an aging parent. If a wife is able to earn more money than her spouse, U.S. males are increasingly likely to become househusbands and take care of the kids. If we are in good health at age 65, and still enjoy our work, we probably won't retire, even if that's what we've been planning to do all our adult lives. If adult children must move back home after graduating from college in order to pay down their tuition debts, most families adapt accordingly.

Each such lifestyle change reflects a personal choice in response to an individual set of circumstances. And, of course, much adaptive behavior is initially undertaken as a temporary measure, to be abandoned when circumstances return to normal. During World War II, millions of women voluntarily entered the industrial workplace in the United States and the United Kingdom, for example, but returned to the domestic sector as soon as the war ended and a prosperous normalcy was restored. But the Information Revolution and the aging of mature industrial societies are scarcely temporary phenomena, suggesting that at least some recent widespread innovations in lifestyle —including delayed retirements and "sandwich households"—are precursors of long-term or even permanent changes in society.

The current propensity to delay retirement in the United States began in the mid-1980s and accelerated in the mid-1990s. Multiple surveys confirm that delayed retirement is much more a result of increased longevity and reduced morbidity than it is the result of financial necessity. A recent AARP survey, for example,

found that more than 75% of baby boomers plan to work into their 70s or 80s, regardless of their economic circumstances. If the baby boomers choose to age on the job, the widely prophesied mass exodus of retirees will not drain the workforce during the coming decade, and Social Security may be actuarially sound for the foreseeable future.

The Industrial Revolution in production technology certainly produced dramatic changes in society. Before the steam engine and electric power, 70% of us lived in rural areas; today 70% of us live in cities and suburbs. Before industrialization, most economic production was home- or family-based; today, economic production takes place in factories and offices. In preindustrial Europe and America, most households included two or three adult generations (plus children), while the great majority of households today are nuclear families with one adult generation and their children.

Current trends in the United States, however, suggest that the three great cultural consequences of industrialization—the urbanization of society, the institutionalization of work, and the atomization of the family—may all be reversing, as people adapt to their changing circumstances. The U.S. Census Bureau reports that, during the 1990s, Americans began to migrate out of cities and suburbs into exurban and rural areas for the first time in the twentieth century. Simultaneously, information work has begun to migrate out of offices and into households. Given the recent accelerated growth of telecommuting, self-employment, and contingent work, one-fourth to one-third of all gainful employment is likely to take place at home within 10 years. Meanwhile, growing numbers of baby boomers find themselves living with both their debt-burdened, underemployed adult

children and their own increasingly dependent aging parents. The recent emergence of the "sandwich household" in America resonates powerfully with the multigenerational, extended families that commonly served as society's safety nets in preindustrial times.

Leadership in Changing Times

The foregoing meta-trends are not the only watershed developments that will predictably reshape daily life in the decades ahead. An untold number of inertial realities inherent in the common human enterprise will inexorably change our collective circumstances—the options and imperatives that confront society and its institutions. Society's adaptation to these new realities will, in turn, create further changes in the institutional operating environment, among customers, competitors, and constituents. There is no reason to believe that the Information Revolution will change us any less than did the Industrial Revolution.

In times like these, the best advice comes from ancient truths that have withstood the test of time. The Greek philosopher-historian Heraclitus observed 2,500 years ago that "nothing about the future is inevitable except change." Two hundred years later, the mythic Chinese general Sun Tzu advised that "the wise leader exploits the inevitable." Their combined message is clear: "The wise leader exploits change."

David Pearce Snyder is the lifestyles editor of THE FUTURIST and principal of The Snyder Family Enterprise, a futures consultancy located at 8628 Garfield Street, Bethesda, Maryland 20817. Telephone 301-530-5807; e-mail davidpearcesnyder@earthlink.net; Web site www.the-futurist.com.

Crimes of (a) Global Nature

**FORGING ENVIRONMENTAL TREATIES
IS DIFFICULT.
ENFORCING THEM IS EVEN TOUGHER.**

by Lisa Mastny and Hilary French

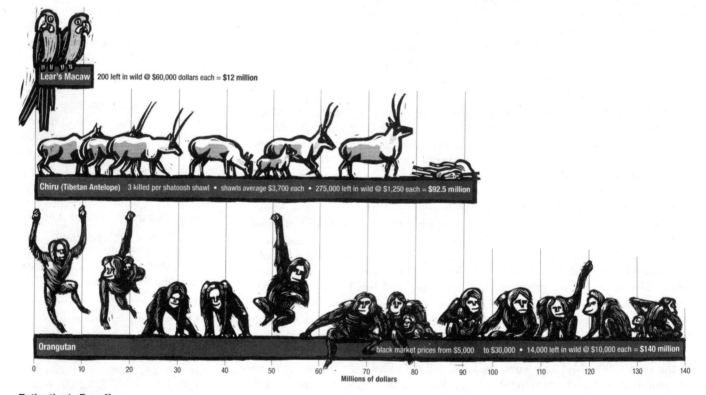

Lear's Macaw 200 left in wild @ $60,000 dollars each = **$12 million**

Chiru (Tibetan Antelope) 3 killed per shatoosh shawl • shawls average $3,700 each • 275,000 left in wild @ $1,250 each = **$92.5 million**

Orangutan black market prices from $5,000 to $30,000 • 14,000 left in wild @ $10,000 each = **$140 million**

0 10 20 30 40 50 60 70 80 90 100 110 120 130 140
Millions of dollars

Extinction's Payoff
Pets, aphrodisiacs, distinctive clothing: these are a few of our favorite things—even if having them drives a species or two to extinction. The staggering prices some threatened animals fetch on the black market create powerful incentives for illegal trafficking and help increase the risk of extinction. The prices in the graph are probably conservative, since prices would tend to rise with increasing scarcity.

Last February, armed troops and fisheries officials on two Australian navy ships and a helicopter boarded and seized the Volga and the Lena, Russian-flagged fishing vessels operating near Heard Island, some 2,200 nautical miles southwest of Perth. The two ships were found to be carrying about 200 tons of illegally caught Patagonian toothfish in their holds. This bounty, valued at an estimated $1.25 million, had been taken in violation of conservation agreements negotiated under the auspices of the Commission for the Conservation of Antarctic Marine Living Resources.

Few casual seafood lovers have heard of Patagonian toothfish, but many are familiar with Chilean sea bass, a different name for the same fish. Chilean sea bass began appearing on menus in the early 1990s, and consumption of the flaky, white fish took off fast, quickly endangering the health of the fishery. Large-scale commercial fishing of the species began only a decade ago, but scientists estimate that at current rates of plunder the fish could become commercially extinct in less than five years.

A few months after the drama in the Southern Ocean, a different front in the same battle opened up thousands of miles away in Washington, D.C., where nearly 60 restaurants and caterers pledged to keep the fish off their menus. More than 90 restaurants in the Los Angeles area did the same a few weeks later, following similar promises by chefs in Northern California, Chicago, and Houston. Thus the fight to save the Patagonian toothfish is beginning to hit close to home. But many Chilean sea bass fans remain unaware that they may be accomplices to a growing phenomenon known as international environmental crime.

Although variously defined, in this article international environmental crime means an activity that violates the letter or the spirit of an international environmental treaty and the implementing national legislation. Trade in endangered species, illegal fishing, CFC smuggling, and the illicit dumping of wastes are all cases in point that are explored below. Illegal logging is another major category of environmental crime, although environmental treaties currently impose few specific constraints on logging (see Box, *Logging Illogic*). The rapidly growing illegal trade in these environmentally sensitive products stems from strong demand, low risk, and other factors (see Box, *Variable Crimes, Constant Incentives*).

The number of international environmental accords has exploded as countries awaken to the seriousness of transboundary and global ecological threats. The UN Environment Programme (UNEP) estimates that there are now more than 500 international treaties and other agreements related to the environment, more than 300 of them negotiated in the last 30 years.

But reaching such agreements is only the first step. The larger challenge is seeing that the ideals expressed in them become reality. What is needed is not necessarily more agreements, but a commitment to breathe life into the hundreds of existing accords by implementing and enforcing them.

Here the genteel world of diplomacy often runs into hard-nosed domestic politics. Countries that ratify treaties are responsible for upholding them by enacting and enforcing the necessary domestic laws. This requires the backing of businesses, consumers, and other constituencies, which may not be easily secured. Countries with strong fossil fuel industries, for instance, may meet staunch resistance to international rules to mitigate climate change. And countries where natural resource industries are politically powerful will probably find it difficult to adequately enforce environmental treaties designed to regulate resource-related activity. The effect has been to expand trafficking in a number of restricted substances, an increasingly urgent problem that is beginning to stimulate a stronger international response.

Trading in Wildlife

Undercover Russian police officers in the port city of Vladivostok recently trailed two investigators from environmental groups posing as eager purchasers of Siberian tiger skins from a corrupt official. When the deal went down, the officers arrested the wildlife trader on the spot. Russian investigators earlier had infiltrated the wildlife trade crime ring and determined that it was raking in some $5 million a year from smuggling wild ginseng, tiger skins, and bear paws and gallbladders across the Russian border.

Trade in these wildlife products is restricted under the terms of the 1973 Convention on International Trade in Endangered Species of Wild Fauna and Flora (CITES), which bans international trade in some 900 animal and plant species in danger of extinction, including all tigers, great apes, and sea turtles, and many species of elephants, orchids, and crocodiles. CITES also restricts trade in some 29,000 additional species that are threatened by commerce; among them birdwing butterflies, parrots, black and stony corals, and some hummingbirds.

CITES has shrunk the trade in many threatened species, including cheetahs, chimpanzees, crocodiles, and elephants. But the trafficking in these and other species continues, earning smugglers profits of $8 billion to $12 billion annually. Among the most coveted black market items are tigers and other large cats, rhinos, reptiles, rare birds, and botanical specimens.

Most illegally traded wildlife originates in developing countries, home to most of the world's biological diversity. Brazil alone supplies some 10 percent of the global black market, and its nonprofit wildlife-trade monitoring body, RENCTAS, estimates that poachers steal some 38 million animals a year from the country's Amazon forests, Pantanal wetlands, and other important habitats, generating annual revenues of $1 billion. Southeast Asian wildlife has also been plundered: the Gibbon Foundation reports that in a single recent year, traders smuggled out some 2,000 orangutans from Indonesia—at an average street price of $10,000 apiece.

The demand for illegal wildlife—for food and medicine, as clothing and ornamentation, for display in zoo collections and horticulture, and as pets—comes primarily from wealthy collectors and other consumers in Europe, North America, Asia, and the Middle East. In the United States (the world's largest market for reptile trafficking) exotic pets such as the Komodo dragon of Indonesia, the plowshare tortoise of northeast Madagascar, and the tuatara (a small lizard-like reptile from New Zealand) reportedly sell for as much as $30,000 each on the black market. Wildlife smuggling is also a growing concern in the United Kingdom, where in 1999 alone, customs officials confiscated some 1,600 live animal and birds, 1,800 plants, 52,000 parts and derivatives of endangered species, and 388,000 grams of smuggled caviar.

The multibillion-dollar traditional Asian medicine industry has also been a strong source of demand for illegal wildlife, with adherents from Beijing to New York purchasing potions made from ground tiger bone, rhino horn, and other wildlife derivatives for their alleged effect on ailments from impotence to asthma. In parts of Asia, bile from bear gall bladders (used to

VARIABLE CRIMES, CONSTANT INCENTIVES

International environmental crime involves many different kinds of activities and contraband, ranging from illegal fishing and trading in endangered wildlife to smuggling ozone-depleting chlorofluorocarbons (CFCs) across borders. But in most cases there are several common elements:

• **Booming demand, minimal investment, high profits.** Such crime is generally very attractive to traders and smugglers. Demand for increasingly rare plants, fish, or CFCs is strong. Investment can be minimal and the rewards great, with prices climbing substantially as the items change hands. For instance, a Senegalese wholesaler may buy a grey parrot in Gabon for $16–20 and sell it to a European wholesaler for $300–360, who then gets $600–1,200 for it.

• **Low risk, low penalties.** Environmental crimes normally carry low risk, with domestic penalties often light or nonexistent. A 2000 study by the secretariat of the Convention on International Trade in Endangered Species of Wild Fauna and Flora found that roughly half of the treaty's 158 parties failed to implement the treaty adequately. A common deficiency is a lack of appropriate penalties to deter treaty violations.

• **Creative smuggling methods.** Smugglers have devised clever ways to evade controls on restricted items, from using unauthorized crossing points to shipping illegal items along with legal consignments.

• **Links to organized crime.** Environmental criminals sometimes launder or funnel the proceeds from their trafficking to other illicit activities, such as buying drugs or weapons. The National Network for Combating the Traffic of Wild Animals (RENCTAS), a Brazilian NGO, reports that as much as 40 percent of the 300 to 400 criminal groups that control that country's wildlife trade also have links with drug trafficking.

• **Weak domestic treaty enforcement.** Because treaty secretariats have little centralized authority, signatory countries are expected to designate their own permitting authorities and train local customs inspectors and police to detect and penalize illegal activity. But many countries simply don't have the resources to do this.

• **Porous borders.** Many countries lack the money, equipment, or political will to effectively monitor illegal activity at border crossings. In the developing world in particular, customs offices are chronically understaffed and underfunded. Agents are often untrained to spot crimes or overwhelmed by the sheer numbers of items they must track.

• **Permitting challenges.** Local authorities may unwittingly issue permits for animals, seafood, chemicals, or other items they don't know are restricted, or let consignments slip through because they can't identify a fake or altered permit. Corrupt local officials may issue false permits or overlook illicit consignments in return for bribes or kickbacks.

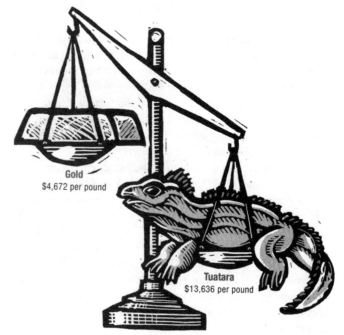

Gold
$4,672 per pound

Tuatara
$13,636 per pound

Scarcity=Value
The tuatara has a third eye on top of its skull and can live to be 100 years old. Its cousins in the reptilian order Rhynchocephalia all died out 139 million years ago. These oddities may help explain why collectors will pay up to $30,000 apiece. At least one species is nominally protected under CITES.

treat cancers, asthma, eye disease, and other afflictions) can be worth more than narcotics. The New York-based Wildlife Conservation Society reports that the illegal hunting and trafficking of animals for medicine, aphrodisiacs, and gourmet food is now the single greatest threat to endangered species in Asia.

The scale of the illegal wildlife trade reflects the serious obstacles to enforcement under CITES. Smugglers may conceal the items on their persons or in vehicles, baggage, or postal and courier shipments, resulting in fatality rates as high as 90 percent for many live species. They may also alter the required CITES permits to indicate a different quantity, type, or destination of species, or to change the appearance of items so they appear ordinary. In February 2000, the U.S. Fish and Wildlife Service arrested a Cote d'Ivoire man for smuggling 72 elephant ivory carvings, valued at $200,000, through New York's John F. Kennedy airport. Many of them had been painted to resemble stone.

As wildlife smuggling grows in sophistication, it accounts for a rapidly rising share of international criminal activity. RENCTAS now ranks the illicit wildlife trade as the third largest illegal cross-border activity, after the arms and drug trades. Wildlife smugglers commonly rely on the same international trafficking routes as dealers of other contraband goods, such as gems and drugs. In the United States, consignments of snakes have been found stuffed with cocaine, and illegally traded turtles have entered on the same boats as marijuana. The U.S.-Mexico border has become a significant transfer point for environmental contraband: in the first eight months of 2001, Mexican authorities reported seizing more than 50,000 smuggled animals en route to the U.S. border.

As the scope of international wildlife trafficking becomes clearer, authorities are beginning to take action on the domestic, regional, and global fronts. In Europe, officials are working together to improve regional cooperation in enforcement of CITES and to strengthen legal mechanisms for prosecuting violators. The United Kingdom has established a new national police unit to combat wildlife crime and announced tougher penalties for persistent offenders. And in the first program of its kind in East Asia, a team at South Korea's Seoul airport now uses specially trained dogs to detect tiger bone, musk, bear gall bladders, and other illegal wildlife derivatives smuggled in luggage and freight.

In general, developing countries have had greater difficulty in controlling wildlife smuggling and implementing CITES. Many lack the political will, money, or equipment to effectively monitor their wildlife populations, much less the wildlife trade. In Kenya, where elephant deaths declined dramatically following a CITES-imposed ban on the international ivory trade in the late 1980s, poaching is again surging as funds to hire additional game wardens have run dry. In two separate incidents in April 2002, poachers armed with automatic weapons slaughtered 25 elephants in the country's wildlife parks and removed the tusks for sale on the black market.

At the international level, one tool that has proved successful in enticing (some would say coercing) countries to uphold their obligations under CITES is the use of trade sanctions. CITES is empowered to recommend that its members temporarily suspend all wildlife trade with noncomplying countries. Within the past year, such sanctions have been levied against the United Arab Emirates for not taking strong enough measures to combat the illicit trade in falcons, against Russia for not cracking down on the illegal caviar trade, and against Fiji and Vietnam for not enacting adequate wildlife trade legislation by the required deadline. In most instances, the governments scrambled to strengthen their legislation and enforcement, and the sanctions were soon lifted.

High Seas, High Crimes

Although CITES focuses mainly on terrestrial animals and plants, it protects several highly endangered fish species as well. In February 2002, the Philippine navy arrested 95 Chinese fishermen near a national marine park in the Sulu Sea and charged them with multiple counts of poaching fish, harvesting endangered species, and using illegal fishing methods like poison and explosives. Among the species found aboard their four vessels were endangered sea turtles and giant clams, both of which are prohibited from trade under CITES.

Countries have negotiated a wide range of other agreements to oversee and regulate the world's fisheries. Like CITES, many of these are poorly enforced, resulting in illegal fishing in all types of fisheries and in all the world's oceans, including national waters, regionally managed fisheries, and the high seas. In some of the world's most important fisheries, as much as 30 percent of the catch is illegal, according to the UN Food and Agriculture Organization (FAO).

As with other forms of wildlife trade, booming consumer demand is an important driver behind the rise in this activity. Big profits can be made selling black market seafood to selective buyers, who are willing to pay a premium for increasingly rare items. In Japan, species such as the threatened Southern bluefin tuna now fetch up to $50,000 per fish.

One major form of illegal fishing occurs when foreign vessels fish without authorization in the waters of other countries, often developing nations unable to patrol their shores adequately. In early 2000, the Tanzanian government estimated that more than 70 vessels, most of them from Mediterranean countries and the Far East, were fishing illegally in its waters. Tanzania has few police boats of its own and has had to rely on assistance from France and the United Kingdom to crack down on offenders. Mozambique, Somalia, and other African countries also report increased illegal fishing, often by heavily armed foreign vessels, and worry that the continued poaching will damage national economies and deprive coastal villagers of the healthy fisheries they depend upon for their livelihoods.

Commercial fishers are also turning to the largely unmonitored high seas, including the Mediterranean Sea and the Indian, South Atlantic, and Southern Oceans. Their ships illegally penetrate the borders of regional fishing grounds that are closed or restricted under international law, and deliberately hide their flags or other markings to avoid recognition. Some offload their illicit catch to other vessels to further disguise its origin and to minimize the penalties if discovered. Often these fishers do not report their activity, or if they do, they may falsify the equipment used, the fishing area frequented, or the species or amount caught.

TRAFFIC, a UK-based nonprofit group that monitors the international wildlife trade, reports that illegal Russian trawlers are accelerating the collapse of once productive fisheries in the Bering Sea. Backed by the Russian mafia, the fishers remove billions of dollars worth of pollack, cod, herring, flounder, halibut, and other species from the ecosystem each year, often trawling in prohibited areas and using illegal nets and other gear. They then transfer the catches to ships bound for ports in the United States, Canada, and Asia, in particular the South Korean port of Pusan, where vessel inspections are rare.

One of the most serious challenges to adequate fisheries enforcement is the rapid rise in so-called flag-of-convenience (FOC) fishing. Increasingly, commercial fishing companies register their ships in countries known to be lax enforcers of international fisheries laws or that are not members of major maritime agreements. By transferring their allegiance to these new "flags," the companies can easily enter the waters of their adopted countries or operate undercover on the high seas where only the country of registry (the flag state) can make an official arrest. Some vessels change their flags frequently in order to hide their origins or identities, making it difficult for enforcement officials to track them down.

An estimated 5 to 10 percent of the vessels in the world fishing fleet now fly flags of convenience. This includes more than 1,300 large industrial fishing vessels registered in such countries as Belize, Honduras, Panama, St. Vincent, and Equatorial Guinea.

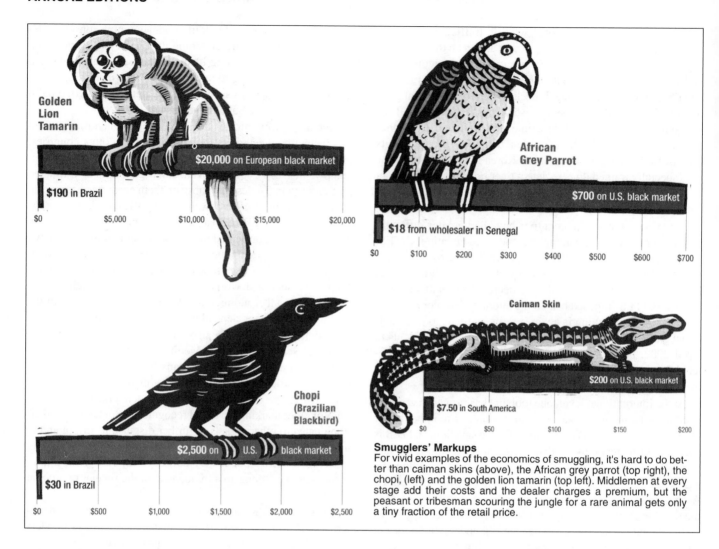

Golden Lion Tamarin
$20,000 on European black market
$190 in Brazil
$0 $5,000 $10,000 $15,000 $20,000

African Grey Parrot
$700 on U.S. black market
$18 from wholesaler in Senegal
$0 $100 $200 $300 $400 $500 $600 $700

Chopi (Brazilian Blackbird)
$2,500 on U.S. black market
$30 in Brazil
$0 $500 $1,000 $1,500 $2,000 $2,500

Caiman Skin
$200 on U.S. black market
$7.50 in South America
$0 $50 $100 $150 $200

Smugglers' Markups
For vivid examples of the economics of smuggling, it's hard to do better than caiman skins (above), the African grey parrot (top right), the chopi, (left) and the golden lion tamarin (top left). Middlemen at every stage add their costs and the dealer charges a premium, but the peasant or tribesman scouring the jungle for a rare animal gets only a tiny fraction of the retail price.

The countries and companies that own the reflagged vessels are also partially to blame for the problem. Fishing companies from Europe (primarily Spain) and Taiwan own the highest numbers of FOC vessels. Many of these firms receive special government subsidies to register their ships abroad.

FOC vessels also pose a serious problem in the cruise and shipping industries. Freight companies may re-flag their vessels in order to avoid higher taxes, labor, and operating costs at home, or to take advantage of the lower safety standards and weaker pollution laws in countries like Liberia, Panama, and Malta. The International Transport Workers' Federation reports that FOC ships accounted for nearly a quarter of all large freight vessels in 2000, and carried 53 percent of the world's gross tonnage. The re-flagged ships account for a disproportionate share of pollution and accidents at sea, as well as detentions in ports for violations of maritime laws.

Some of the most serious offenses have been breaches of MARPOL, the international treaty regulating the disposal of garbage and other pollutants at sea. Vessel owners may decide to dump their waste overboard illegally either because they lack the storage or incineration capacity onboard or because they wish to avoid the high costs of eventual disposal on land. Often this at-sea discharge occurs in the waters of countries where monitoring and enforcement are minimal and the activity can go undetected. In 1999, however, U.S. officials fined Royal Caribbean Cruises Ltd. a record $18 million for releasing oil waste from its ships into U.S. waters, among other violations. In April 2002, rival Carnival Corporation was also fined $18 million for similar offenses.

As the scope of maritime violations becomes increasingly evident, many countries are redoubling their efforts to combat illegal fishing, including by exchanging information about illicit activity, making vessel registries more transparent, increasing inspections at sea and in ports, denying landing and trans-shipment rights for illegally caught fish, and improving monitoring and surveillance of fleets. These steps appear to be working, at least in some places: in the North Pacific, Canadian officials attribute the recent drop in the number of vessels using illegal driftnets to stepped-up air patrols by Canada and its partners (the United States, Russia, and China) in regional enforcement.

Significant effort has also been made at the global level to crack down on illegal fishing. In March of last year, 114 countries agreed to a non-binding plan to combat illicit activity, including that by flag-of-convenience vessels. Developed under the auspices of the UN Food and Agriculture Organization, the plan calls for improved oversight of vessels and coastal waters by flag states, better coordination and sharing of information

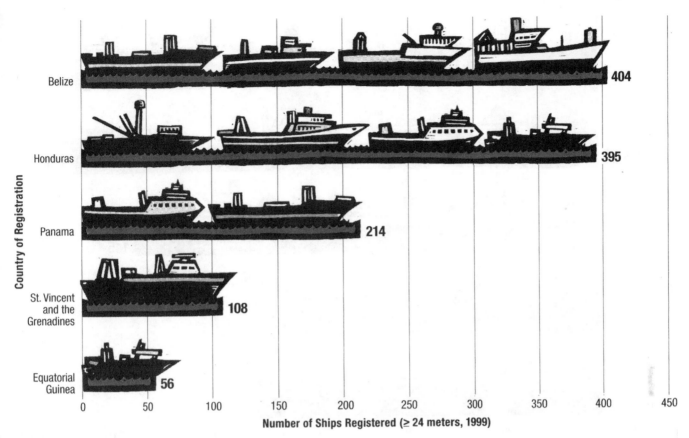

Country of Registration (y-axis)

- Belize — 404
- Honduras — 395
- Panama — 214
- St. Vincent and the Grenadines — 108
- Equatorial Guinea — 56

Number of Ships Registered (≥ 24 meters, 1999) (x-axis: 0, 50, 100, 150, 200, 250, 300, 350, 400, 450)

False Colors
Treaties to control overfishing and other marine crimes generally require the nation where a ship is registered to do the policing. Commercial fishers often register their ships with countries unable, or deliberately reluctant, to enforce the treaties. Such flag-of-convenience (FOC) vessels can operate with little risk of prosecution, which leads to overfishing and depletion of fish stocks. As much as one-tenth of the world's fishing vessels fly flags of convenience. The top five FOC registration countries are shown here, and the opposite page shows the top five countries where firms owning the vessels are based.

among countries, and stronger efforts to ratify, implement, and enforce existing fisheries accords. The plan is expected to work in conjunction with other international efforts to protect global fisheries, including a 1993 pact known as the Compliance Agreement and the 1995 UN Agreement Relating to the Conservation and Management of Straddling Fish Stocks and Highly Migratory Fish Stocks.

Dumping on Land

A Japanese district court last March sentenced Hiromi Ito, the president of a waste disposal company, to four years in prison and slapped him with a 5 million yen ($40,000) fine. Ito and his accomplices had masterminded an elaborate scheme to illegally dump some 2,700 tons of industrial and medical waste in the Philippines, in containers marked 'paper for recycling.' When the Filipino importer opened the 122 shipping containers, each 40 feet long, he found not just paper but also hazardous materials, including contaminated hypodermic needles and bandages, used plastic sheeting, and old electronics equipment.

With Ito's conviction, the Japanese government closed the books on an embarrassing international incident. But the episode points to a much larger global challenge: effectively regulating the 300 to 500 million tons of hazardous waste generated

worldwide each year. This vast waste mountain includes everything from used batteries, electronic wastes, old ships, and toxic incinerator ash to industrial sludge and contaminated medical and military equipment.

Roughly 10 percent of this waste is shipped legally across international borders, under the terms of the 1989 Basel Convention on the Control of Transboundary Movements of Hazardous Wastes and Their Disposal. The agreement requires member countries to obtain prior consent from the importing country for waste shipments and uses a system of permits to track the pathway to disposal. Most of the waste originates in and moves among industrial countries, but some also travels to and within the developing world.

As waste disposal problems mount in countries like Japan, so does the likelihood of a lucrative illegal trade in hazardous wastes. Asia is thought to be one of the biggest destinations for the illicit waste. Greenpeace India reports that more than 100,000 tons of unauthorized wastes entered India in 1998 and 1999, including toxic zinc ash and residues, lead waste, used batteries, and scraps of chromium, cadmium, thallium, and other heavy metals. The illicit imports, originating in places like Australia, Belgium, Germany, Norway, and the United States, violated both the Basel treaty's notification rules and a 1997 Indian government ban on waste imports.

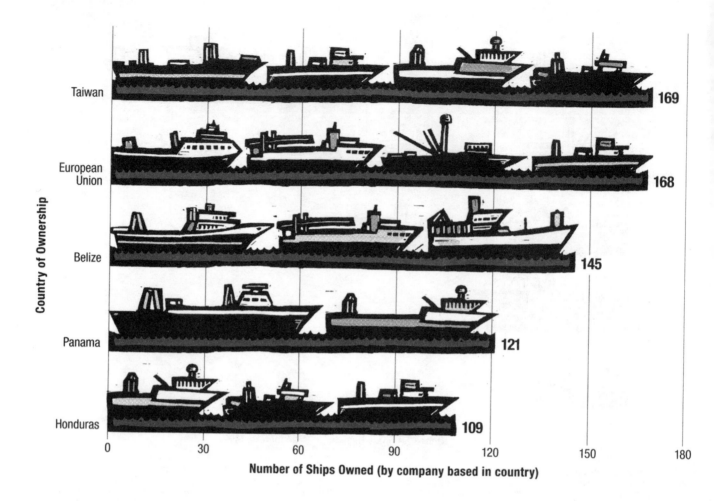

Number of Ships Owned (by company based in country)

China receives a massive flow of illegally imported hazardous electronic waste each year. A recent study by the Basel Action Network (BAN), a watchdog group that monitors implementation of the Basel treaty, and the California-based Silicon Valley Toxics Coalition reported that workers in Chinese recycling factories risk serious exposure to heavy metals and other poisonous chemicals as they salvage components from old computer circuitboards, monitors, batteries, and other equipment. This toxic trade continues despite a Chinese import ban on the material, and hence violates Basel rules that forbid the export of wastes to countries that have banned their import.

Roughly half of this "e-waste" originates in the United States, where an estimated 20 million computers become obsolete each year. Yet the U.S. government doesn't consider the high-tech shipments to be illegal because the waste isn't technically classified as hazardous. The United States is also the only industrial country that hasn't ratified the Basel Convention.

Developing countries are particularly vulnerable to the health and environmental effects associated with the illegal waste trade. Many governments lack the infrastructure or equipment to dispose of or recycle waste safely, prevent exposure to workers and communities, clean up dumpsites, or monitor waste movements. For this reason, a bloc of developing countries secured passage in 1995 of a far-reaching amendment to the Basel Convention, known as the Basel Ban, which would

outlaw all transfers of hazardous wastes from industrial countries to the developing world. Though many countries already observe its terms voluntarily and the European Union has already implemented it, the Ban is not yet in strict legal force and still faces serious opposition from a few industrialized countries, including the United States.

Even so, the Basel Ban has helped slow the flow of hazardous wastes from industrial countries to the developing world, says Jim Puckett, coordinator of BAN. And once the amendment enters into legal force (it needs 32 more ratifications), it should be even harder for waste traders to operate, as violators would face strict criminal penalties. If the ban fails to enter into force, however, or if it is weakened through continued opposition, the waste flood could resume.

But the ban alone would likely fail to wipe out the illegal waste trade. Smugglers rely on a wide range of tactics, including false permits, bribes, and mislabeling of wastes as raw materials, less dangerous substances, or other products, to evade the laws. Moreover, no port in the world can check all sea-going containers, let alone developing-country ports.

One growing trend is the export of hazardous wastes under the pretext of "recycling." Like Hiromi Ito, illicit waste dealers increasingly pass off waste as recyclable material, which in many cases frees them from strict government oversight. Greenpeace estimates that as much as 90 percent of waste shipped to the developing world—particularly plastics and

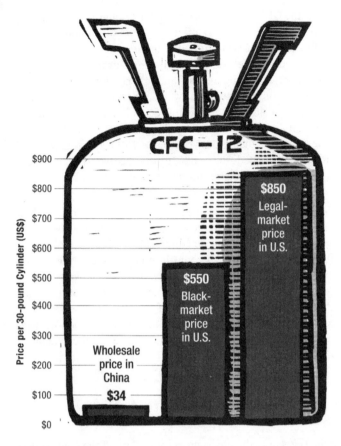

The Price of Cool
The Montreal Protocol ended CFC manufacture in the industrialized world, but developing nations have until 2010 to phase it out. Some of their output feeds illicit demand in the United States and Europe.

countries by agreeing to set up a centralized system for reporting suspect activity. The treaty also requires members to pass laws to curb and punish illegal waste traffic and outlines how to handle illegally traded waste once it is discovered.

A new liability protocol to the convention, negotiated in December 1999, could further discourage illegal activity—if it ever comes into force. It makes exporters and disposers of hazardous waste liable for any harm that might occur during transport, both legal and illegal. It also requires dealers to be insured against the damage and to provide financial compensation to those affected. But the protocol will not be legally binding until 20 ratifications are received (none have been registered so far). Moreover, it still only covers harm that occurs in transit, not after disposal, and only applies to damage suffered in the jurisdiction of a treaty member.

CFCs on the Loose

The landmark Montreal Protocol on Substances That Deplete the Ozone Layer, adopted in 1987, mandated far-reaching restrictions in the use of certain chemicals that damage the thin, vital veil of stratospheric ozone that protects the earth and its inhabitants from excessive ultraviolet radiation. The Protocol and its later amendments set target dates for the phaseout of 96 different ozone-depleting substances, most notably CFCs and halons, chemicals once widely used in a range of industrial applications. Industrialized countries were required to halt production and import of CFCs in 1996, while developing countries have until 2010 to complete the phaseout.

Considered one of the world's most successful environmental treaties, the Protocol has resulted in a dramatic decline in the overall use of ozone-depleting substances. But starting in the mid-1990, the different phaseout schedules for industrial and developing countries helped stoke a flourishing illegal trade in the banned chemicals. CFCs that were still legally produced in the developing world began to make their way to lucrative black markets in the United States and Europe, where demand for substances like Freon (used in older-model auto air conditioners) remained high.

Government and industry reports suggest that in the mid-1990s, as much as 15 percent of global annual production of the chemicals, or 38,000 tons, was smuggled into industrial countries. Early shipments originated in Russia, but today the bulk of the smuggled CFC supply is thought to originate in China and India, which together account for more than half of the world's remaining CFC production.

CFC importers resort to fraud and other evasive tactics to smuggle the banned chemicals. For instance, traders abuse existing loopholes in the Montreal Protocol and domestic laws to pass off shipments of new CFCs as recycled material or as CFC replacements, neither of which are restricted under the treaty. The chemicals are typically colorless and odorless, making them easy to disguise and virtually impossible to differentiate without chemical analysis.

In the United States, the illegal CFC trade is believed to have peaked in the mid-1990s, just after the initial phaseout. At that

heavy metals—is now labeled as destined for recycling. Much of this waste, however, is never recycled—or, as with e-waste in China, is recycled in highly polluting operations that are little better than dumping. Typically, the end result is the same: the export of a serious pollution problem from a rich country to a poor one.

In many cases, organized crime is thought to be behind large-scale waste trading, which can be closely linked with money laundering, the illegal arms trade, and other criminal activities. The Italian mafia is reportedly a key player in the robust trade in radioactive metal waste from Eastern Europe and the former Soviet Union, which it re-sells to smelters as "safe" scrap metal.

Efforts to combat the illegal waste trade face serious challenges. A recent survey by the Basel secretariat revealed that many countries lack adequate—or any—legislation for preventing and punishing illegal waste traffic. On a global scale, the absence of uniform definitions of hazardous waste and of coordinated enforcement efforts among customs officers and port authorities has contributed to the spread of illegal waste trading. Even for countries that do have the resources, the lack of hard data on the extent or geographic flow of this trade makes it hard for officials to know how to allocate the resources properly.

The member countries of the Basel Convention have taken some steps to give the treaty sharper teeth. In 1994, the Basel secretariat strengthened information sharing among member time, as much as 10,000 tons of the chemicals entered the country each year. By 1995, CFCs were considered the most valuable contraband entering Miami, after cocaine. Following a crackdown on large consignments through East Coast ports, much of the illegal trade shifted to the Mexican border.

The U.S. Department of Justice estimates that in total, some 10,000–20,000 tons of CFCs have been smuggled into the United States since 1992. Despite stronger enforcement efforts, officials have recovered only a fraction of this contraband. So far, activities under the North American CFC-Anti-Smuggling Initiative, an interagency task force established in 1994, have led to 114 convictions and the seizure of some 1,125 tons of smuggled CFCs.

Today, a second, smaller, spike in black market activity is occurring as the remaining U.S. stockpiles of legal CFC-12 are depleted. Traders can once again earn a high profit on the contraband supplies: a 30-pound cylinder of CFC-12, bought in China for as little as $30 to $60, can be resold on the U.S. black market for as much as $600. For auto repair shops and other end users, obtaining this illegal product can still be cheaper than buying the legal supplies, which now cost as much as $1,000 per cylinder.

Europe has been another significant market for illegal CFCs. In the mid 1990s, researchers with the London-based Environmental Investigation Agency (EIA) uncovered a thriving regional trade amounting to between 6,000 and 20,000 tons of the chemicals annually. Well after the phaseout deadline, supplies were still abundant and prices disproportionately low, suggesting that the market was being swamped with illegal imports. Meanwhile, regional sales of CFC replacements were slower than expected.

The European black market has thrived in part because regional refrigerant management programs have been poorly organized, and because consumers perceive alternatives to CFCs to be too costly and less efficient. In Central and Eastern Europe, where the illegal CFC trade is thought to be increasing, a major problem is that border officials are typically untrained in identifying the chemicals and have difficulty deciphering their often vague customs codes.

European efforts to control CFC smuggling have generally lagged behind those of the United States, but there are signs that European enforcement is improving. In 1997, authorities in Belgium, Germany, the Netherlands, and the United Kingdom jointly nabbed a multimillion-dollar crime ring that had illegally imported more than 1,000 tons of Chinese-made CFCs for redistribution in Europe and the United States. And in an unprecedented move, in September 2000 the European Union adopted a regional ban on CFC sales and use.

Demand for contraband CFCs has also been high in Japan, especially as retail prices for the chemicals have skyrocketed. Although the Japanese government banned the use of Freon in new cars in 1994, 15 to 20 million vehicles in the country still use the refrigerant. The *Japan Times* reports that auto repair shops in the country circulated more than 100,000 canisters of illicit CFC-12 in 2001, most of them originating in China and other developing countries.

As the CFC phaseout begins to take hold in the developing world (countries were required to freeze consumption in July 1999), black markets are beginning to emerge in places like Asia, where there are still millions of users of CFC-based equipment, including old cars and refrigerators that have been exported from the industrial world. In October 2001, EIA reported on a growing multi-million dollar market for illegal CFCs in India, Pakistan, Bangladesh, Malaysia, the Philippines, and Vietnam. Between early 1999 and March 2000, smugglers slipped some 880 tons of ozone-depleting substances into India, representing an estimated 12 percent of national consumption.

To improve monitoring of the CFC trade and head off future black markets, parties to the Montreal Protocol recently adopted a new licensing amendment that entered into force in 1999. It requires member countries to issue licenses or permits for the import and export of all new, used, and recycled ozone-depleting substances and to exchange information regularly about these activities. By identifying who is and is not licensed to trade, the system should make it easier for police and customs officials to track the movement of the chemicals worldwide.

From Words to Action

Four years ago, the environment ministers of the leading economic powers expressed "grave concern about the ever-growing evidence of violations of international environmental agreements," and called for a range of cooperative actions aimed at stepped-up enforcement. This initiative was followed by the adoption last February of international guidelines to promote compliance with multilateral environmental agreements and prevent cross-border environmental crime. This August's World Summit on Sustainable Development will focus renewed international attention on the importance of adequately implementing and enforcing international environmental treaties and other agreements.

In other promising developments, the World Customs Organization is working with governments to harmonize classification systems for waste and other environmental contraband. The international police organization INTERPOL is training national enforcement officers and customs agents to identify illicitly traded goods more easily, and is also working with national police forces to bring international environmental criminals to justice. Both institutions have established close working relations with UNEP and with the CITES and Basel Convention secretariats.

NGOs are playing a strong role as well. Brazil's RENCTAS is cooperating with the police and the federal environment ministry to train officers in wildlife inspection and is investigating anonymous tips about wildlife smuggling left on its Web site. At the international level, TRAFFIC is tracking national customs enforcement efforts, documenting areas of unsustainable wildlife trade, identifying trade routes for wildlife commodities, and investigating smuggling allegations.

FOR FURTHER INFORMATION

Wildlife Trade
CITES: www.cites.org
TRAFFIC: www.traffic.org
RENCTAS: www.renctas.org.br

Illegal Fishing
Take a Pass on Chilean Sea Bass campaign: http://environet.policy.net/marine/csb

International Transport Workers'
  Federation FOC campaign:
  www.itf.org.uk/seafarers/foc/foc.htm
FAO Fisheries Department: www.fao.org/fi

Hazardous Waste Trade
Basel Convention: www.basel.int
  Basel Action Network: www.ban.org

CFC Smuggling
  Montreal Protocol: www.unep.org/ozone

Miscellany
Environmental Investigation Agency:
  www.eia-international.org
UN Environment Programme: www.
  unep.org
World Summit on Sustainable Development: www.johannesburgsummit.org
INTERPOL: www.interpol.int
World Customs Organization:
  www.wcoomd.org

New technologies are also being deployed against international environmental crime. Remote sensing, for instance, was used by the U.S. Coast Guard in the late 1990s to gather the evidence of illegal dumping of oil in international waters that helped to bring Royal Carribean to justice. Satellite-linked Vessel Monitoring Systems are also increasingly being used to monitor the movement of fishing boats in order to detect illegal harvesting. DNA tracing is being used to monitor both fishing and wildlife trade by enabling researchers to link seafood items and wildlife parts back to the species or even the animal of origin.

Targeted industries and countries have bowed to the combined might of NGOs and consumers in several cases. In 1999, pressure from the World Wide Fund for Nature and increased public awareness of the threats that traditional medicine usage poses to wildlife caused leading practitioners and retailers in China to pledge not to prescribe or promote medicines containing parts from tigers, rhinos, bears, and other endangered species. Late last year, Belize, which Greenpeace calls "the world's most fish pirate-friendly country," struck five notorious pirate vessels from its shipping register after Greenpeace showcased the plight of the Patagonian toothfish.

Although so far they are exceptions rather than the rule, these examples demonstrate that international environmental crime can be controlled through the combined efforts of governments, international institutions, businesses, NGOs, and ordinary citizens. That's good news for the integrity of the Earth's protective ozone layer, numerous threatened species, and the health of communities poisoned by hazardous wastes.

Lisa Mastny is a research associate at Worldwatch Institute. Hilary French is director of the Institute's Global Governance Project.

ADVOCATING FOR THE ENVIRONMENT

LOCAL DIMENSIONS OF TRANSNATIONAL NETWORKS

By Maria Guadalupe Moog Rodrigues

"Never doubt that a small group of thoughtful, committed citizens can change the world. Indeed, it is the only thing that ever has."

—*Margaret Mead*

Transnational environmental advocacy networks have been responsible for many achievements of the global environmental movement to date. Among such achievements are reforms leading to the "greening" of multilateral development banks,[1] international treaties for the conservation of biodiversity and ecosystems,[2] and seminal studies and policy orientations, among them the *Report of the World Commission on Dams*.[3] Less evident, however, is the effectiveness of this type of activism locally, both in terms of protecting specific ecosystems and institutionalizing environmental activism.

Political scientists Margaret E. Keck and Kathryn Sikkink define transnational advocacy networks as networks of "relevant actors working internationally on an issue, who are bound together by shared values, a common discourse, and dense exchange of information and services." The organizational flexibility, capacity to produce and disseminate information, and ability to operate across national borders are major assets for

such networks in influencing international environmental politics.[4] Besides environmental issues, transnational advocacy networks have mobilized around issues related to human rights, women's rights, and most recently, free trade.[5] Although such networks have generated great interest among students of international relations and global governance, it can be argued that knowledge about their local impacts has been constrained by limited challenges against two established assumptions: first, that international nongovernmental organizations (NGOs) are the engines behind transnational activism[6] and second, that the participation of local actors (individuals, civil society organizations, and grassroots groups) in transnational advocacy networks leads to their political empowerment.[7] A corollary of the latter assumption is that, once empowered, local actors committed to environmental protection will maintain and deepen their activism, even in the absence of continued transnational mobilization.

A comparison among three major transnational environmental advocacy networks mobilized during the 1980s and 1990s in Brazil, India, and Ecuador shows that both assumptions require qualification. Evidence suggests that while the actions of international NGOs have been vital to initiate transnational

environmental mobilization, it is the role played by the local membership base of transnational environmental advocacy networks that determines the effectiveness of a network's efforts. In addition, while it is true that participation in transnational activism tends to empower local actors—temporarily and for the duration of the mobilization—the failure to institutionalize transnational environmental cooperation often compromises such a process.

To better account for the effectiveness of transnational environmental advocacy networks at the local level, it is important to focus the analysis on the networks' local members, particularly with regards to two key sets of factors: those that foster—or hinder—local organizations' proactive role within the networks and those that affect the institutionalization of transnational activism at the local level (consequently affecting the long-term empowerment of local actors committed to environmental sustainability).[8]

What Do Local Groups Want and Why Should We Ask?

One of the major challenges of transnational environmental advocacy networks is to create a working consensus among members about the objectives of their activism. Original research on transnational advocacy networks found that network members were driven to coordinated action by common principles and ideals.[9] Yet, as knowledge on the topic has advanced, this assumption has been questioned.[10] In fact, individuals and organizations may choose to participate in a transnational advocacy effort for a variety of reasons. In the case of transnational environmental advocacy networks, even when a common goal exists—that of promoting environmentally sustainable development—it may be approached by different network members in different ways. Accounting for the interests and priorities of local network members is particularly difficult for a variety of reasons: Local groups may be located in remote areas, may not have easy access to communication devices (including telephone, email, or fax), may have low levels of technical capacity to formulate an agenda, and may lack adequate budgets to participate in network consensus-building initiatives (such as meetings, conference calls, or workshops). In light of these difficulties, is not uncommon that the interests and goals of more resourceful network members end up being perceived as those of the network as a whole.

Engaging Local Groups in Rondônia

Local civil society organizations were practically nonexistent in the Brazilian state of Rondônia—which borders Bolivia and sits in a remote corner of the Amazon—in the early 1980s. This situation reflected the transient nature of the state's civil society, which was at the time composed mostly of recently arrived migrants with no roots in or understanding of the region's sociopolitical and ecological dynamics. Traditional populations in the area, such as indigenous groups and rubber tappers, were essentially non-entities in the local political arena due to their low levels of organization. In this context, transnational mobilization against the environmental impact of the massive development project known as Polonoroeste[11] unfolded, geographically and politically, outside Rondônia.

The absence of institutionalized local interlocutors aggravated certain cleavages that plagued the Rondônia network in its formative years. The most significant of such cleavages referred to the network's main goals and arenas of activism. International groups, organized within the framework of a campaign to reform unsustainable practices of multilateral development banks (the MDB Campaign),[12] used Brazil's Polonoroeste project as a showcase of bad development practices. They focused the network's leveraging capabilities on reforming the multilateral development banks toward increased environmental accountability. Brazilians, however, particularly environmentalists and anthropologists consulting for Polonoroeste, resented the network's international focus. They attributed to it the loss of many opportunities to challenge the Brazilian military regime and its development strategy for Amazonia.

It is not surprising that the Rondônia network's most significant impact during the 1980s was on the World Bank. Activism against Polonoroeste became a cause célèbre within the MDB Campaign and was instrumental in leading the World Bank to reformulate some of its policies and lending priorities. In Rondônia, however, deforestation, unsustainable agriculture, and invasion of Amerindian lands and conservation units remained unabated.[13]

Mindful of the constraints that the absence of a local membership base imposed on network activism, the members of the Rondônia network invested significant resources in building such a base in the late 1980s, culminating in the creation of the Rondônia Forum in 1991. The forum's existence was, in itself, an asset to the Rondônia network. It constituted a valuable formal interlocutor for local grassroots groups—such as rubber tappers and indigenous peoples' organizations and rural workers' unions—to voice their needs and expectations with regards to the network's initiatives. Yet during the early 1990s, despite the claims of the Rondônia Forum's leadership about its primary commitment to the interests of its constituent members, it remained mostly responsive to the agendas of its national and international partners. The low levels of technical and political capacity of local groups limited their contribution to the network's activism.

Under the leadership of international environmental NGOs such as the Environmental Defense Fund, and strengthened in its legitimacy due to the Rondônia Forum's existence, the Rondônia network was able to achieve some of its goals. The network's successes included the redesign of the World Bank-sponsored natural resource management project known as Planafloro[14] in more environmentally sustainable terms, and the selective incorporation of civil society organizations in the project's decision-making and implementing institutions.

As important as these conquests were in conceptual terms, they remained mainly on paper and were not effective in practice. As a result, the Rondônia Forum suffered a legitimacy crisis in 1994, which became a turning point in the evolution of the entire Rondônia network. It generated opportunities for revision of network strategies and priorities. It also forced the forum to reassess its institutional identity and commitments, a process that brought the organization's leadership closer to its members and to the populations it represented. An unprecedented level of political cohesion among local civil society organizations emerged from these processes, leading to a natural rise in their political assertiveness in relation to other network members, the Rondonian government, and World Bank officials. In this atmosphere, the proposal by international environmental NGOs of taking the Planafloro project to a newly created (in 1992) international grievances mechanism, the World Bank Inspection Panel, fell on fertile ground.

One measure of success of the Inspection Panel strategy was the empowerment experienced by Rondonian groups in the process of restructuring of the Planafloro project. Local groups demanded that one-third of remaining project funds be allocated to the Program for the Support of Community Initiatives (Programa de Apoio a Iniciativas Comunitárias, PAIC). One of the most interesting aspects of the Rondônia network's story starts where many assumed it was close to the end. The Planafloro restructuring process and the "upper hand" that local civil society organizations had in its outcome were clear evidence of their increased political empowerment. But

THE RONDÔNIA NETWORK: RONDÔNIA, BRAZIL

The Rondônia network began to mobilize in the early 1980s motivated by concerns with the environmental degradation caused by the Polonoroeste development project.[1] The project, a settlement and road-paying initiative cofunded by the World Bank, was part of a strategy to develop the Amazon frontier. The Brazilian military, which ruled the country at the time, considered the integration of Amazonia to the national economy a strategic priority. Polonoroeste quickly intensified deforestation and contributed to the deterioration of indigenous populations health.[2] International environment NGOs effectively turned the project into the main case study of their campaign against multilateral development banks and their environmentally sustainable funding practices (known as the MDB Campaign). Network partners in Brazil (scientists and research and professional organizations) contributed to the campaign by providing information, and informally working with the Brazilian Indian Agency (Fundação Nacional do Índio, FUNAI) to ameliorate the plight of indigenous peoples. Campaigners in Brazil were politically constrained and limited by distance and lack of resources. Most domestic activists were based in Brazil's largest cities—particularly in São Paulo and Rio de Janeiro, both of which lie more than 1,500 miles to the southeast on the Atlantic coast—and were rarely able to visit Rondônia. Outside Brazil, activism thrived, fueled by increased global awareness of environmental problems. In 1985, MDB campaigners forced the World Bank to interrupt

funding for Polonoroeste, creating a major precedent (and a serious diplomatic crisis).

As a follow-up to Polonoroeste, the World Bank encouraged the state of Rondônia to formulate the Planafloro project.[3] Its goal was to map and sustainably manage the state's natural resources, prioritizing the social and environmental needs of small farmers, indigenous peoples, and rubber tappers. Planafloro's original design was scrutinized by members of the Rondônia network, who demanded a higher degree of beneficiary participation in its formulation, implementation, and monitoring. However, participation required the organization of project beneficiaries. In 1991, with the support of international and domestic NGOs, significant sectors of the Rondônia civil society (small environmental and indigenous rights organizations and recently formed grassroots groups, such as the rubber tappers' organization and rural workers' unions) came together under the Rondônia Forum. An umbrella organization, the Rondônia Forum quickly became the catalyst of network activism and the main interlocutor of local groups vis-à-vis the Rondônia government and the World Bank. It eventually added co-implementation responsibilities to its project monitoring goals. By the mid-1990s, however, Planafloro's environmental performance was abysmal. The Rondônia Forum's association with the project challenged its legitimacy among project beneficiaries. In a radical

change of course, the forum, supported by its international partners, took Planafloro to the World Bank's recently inaugurated grievances mechanism, the Inspection Panel. The strategy sent shock waves throughout the World Bank and the Brazilian federal and Rondônia state governments. Although Planafloro was not fully investigated, it underwent significant restructuring in 1995-1996. One-third of remaining project funds were channeled to community initiatives, over which local civil society groups had the upper hand.

---

1. "Polonoroeste" is the acronym for the project's title in Portuguese: Programa de Desenvolvimento Integrado do Noroeste do Brasil (Northwest Brazil Integrated Development Program).

2. In 1980, levels of deforestation in the states affected by Polonoroeste were, for Rondônia, 1,680 square miles and for Mato Grosso, 8,000 square miles. In 1990 (three years after the completion of the project), deforestation rates had reached 13,400 square miles in Rondônia, and 33,440 square miles in Mato Grosso. See Instituto Brasileiro de Geografia e Estatistica (Brazilian Institute of Geography and Statistics). *Anuário Estatístico de Brasil* (Brazil's Annual Statistics Report) (Rio de Janeiro: 1993); and Instituto Nacional de Pesquisas Espaciais (National Institute of Space Research), *Deforestation in Brazilian Amazon* (São José Dos Campos; May 1992).

3. "Planafloro" is the acronym for the project's title in Portuguese: Plano Agropecuário e Florestal de Rondônia (Rondônia Natural Resources Management Project).

project restructuring and the (temporary) empowerment of local groups was not the end of the story. The process shifted the balance of forces within Rondonian politics and among the actors that participate in local environmental and development policymaking.

The consequences of such a shift were not fully appreciated or accounted for by network members, and they failed to devise strategies to address them in the long term. In fact, the case of Rondônia demonstrates that the very success of the net-

work's activism brought upon local groups new responsibilities for the formulation, implementation, and monitoring of policies. Local groups, however, were not always prepared, technically or politically, to deliver on such commitments. PAIC projects were plagued by bureaucratic delays, formulation inconsistencies, and accounting errors. In addition, the concrete environmental gains achieved by the network in the context of Planafloro generated a political backlash from the part of sectors of the government and of

local economic and political elites.[15] In recent years, opposition forces have reiterated efforts to roll back some of the environmental gains obtained by the Rondônia network. It is not clear how local civil society will continue to respond to such a backlash in the absence of ongoing network mobilization.

Activism at India's Narmada River

In India, the backlash against local groups has been even more pronounced than in Rondônia. As in Rondônia, one could argue that the Narmada network—which was formed to call attention to the human and environmental impacts of a large hydroelectric project on central India's Narmada River—was ineffective at the local level in part because of its over-reliance on strategies and priorities that were selected by its international members and that unfolded in international arenas. As a corollary, it was also a result of the incapacity of local groups to assert their priorities within the scope of the network itself.

It is undeniable that the Narmada network has obtained an unprecedented level of success internationally, influencing changes in international organizations (the creation of the World Bank's Inspection Panel is attributed to the network's activism)[16] and contributing to the establishment of the World Commission on Dams. Even at the national level, the network has had important impacts, formulating a solid critique of India's democracy and development model and leading discussions on the need for a national policy for resettlement and rehabilitation of populations affected by large dams. Yet network activism has been unable to successfully address the immediate interests of its local constituency. It has not been able to prevent the flooding of villages in the Narmada River's Nimar Valley, nor has it been able to shelter villagers or displaced populations from state repression.

In part, these difficulties can be attributed to the incapacity of the Narmada Bachao Andolan (NBA)—or "Save the Narmada Movement"—the main catalyst for local activism, to cope with the demands and expectations placed upon it (simultaneously at local, national, and international levels). Such demands have increased in proportion to the organization's various achievements and enhanced political visibility. In light of disparate demands on its resources and confronted with the technical and political weakness of local groups, NBA seemed to have opted for keeping the struggle of the Narmada network alive through a focus on its national and international dimensions. A vicious cycle was thus created: The network's effectiveness at national and international levels encouraged the continuing allocation

of network resources toward actions in these spheres, whereas the weakness of local groups and limited local victories further discouraged the institutionalization of activism at that level.

It is true that local villagers and grass-roots organizations have actively participated in many actions organized by NBA and thus demonstrated their continued commitment to the struggle against the Sardar Sarovar Project (SSP)—a hydroelectric dam, reservoir, and irrigation project. Among such actions are *satyagraha*—nonviolent marches or rallies by affected people—and "drought squads"—activists willing to drown if/when reservoir waters flood villages. Research has revealed, however, that there was an absence of channels within the Narmada network through which local voices could discuss pragmatic alternatives to the network's main strategies (which consisted of opposing SSP, challenging India's development model and democratic structure, and criticizing large dams). It is not entirely clear what factors have contributed to this gap. Certainly, NBA has stretched itself too thin answering to the many international and national demands that the network's initial successes imposed on it. But it is also probable that limited capacity among grassroots groups in the Nimar Valley has constrained their autonomy vis-à-vis the NBA. As a result, local grassroots groups may have missed opportunities to use the organization as an effective channel for bringing their priorities to the forefront of the network's agenda.

Local Voices in Ecuador's Oriente

In Ecuador, the challenges to the process of listening to local voices had specific nuances. Network activism, combined with the growing assertiveness of the country's indigenous movement throughout the 1990s, has created several arenas for local groups' participation in decisions about oil development in the Oriente—a relatively remote region in eastern Ecuador that lies in the Amazon basin (literally translated, "the East"). Yet the level of capacity for a meaningful participation in these arenas varies greatly among local groups, with direct consequences for the network as a whole. For instance, the anti-oil network

attempted to form a united front to demand clean-up operations and reparations from Texaco (la Campaña Amazonia por la Vida, or the "Amazonia for Life Campaign"). Its most visible—and at the time most effective—strategy was a lawsuit against the company, pursued in a New York District Court. Yet despite the potential gains of a successful trial and the immediate political leverage that the network derived from such a strategy,[17] a number of indigenous groups opted out: In 1995, an agreement brokered by the Ecuadorian government brought clean-up initiatives and one million dollars in "compensation" to selected Quichua communities. In return, the indigenous communities refrained from participating in any political or legal action against the company.[18] While many would argue that the Quichua leaders were co-opted by Texaco, the charge is harder to make when local groups represented by the Frente de Defensa de la Amazonia, the "Amazon Defense Front," asserted their fight to consider a settlement with the company.[19] Regardless of the priority of national and international partners in creating a precedent in international law, local groups still held the upper hand in determining the network's approach to the lawsuit.

How Do Local Groups Communicate What They Want?

The experience of these three transnational efforts to promote environmentally sustainable development at the local level indicates that it is vital that local groups are able to formulate their own approach to the concept. Such an approach, it can be argued, should determine a network's priorities and selection of strategies. Yet as the comparison among the case studies demonstrates, local groups often have difficulties formulating a coherent vision of sustainable development. Even when such difficulties are overcome, the mediation between local groups' approach to sustainable development and those of other network members may be problematic.[20]

Indigenous and settlers' populations in Ecuador's Oriente have attempted to formulate a vision of what environmentally sustainable development means for

THE NARMADA NETWORK: NIMAR VLLEY, CENTRAL INDIA

The Narmada network came together through the work of Indian activists concerned with the fate of populations to be displaced by the Sardar Sarovar Project (SSP). The hydroelectric dam, reservoir, and irrigation canal network to be built in the Nimar Valley (which is crossed by the Narmada River) would displace villagers in three Indian states (Madhya Pradesh, Maharashtra, and Gujarat). The project contained only vague provisions for resettlement and rehabilitation and proceeded in complete disregard for the environment. Yet it received significant funding from the World Bank and the Japanese Overseas Economic Cooperation Fund (OECF). From the mid-1980s on, opposition to SSP was channeled through the Narmada Bachao Andolan (NBA), the "Save the Narmada Movement." The organization brought together affected populations and national activists' organizations. Through the networking efforts of one of its key activists, Metha Patkar, it obtained the support of international environmental groups and placed the campaign against SSP on the radar screen of the Multilateral Development Bank Campaign. The activism of the Narmada campaign during the 1980s and early 1990s led to the withdrawal of funds from the Japanese government and the World Bank and to the commissioning of landmark studies about the viability of SSP and other large dams. In India, NBA initiated a lawsuit against the project and coordinated an ongoing critique of the country's democracy and development model. In the Nimar Valley, however, repression of activists and villagers steadily increased during the 1990s. In 2002, the Indian Supreme Court reversed previous stays and authorized the continuation of reservoir works. Although the Narmada network remains mobilized, it has redirected most of its activist efforts to other areas of the Narmada River, where projects similar to SSP are either being planned or already underway.

a region endowed with significant natural resources, crude oil among them. This vision stresses the need for political participation of grassroots groups in larger development policies for the Oriente. Thanks to the anti-oil network's activism, this notion has become accepted (if not always implemented) by most sectors of Ecuadorian politics. Dialogue and consultations among government officials, oil companies, and affected populations have sometimes created opportunities for the formulation of instruments, for the compensation of affected populations, and for further research on the area's ecological characteristics, among others.

Yet difficulties in formulating a consensual approach to sustainable development still remain. Negotiations between indigenous groups and the oil company ARCO best illustrate this point: Despite years of dialogue and joint initiatives, indigenous organizations have yet to come together around a common socioeconomic development plan for the region to be sponsored by the company. ARCO has used such lack of unity as an excuse to delay important initiatives on behalf of local populations.

On a brighter note, participation in the Rondônia network has created conditions for local groups to formulate their own approach to environmentally sustainable development. Such an approach was fully developed—and made operational—in the context of PAIC. Based on that experience, it is not farfetched to predict that PAIC's approach to sustainable development is likely to prevail in future policy initiatives in Rondônia. The lessons from PAIC have made clear that, for locals, sustainable development is a process that integrates environmental protection with the improvement of communities' socioeconomic well-being and their participation in the policymaking process.

Although local groups' definitions of environmentally sustainable development—and of criteria to make it attainable—is vital for the success of a network's initiatives, their vision must be communicated to other network partners. Thus the mediating role of catalyst organizations deserves close attention. In Rondônia, mediation among interests and visions of network members was primarily performed by the Rondônia Forum. In Ecuador, regional and national indigenous federations and confederations, as well as national environmental groups, were instrumental. In the case of the Narmada network, NBA was a key catalyst for resources and strategic planning. Yet the nature of its mediating role remains unclear, in part due to problems related to the organization's identity.

The Rondônia Forum was essentially conceived and made operational by international and Brazilian nongovernmental organizations involved in the MDB Campaign. While local groups welcomed the initiative, they took a back seat during the forum's formative years. It was the Rondônia Forum's leadership—individuals and local NGOs closer to international network members—that played a proactive role in the Rondônia network in the early 1990s. Such a role, however, did not always reflect the level of commitment or engagement of all the forum's affiliated organizations to the network's priorities and strategies. Inevitably, this situation led the forum to an identity crisis that affected its legitimacy as a valid interlocutor of local groups' interests within and outside the Rondônia network.

As Rondonian civil society groups resolved the forum's legitimacy issues and reassessed its mission, they changed the nature of their engagement in the Rondônia network. In becoming proactive players within the Rondônia Forum, grassroots groups such as the Organization of Rondonian Rubber Tappers (Organização dos Seringueiros de Rondônia, OSR) and the Federation of Rondonian Rural Workers (Federação dos Trabalhadores Agrícolas de Rondônia, FETAGRO), came to have a decisive voice within the network as a whole. This process reflects a dual dynamic: participation in the Rondônia Forum led to a gradual increase in grassroots organizations' capacity, and their increased capacity eventually constrained the forum's role. At the turn of the millennium, the Rondônia Forum has become less a catalyst organization for local groups' activism and more a source of resources for the support of grassroots initiatives. This change is significant in that it allows a clear assessment of the role of local groups in transnational advocacy networks.

The role played by grassroots groups in shaping strategies and defining the priorities of the Narmada network, on the contrary, is less evident. The extraordinary success of NBA as a catalyst for local groups' activism may have hindered these groups' ability to develop a more proactive (and autonomous) behavior within the network. As a consequence, it is difficult to distinguish between the contribution of NBA's leadership and that of its individual affiliated organizations. This difficulty—at least for outside observers—in identifying a pluralist pattern of deliberation within the Narmada network should not lead one to infer that NBA is any less committed and accountable to the interests of its affiliated groups than, for instance, the Rondônia Forum. The issue here is how the organization has mediated between the priorities of its very diverse constituency and how this process has impacted on the evolution of the Narmada network. As NBA keeps the focus of its activism on a larger critique of India's development model, we are left wondering: How do populations in the Nimar Valley define environmentally sustainable development in light of the constraints imposed by the Sardar Sarovar Project (and other issues)? Is it correct to assume that their vision is entirely subsumed under NBA's critique of national institutions and processes?

The nature of NBA's mediation between local grassroots groups and other members of the Narmada network is further illuminated by contrasting it to the mediation performed by Ecuadorian national organizations in the context of the anti-oil network. In Ecuador, the relationship between local grassroots groups, national organizations (indigenous federations and environmental

NGOs), and their international partners has unfolded according to clearly defined patterns of interaction at all levels. Local indigenous groups, for instance, formally declare their affiliation to regional and national indigenous confederations and give them a mandate to represent their interests at the national and international levels. In local matters, however, these groups tend to assert their autonomy. The nature of relations between groups at different levels is even better defined in the case of the Oriente's indigenous and settlers' groups and their links with Ecuadorian environmental NGOs. Partnership for these groups is defined in very specific terms, and environmental groups were never given a mandate to represent local populations.

National NGOs and indigenous confederations in Ecuador do assume the largest part of the responsibility for mediation between local and international interests, with different degrees of success. National organizations are thus key elements in the process of resource-sharing that characterizes transnational networks. For instance, in the anti-oil network, they were the main channels of articulation between local and international actors within the Amazon Coalition, a transnational umbrella organization whose mandate is to foster cooperation among forest peoples in the Amazon and their supporters. One result of this mediation was that the latter became particularly skillful in avoiding challenges to the legitimacy of its initiatives, regardless of the geographic and political arenas in which they were pursued.

Similar to the anti-oil network, the relationship between NBA and the Narmada network's international partners was clearly defined. From the onset of mobilization, NBA was able to assert its financial and political autonomy and its leadership role in determining the network's priorities and strategies. In contrast with the anti-oil network in Ecuador, however, the nature of NBA's relations with its local partners is less evident. This is in part due to NBA's own ambivalence regarding its identity: Is it a grassroots organization—representing the interests of specific populations—-or a coalition/umbrella organization that

mediates the interests of different groups and facilitates their activism?[21]

However, the members of the Narmada network, more specifically NBA, are not the only ones plagued by the porous boundaries of multilevel activism. Compared to the success of Ecuador's anti-oil network in this area, the shortcomings of the Rondônia network are even more striking. In the approximately 20 years in which the Rondônia network remained mobilized, there were few occasions when national actors actively participated in its efforts. For the most part, the Rondônia Forum played the role of mediator between the interests of international and grassroots groups, not always successfully. From the early 1990s on, activism in Rondônia was characterized by direct interactions between international and local organizations—in complete disregard for the potential advantages of mediation by national advocacy organizations. As a result, the process of resource-sharing among network members was constrained. This has had negative consequences for local groups, particularly in the years following the restructuring of the Planafloro project. As the network's priorities became essentially local, international interest in it decreased. The resources once available to Rondonian groups from their international partners have since diminished.

Historically, the lack of systematic participation of national organizations has prevented the Rondônia network from institutionalizing avenues for resource sharing between local and national activist groups. This partially explains the difficulties that local groups have had in addressing issues related to the low technical capacity of their cadres. Yet the failure of local groups to reach out to and participate in national arenas for civil society environmental and development activism has perpetuated this problem. The irony is that part of the explanation for why local groups have failed to strengthen links with their national partners in Brazil is their limited capacity. They have lacked human and financial resources (such as competent cadres who understand issues debated in national meetings and financial resources to send representatives to conferences and rallies) that could enable

THE ANTI-OIL NETWORK IN ECUADOR'S ORIENTE

The anti-oil network in Ecuador mobilized as a result of grassroots activism among indigenous populations and the awareness raised by descriptions of the environmental impact of Texaco's operation in Ecuador's Amazonia.[1] The initial phase of the campaign targeted the Ecuadorian government. Activists demanded that it evaluate the environmental impact of Texaco's operations and mandate clean-up intiatives. La Campaña Amazonia por la Vida (the Amazonia for Life Campaign) was led by Ecuadorian environmental organizations such as COR-DAVI (Corporation de Defensa de la Vida, or "Corporation for the Defense of Life") and Acción Ecológica (Ecological Action), and it counted on a key ally, the Amazon Coalition. Formed in 1990, the Amazon Coalition aimed at institutionalizing processes of sharing and exchanging political, technical, and material resources among local and international actors concerned with environmental and indigenous issues in the entire Amazon region. It played a key role in bringing together indigenous groups and settlers' organizations, which had traditionally competed for the same resource—land—in the Oriente region in eastern Ecuador.

Pressures on the Ecuadorian government to increase its oversight of oil operations led to growing national and global awareness about the problem but no concrete action. In 1993, the network resorted to a different strategy. It initiated a lawsuit against Texaco in a New York District Court.[2] The goal was to hold the company's international operations to U.S. environmental standards. To best coordinate the interests of affected populations, the network encouraged the formation of the Frente de Defensa de la Amazonia (Amazon Defense Front) in 1994. The Frente lobbied Ecuadorian authorities to cooperate with the lawsuit, disseminated information among interested parties, and maintained a flow of information between affected communities and their lawyers in the United States. But despite the network's efforts, the lawsuit was dismissed in 2002. (It is currently being pursued in Ecuador.)

The Texaco experience has nonetheless continued to illuminate the actions of the anti-oil network. Throughout the early 1990s, indigenous groups in the Pastaza region (a large section of central Oriente) and their allies pressured ARCO (a subsidiary of the U.S.-based Atlantic Richfield Company) to become accountable for the socioenvironmental impacts of its operations. Eventually the network succeeded in forcing the company into a twofold agreement: It would fund a regional development plan to be drafted by indigenous groups, and it would sponsor environmental impact studies to be conducted with the participation of indigenous representatives. The implementation of these initiatives, however, has encountered a series of difficulties, ranging from a lack of interest by company officials to indigenous peoples' conflicts regarding the priorities of a regional development plan.

1. See J. Kimerling, *Amazon Crude* (New York: Natural Resources Defense Council, 1991).

2. Maria Aguida et al., vs. Texaco, filed on 3 November 1993 in the U.S. District Court for the Southern District of New York.

them to fully engage in national activism. The consequence is the perpetuation of a vicious cycle in which lack of engagement and closer cooperation with national groups prevents gains in local groups' technical capacity, and limited capacity constrains local-national cooperation.

Lessons and Recommendations

It is clear from the three case studies that many transnational environmental advocacy networks have attained major victories at international and national levels, altering policies of multilateral organizations, changing priorities and practices of corporations, and even affecting the rhetoric of nation-states regarding their environmental goals. Yet the networks' record at the local level has been less striking. Thus it is key to reflect on networks' successes and failures with the goal of improving their local performance in future initiatives. It is in this spirit that the recommendations below were formulated. In essence, the strategies suggested aim at facilitating the incorporation of local voices in networks' efforts toward pluralist deliberation and consensus building.

One key priority is to establish or identify local umbrella organizations or clearinghouses that support the organization of local groups. This would facilitate mediation between the interests of these local groups and those of their national and international partners. It is also important that efforts be made to guarantee that umbrella organizations remain a participatory forum where different local groups may reconcile their differences and reach consensus regarding their goals and expectations as network members. In addition, as the experience of the Rondônia network illustrates, it is crucial that the leaders or members of an umbrella organization's executive secretariat do not overshadow the autonomy of local groups. Umbrella organizations work best as arenas for resource-sharing between network members and for the facilitation of members' initiatives, not as the sole catalyst or initiator of activism.

As the case of Ecuador shows, the existence of local umbrella organizations, such as the Amazon Coalition, are no substitute for well-established lines of cooperation between network members at all levels (particularly between local and regional/national indigenous organizations). A well-structured transnational advocacy network engages partners at local, national, and international levels, and it is vital that members at each level are mindful of the political boundaries of their activism.

The establishment of local umbrella organizations and the processes of resource sharing and information diffusion typical of transnational activism are likely to affect the balance of power among local political actors. To avoid possible backlashes from established forces against local groups circumstantially empowered from their participation in transnational activism, it would be helpful for network members at all levels to be aware of—and prepared for—such possibilities. This issue is of particular concern for international network members. Transnational activism inevitably affects local politics; thus responsible activism requires consideration to the following questions: To what extent are international network members willing or able to become (even if indirectly) involved in domestic and local politics? What are the consequences for such actors' legitimacy and accountability to their global constituencies? To what extent do their resources allow them to commit to struggles that are inherently long-term, since they involve structural change? How may international network members best cope with the political responsibility of committing to transnational socioenvironmental activism and campaigns?

Another challenge to the relationship between international and national/local network members refers to their choice of strategies. Given the superior resources of international groups, transnational networks have had a tendency to concentrate advocacy efforts on international arenas. Yet this practice generates at least two risks for a network's effectiveness. First, successful strategies, conceived and implemented by international NGOs in industrialized countries, do not necessarily fare well when reproduced by activist groups in developing countries. A second risk of an excessive reliance on international strategies is the possibility that it will constrain the emergence of endogenous and often innovative channels for activism.

The priority to locally devised strategies within the context of transnational activism is inherently linked to another key process, that of defining environmentally sustainable development. To this end, it is necessary that network members' encourage the formulation of an approach to environmentally sustainable development that truly represents the needs and expectations of local network members. A network's strategies (legal, educational, and political) and goals will likely be most effective if they remain faithful to such an approach. The essence of this process is to define what specific actions, projects, and policies must be pursued—at different levels—o foster environmentally sustainable development in a given region. The contribution of such actions and policies to national and global environmental sustainability, while a desirable outcome, should remain secondary goals on the agendas of network members.

Legitimacy problems may plague local network members if expectations raised by network activism remain unfulfilled. Thus, it is advisable that local groups prepare to respond to new responsibilities that may emerge as a consequence of successful activism. For instance, technical and material demands on network members are likely to increase as a result of a network's enhanced political assertiveness. Network members may be called upon to participate in the design of socio-environmental policies (PAIC in Rondônia is an example of this). They may also be requested to cooperate in environmental impact assessments, environmental monitoring missions, and community development plans (as happened to indigenous groups in Ecuador's Oriente). It is important for all network members to devise policies for their long-term engagement in decisionmaking arenas, which may become available to them—particularly local groups—as a result of successful activism. Conversely, network members may consider the costs of embracing roles beyond their capacities, as well as of assuming responsibility for tasks that traditionally fall under the competence of national, regional, or local governments or large businesses such as oil companies. Although NGOs (domestic and international), research institutes, and grassroots groups may be open to cooperate with state agencies and corporations, they may best defray legitimacy and credibility challenges by limiting their role to independent advising. In very well-defined and limited cases, network members may assume implementation responsibilities for small, locally based initiatives. By assuming large-scale executive roles in environmental and development projects and policies, network members may be misidentified by their constituencies as being in charge of delivering goods and services. When such goods and services are not provided, for reasons often outside the control of network members, they may lose appeal among their constituencies.

NOTES

1. B. Rich, "The Emperor's New Clothes: The World Bank and Environmental Reform," *World Policy Journal* 7, no. 2 (1990): 305-29.

2. T. Princen and M. Finger, *Environmental NGOs in World Politics: Linking the Local and the Global* (London and New York: Routledge, 1994).

3. The World Commission on Dams, ed., *Dams and Development: A New Framework for Decision-Making, The Report of the World Commission on Dams* (London: Earthscan, 2000). Also see R. Bissell, "A Participatory Approach to Strategic Planning," *Environment*, September 2001, 37-40.

4. M. E. Keck and K. Sikkink, *Activists Beyond Borders* (Ithaca, NY: Cornell University Press, 1998), 2-4.

5. Well known examples are, respectively, transnational activism against human rights violations in Chile and Argentina during military rule in the 1970s, against female genital mutilation in African countries, and against the World Trade Organization.

6. For instance, see Keck and Sikkink, note 4 above; and Princen and Finger, note 2 above.

7. L. Jordan and P. Van Tuijl, "Political Responsibility in Transnational NGO Advocacy," in *World Development* 28, no. 12 (2000): 2051-65; and T. Jezic, "Ecuador: The Campaign Against Texaco Oil," in D. Cohen, R. Vega, and G. Watson, eds., *Advocacy for Social Justice: A Global Action and Reflection Guide* (Bloomfield, CT: Kumarian Press, 2001).

8. Data were obtained through open-ended interviews with activists, and analysis of primary documents (technical reports, correspondence, and legal documents) and secondary literature.

9. Keck and Sikkink, note 4 above.

10. Jordan and Van Tuijl, note 7 above; M. Rodrigues: "Environmental Protection Issue Networks in Amazonia," *Latin American Research Review* 35, no. 3 (2000): 125-53.

11. "Polonoroeste" is the acronym for the project's title in Portuguese: Programa de Desenvolvimento Integrado do Noroeste do Brasil (Northwest Brazil Integrated Development Program).

12. For details on the campaign, see B. Rich, *Mortgaging the Earth: The World Bank, Environmental Impoverishment, and the Crisis of Development* (Boston: Beacon Press, 1994);

and B. Bramble and G. Porter, "Non-Governmental Organizations and the Making of US International Environmental Policy," in A. Hurrell and B. Kingsbury, eds., *The International Politics of the Environment* (Oxford, UK: Clarendon Press, 1992).

13. Evidently, these outcomes cannot be blamed entirely on the network's cleavages. The point here is that the network's international focus led activists to overlook opportunities for action at the local level.

14. "Planafloro" is the acronym for Piano Agropecuário e Florestal de Rondônia (Rondônia Natural Resource Management Project).

15. In the wake of Planafloro restructuring, implementing agencies finally acted to demarcate rubber tappers' and indigenous reserves and to create protected areas that the project had mandated since its inception.

16. P. McCully, *Silenced Rivers: The Ecology and Politics of Large Dams* (London: Zed Books, 1996).

17. If successful, the lawsuit would create a precedent in international law, whereby multinational corporations responsible for environmental damage abroad could be sued in their countries of origin and held to those countries' environmental standards.

18. It should be noted, however, that Texaco's money never benefited indigenous communities. It was squandered by the leadership, and some leaders eventually had to respond to charges of corruption.

19. Luis Yanza, president of the Frente de Defensa de la Amazonia (Amazon Defense Front), stated in a 16 December 1999 letter to Ivonne Ramos of Acción Ecológica (Ecological Action): "In effect, in the face of a possible proposal for dialogue from Texaco, we have initiated a process of information and reflection that will culminate in a workshop with community leaders, under the coordination of an independent international expert, with the goal of deepening our understanding of US legislation. After receiving all this information, the leaders will be able to guide their bases along the path that they (the grassroots) find most convenient" (Translation by M. G. M. Rodrigues).

20. As it has been stated by Sharachchandra Lélé and other experts on sustainable development, the "fuzziness" or "vagueness" of the concept of environmentally sustainable development has made its operationalization a difficult task. A key priority of network members is to formulate a unified vision of environmentally sustainable development and determine how it can be (locally) implemented. See S. Lélé, "Sustainable Development: A Critical Review," *World Development* 19, no. 6 (1991): 607-21.

21. Narmada Bachao Andolan's (NBA) own activists have described the nature of the organization in contradictory terms. For instance, Medha Patkar—one of the central organizers of the group—draws a distinction between NBA as a "support" organization and the specific groups that represent local populations. She states that "a basic principle was that the people's representatives should accompany the [NBA] activists every time we met officials." M. Patkar, "The Struggle for Participation and Justice: a Historical Narrative," in W. Fisher, ed., *Toward Sustainable Development: Struggling Over India's Narmada River* (Armonk, NY, and London, UK: M.E. Sharpe, 1995), 158. In the words of another NBA activist, the organization is not one working "on behalf of the local people, (but one that is) primarily a body of the affected people, with some—very few—activists from outside—these too living in the valley itself among the people...," Shripad Dharmadhikary, NBA activist, in letter to Suzanne Moxon, writer for *International Water Power and Dam Construction*, 4 April 1998.

Maria Guadalupe Moog Rodrigues is assistant professor of political science at the College of the Holy Cross in Worcester, Massachusetts. Previous to obtaining her Ph.D. in Political Science from Boston University, she was an intern and consultant for the National Wildlife Federation in Washington, DC. she was recently awarded a fellowship from the Carnegie Council on Ethics and International Affairs' 2002-2003 Fellows Program. Her publications related to the topic of this article include "Environmental Protection Issue Networks in Amazonia," in *Latin American Research Review* 35, no. 3 (2000), and "The Planafloro Inspection Panel Claim," in D. Clark, J. Fox, and K. Treakle, eds., *Demanding Accountability: Civil Society Claims and the World Bank Inspection Panel* (Lanham, MD: Rowman & Littlefield Publishers, Inc., 2003). Maria Guadalupe Moog Rodrigues can be contacted via e-mail at mrodrigu@holycross.edu. This article is a revised and edited version of the concluding chapter of Rodrigues's book, *Global Environmentalism and Local Politics*, (Albany, NY: State University of New York Press, 2003).

Rescuing a planet under stress

Lester R. Brown

Understanding the Problem

As world population has doubled and as the global economy has expanded sevenfold over the last half-century, our claims on the Earth have become excessive. We are asking more of the Earth than it can give on an ongoing basis.

We are harvesting trees faster than they can regenerate, overgrazing rangelands and converting them into deserts, overpumping aquifers, and draining rivers dry. On our cropland, soil erosion exceeds new soil formation, slowly depriving the soil of its inherent fertility. We are taking fish from the ocean faster than they can reproduce.

We are releasing carbon dioxide into the atmosphere faster than nature can absorb it, creating a greenhouse effect. As atmospheric carbon dioxide levels rise, so does the earth's temperature. Habitat destruction and climate change are destroying plant and animal species far faster than new species can evolve, launching the first mass extinction since the one that eradicated the dinosaurs sixty-five million years ago.

Throughout history, humans have lived on the Earth's sustainable yield—the interest from its natural endowment. But now we are consuming the endowment itself. In ecology, as in economics, we can consume principal along with interest in the short run but in the long run it leads to bankruptcy.

In 2002 a team of scientists led by Mathis Wackernagel, an analyst at Redefining Progress, concluded that humanity's collective demands first surpassed the Earth's regenerative capacity around 1980. Their study, published by the U.S. National Academy of Sciences, estimated that our demands in 1999 exceeded that capacity by 20 percent. We are satisfying our excessive demands by consuming the Earth's natural assets, in effect creating a global bubble economy.

Bubble economies aren't new. U.S. investors got an up-close view of this when the bubble in high-tech stocks burst in 2000 and the NASDAQ, an indicator of the value of these stocks, declined by some 75 percent. According to the *Washington Post*, Japan had a similar experience in 1989 when the real estate bubble burst, depreciating stock and real estate assets by 60 percent. The bad-debt fallout and other effects of this collapse have left the once-dynamic Japanese economy dead in the water ever since.

The bursting of these two bubbles affected primarily people living in the United States and Japan but the global bubble economy that is based on the overconsumption of the Earth's natural capital assets will affect the entire world. When the food bubble economy, inflated by the overpumping of aquifers, bursts, it will raise food prices worldwide. The challenge for our generation is to deflate the economic bubble before it bursts.

Unfortunately, since September 11, 2001, political leaders, diplomats, and the media worldwide have been preoccupied with terrorism and, more recently, the occupation of Iraq. Terrorism is certainly a matter of concern, but if it diverts us from the environmental trends that are undermining our future until it is too late to reverse them, Osama bin Laden and his followers will have achieved their goal in a way they couldn't have imagined.

In February 2003, United Nations demographers made an announcement that was in some ways more shocking than the 9/11 attack: the worldwide rise in life expectancy has been dramatically reversed for a large segment of humanity—the seven hundred million people living in sub-Saharan Africa. The HIV epidemic has reduced life expectancy among this region's people from sixty-two to forty-seven years. The epidemic may soon claim more lives than all the wars of the twentieth century. If this teaches us anything, it is the high cost of neglecting newly emerging threats.

The HIV epidemic isn't the only emerging mega-threat. Numerous nations are feeding their growing populations by overpumping their aquifers—a measure that virtually guarantees a future drop in food production when the aquifers are depleted. In effect, these nations are creating a food bubble economy—one where food production is artificially inflated by the unsustainable use of groundwater.

Another mega-threat—climate change—isn't getting the attention it deserves from most governments, particularly that of the United States, the nation responsible for one-fourth of all carbon emissions. Washington, D.C., wants to wait until all the evidence on climate change is in, by which time it will be too late to prevent a wholesale warming of the planet. Just as governments in Africa watched HIV infection rates rise and did little about it, the United States is watching atmospheric carbon dioxide levels rise and doing little to check the increase.

Other mega-threats being neglected include eroding soils and expanding deserts, which jeopardize the livelihood and food supply of hundreds of millions of the world's people. These issues don't even appear on the radar screen of many national governments.

Thus far, most of the environmental damage has been local: the death of the Aral Sea, the burning rainforests of Indonesia, the collapse of the Canadian cod fishery, the melting of the glaciers that supply Andean cities with water, the dust bowl forming in northwestern China, and the depletion of the U.S. great plains aquifer. But as these local environmental events expand and multiply, they will progressively weaken the global economy, bringing closer the day when the economic bubble will burst.

Humanity's demands on the Earth have multiplied over the last half-century as our numbers have increased and our incomes have risen. World population grew from 2.5 billion in 1950 to 6.1 billion in 2000. The growth during those fifty years exceeded that during the four million years since our ancestors first emerged from Africa.

Incomes have risen even faster than population. According to Erik Assadourian's *Vital Signs* 2003 article, "Economic Growth Inches Up," income per person worldwide nearly tripled from 1950 to 2000. Growth in population and the rise in incomes together expanded global economic output from just under $7 trillion (in 2001 dollars) of goods and services in 1950 to $46 trillion in 2000—a gain of nearly sevenfold.

Population growth and rising incomes together have tripled world grain demand over the last half-century, pushing it from 640 million tons in 1950 to 1,855 million tons in 2000, according to the U.S. Department of Agriculture (USDA). To satisfy this swelling demand, farmers have plowed land that was highly erodible—land that was too dry or too steeply sloping to sustain cultivation. Each year billions of tons of topsoil are being blown away in dust storms or washed away in rainstorms, leaving farmers to try to feed some seventy million additional people but with less topsoil than the year before.

Demand for water also tripled as agricultural, industrial, and residential uses increased, outstripping the sustainable supply in many nations. As a result, water tables are falling and wells are going dry. Rivers are also being drained dry, to the detriment of wildlife and ecosystems.

Fossil fuel use quadrupled, setting in motion a rise in carbon emissions that is overwhelming nature's capacity to fix carbon dioxide. As a result of this carbon-fixing deficit, atmospheric carbon dioxide concentrations climbed from 316 parts per million (ppm) in 1959, when official measurement began, to 369 ppm in 2000, according to a report issued by the Scripps Institution of Oceanography at the University of California.

The sector of the economy that seems likely to unravel first is food. Eroding soils, deteriorating rangelands, collapsing fisheries, falling water tables, and rising temperatures are converging to make it more difficult to expand food production fast enough to keep up with demand. According to the USDA, in 2002 the world grain harvest of 1,807 million tons fell short of

world grain consumption by 100 million tons, or 5 percent. This shortfall, the largest on record, marked the third consecutive year of grain deficits, dropping stocks to the lowest level in a generation.

Now the question is: can the world's farmers bounce back and expand production enough to fill the hundred-million-ton shortfall, provide for the more than seventy million people added each year, and rebuild stocks to a more secure level? In the past, farmers responded to short supplies and higher grain prices by planting more land and using more irrigation water and fertilizer. Now it is doubtful that farmers can fill this gap without further depleting aquifers and jeopardizing future harvests.

At the 1996 World Food Summit in Rome, Italy, hosted by the UN Food and Agriculture Organization (FAO), 185 nations plus the European community agreed to reduce hunger by half by 2015. Using 1990–1992 as a base, governments set the goal of cutting the number of people who were hungry—860 million—by roughly 20 million per year. It was an exciting and worthy goal, one that later became one of the UN Millennium Development Goals.

But in its late 2002 review of food security, the UN issued a discouraging report:

> This year we must report that progress has virtually ground to a halt. Our latest estimates, based on data from the years 1998-2000, put the number of undernourished people in the world at 840 million.... a decrease of barely 2.5 million per year over the eight years since 1990–92.

Since 1998–2000, world grain production per person has fallen 5 percent, suggesting that the ranks of the hungry are now expanding. As noted earlier, life expectancy is plummeting in sub-Saharan Africa. If the number of hungry people worldwide is also increasing, then two key social indicators are showing widespread deterioration in the human condition.

The ecological deficits just described are converging on the farm sector, making it more difficult to sustain rapid growth in world food output. No one knows when the growth in food production will fall behind that of demand, driving up prices, but it may be much sooner than we think. The triggering events that will precipitate future food shortages are likely to be spreading water shortages interacting with crop-withering heat waves in key food-producing regions. The economic indicator most likely to signal serious trouble in the deteriorating relationship between the global economy and the Earth's ecosystem is grain prices.

Food is fast becoming a national security issue as growth in the world harvest slows and as falling water tables and rising temperatures hint at future shortages. According to the USDA more than one hundred nations import part of the wheat they consume. Some forty import rice. While some nations are only marginally dependent on imports, others couldn't survive without them. Egypt and Iran, for example, rely on imports for 40 percent of their grain supply. For Algeria, Japan, South Korea, and Taiwan, among others, it is 70 percent or more. For Israel and Yemen, over 90 percent. Just six nations—Argentina,

Australia, Canada, France, Thailand, and the United States—supply 90 percent of grain exports. The United States alone controls close to half of world grain exports, a larger share than Saudi Arabia does of oil.

Thus far the nations that import heavily are small and middle-sized ones. But now China, the world's most populous nation, is soon likely to turn to world markets in a major way. As reported by the International Monetary Fund, when the former Soviet Union unexpectedly turned to the world market in 1972 for roughly a tenth of its grain supply following a weather-reduced harvest, world wheat prices climbed from $1.90 to $4.89 a bushel. Bread prices soon rose, too.

If China depletes its grain reserves and turns to the world grain market to cover its shortfall—now forty million tons per year—it could destabilize world grain markets overnight. Turning to the world market means turning to the United States, presenting a potentially delicate geopolitical situation in which 1.3 billion Chinese consumers with a $100-billion trade surplus with the United States will be competing with U.S. consumers for U.S. grain. If this leads to rising food prices in the United States, how will the government respond? In times past, it could have restricted exports, even imposing an export embargo, as it did with soybeans to Japan in 1974. But today the United States has a stake in a politically stable China. With an economy growing at 7 to 8 percent a year, China is the engine that is powering not only the Asian economy but, to some degree, the world economy.

For China, becoming dependent on other nations for food would end its history of food self-sufficiency, leaving it vulnerable to world market uncertainties. For Americans, rising food prices would be the first indication that the world has changed fundamentally and that they are being directly affected by the growing grain deficit in China. If it seems likely that rising food prices are being driven in part by crop-withering temperature rises, pressure will mount for the United States to reduce oil and coal use.

For the world's poor—the millions living in cities on $1 per day or less and already spending 70 percent of their income on food—rising grain prices would be life threatening. A doubling of world grain prices today could impoverish more people in a shorter period of time than any event in history. With desperate people holding their governments responsible, such a price rise could also destabilize governments of low-income, grain-importing nations.

Food security has changed in other ways. Traditionally it was largely an agricultural matter. But now it is something that our entire society is responsible for. National population and energy policies may have a greater effect on food security than agricultural policies do. With most of the three billion people to be added to world population by 2050 (as estimated by the UN) being born in nations already facing water shortages, childbearing decisions may have a greater effect on food security than crop planting decisions. Achieving an acceptable balance between food and people today depends on family planners and farmers working together.

Climate change is the wild card in the food security deck. The effect of population and energy policies on food security differ from climate in one important respect: population stability can be achieved by a nation acting unilaterally. Climate stability cannot.

Instituting the Solution

Business as usual—Plan A—clearly isn't working. The stakes are high, and time isn't on our side. The good news is that there are solutions to the problems we are facing. The bad news is that if we continue to rely on timid, incremental responses our bubble economy will continue to grow until eventually it bursts. A new approach is necessary—a Plan B—an urgent reordering of priorities and a restructuring of the global economy in order to prevent that from happening.

Plan B is a massive mobilization to deflate the global economic bubble before it reaches the bursting point. Keeping the bubble from bursting will require an unprecedented degree of international cooperation to stabilize population, climate, water tables, and soils—and at wartime speed. Indeed, in both scale and urgency the effort required is comparable to the U.S. mobilization during World War II.

Our only hope now is rapid systemic change—change based on market signals that tell the ecological truth. This means restructuring the tax system by lowering income taxes and raising taxes on environmentally destructive activities, such as fossil fuel burning, to incorporate the ecological costs. Unless we can get the market to send signals that reflect reality, we will continue making faulty decisions as consumers, corporate planners, and government policymakers. Ill-informed economic decisions and the economic distortions they create can lead to economic decline.

Stabilizing the world population at 7.5 billion or so is central to avoiding economic breakdown in nations with large projected population increases that are already overconsuming their natural capital assets. According to the Population Reference Bureau, some thirty-six nations, all in Europe except Japan, have essentially stabilized their populations. The challenge now is to create the economic and social conditions and to adopt the priorities that will lead to population stability in all remaining nations. The keys here are extending primary education to all children, providing vaccinations and basic health care, and offering reproductive health care and family planning services in all nations.

Shifting from a carbon-based to a hydrogen-based energy economy to stabilize climate is now technologically possible. Advances in wind turbine design and in solar cell manufacturing, the availability of hydrogen generators, and the evolution of fuel cells provide the technologies needed to build a climate-benign hydrogen economy. Moving quickly from a carbon-based to a hydrogen-based energy economy depends on getting the price right, on incorporating the indirect costs of burning fossil fuels into the market price.

On the energy front, Iceland is the first nation to adopt a national plan to convert its carbon-based energy economy to one based on hydrogen. Denmark and Germany are leading the world into the age of wind. Japan has emerged as the world's leading manufacturer and user of solar cells. With its commercialization of a solar roofing material, it leads the world in electricity generation from solar cells and is well positioned to assist in the electrification of villages in the developing world. The Netherlands leads the industrial world in exploiting the bicycle as an alternative to the automobile. And the Canadian province of Ontario is emerging as a leader in phasing out coal. It plans to replace its five coal-fired power plants with gas-fired plants, wind farms, and efficiency gains.

Stabilizing water tables is particularly difficult because the forces triggering the fall have their own momentum, which must be reversed. Arresting the fall depends on quickly raising water productivity. In pioneering drip irrigation technology, Israel has become the world leader in the efficient use of agricultural water. This unusually labor-intensive irrigation practice, now being used to produce high-value crops in many nations, is ideally suited where water is scarce and labor is abundant.

In stabilizing soils, South Korea and the United States stand out. South Korea, with once denuded mountainsides and hills now covered with trees, has achieved a level of flood control, water storage, and hydrological stability that is a model for other nations. Beginning in the late 1980s, U.S. farmers systematically retired roughly 10 percent of the most erodible cropland, planting the bulk of it to grass, according to the USDA. In addition, they lead the world in adopting minimum-till, no-till, and other soil-conserving practices. With this combination of programs and practices, the United States has reduced soil erosion by nearly 40 percent in less than two decades.

Thus all the things we need to do to keep the bubble from bursting are now being done in at least a few nations. If these highly successful initiatives are adopted worldwide, and quickly, we can deflate the bubble before it bursts.

Yet adopting Plan B is unlikely unless the United States assumes a leadership position, much as it belatedly did in World War II. The nation responded to the aggression of Germany and Japan only after it was directly attacked at Pearl Harbor on December 7, 1941. But respond it did. After an all-out mobilization, the U.S. engagement helped turn the tide, leading the Allied Forces to victory within three and a half years.

This mobilization of resources within a matter of months demonstrates that a nation and, indeed, the world can restructure its economy quickly if it is convinced of the need to do so. Many people—although not yet the majority—are already convinced of the need for a wholesale restructuring of the economy. The issue isn't whether most people will eventually be won over but whether they will be convinced before the bubble economy collapses.

History judges political leaders by whether they respond to the great issues of their time. For today's leaders, that issue is how to deflate the world's bubble economy before it bursts. This bubble threatens the future of everyone, rich and poor alike. It challenges us to restructure the global economy, to build an eco-economy.

We now have some idea of what needs to be done and how to do it. The UN has set social goals for education, health, and the reduction of hunger and poverty in its Millennium Development Goals. My latest book, *Plan B*, offers a sketch for the re-

structuring of the energy economy to stabilize atmospheric carbon dioxide levels, a plan to stabilize population, a strategy for raising land productivity and restoring the earth's vegetation, and a plan to raise water productivity worldwide. The goals are essential and the technologies are available.

We have the wealth to achieve these goals. What we don't yet have is the leadership. And if the past is any guide to the future, that leadership can only come from the United States. By far the wealthiest society that has ever existed, the United States has the resources to lead this effort.

Yet the additional external funding needed to achieve universal primary education in the eighty-eight developing nations that require help is conservatively estimated by the World Bank at $15 billion per year. Funding for an adult literacy program based largely on volunteers is estimated at $4 billion. Providing for the most basic health care is estimated at $21 billion by the World Health Organization. The additional funding needed to provide reproductive health and family planning services to all women in developing nations is $10 billion a year.

Closing the condom gap and providing the additional nine billion condoms needed to control the spread of HIV in the developing world and Eastern Europe requires $2.2 billion—$270 million for condoms and $1.9 billion for AIDS prevention education and condom distribution. The cost per year of extending school lunch programs to the forty-four poorest nations is $6 billion per year. An additional $4 billion per year would cover the cost of assistance to preschool children and pregnant women in these nations.

In total, this comes to $62 billion. If the United States offered to cover one-third of this additional funding, the other industrial nations would almost certainly be willing to provide the remainder, and the worldwide effort to eradicate hunger, illiteracy, disease, and poverty would be under way.

The challenge isn't just to alleviate poverty, but in doing so to build an economy that is compatible with the Earth's natural systems—an eco-economy, an economy that can sustain progress. This means a fundamental restructuring of the energy economy and a substantial modification of the food economy. It also means raising the productivity of energy and shifting from fossil fuels to renewables. It means raising water productivity over the next half-century, much as we did land productivity over the last one.

It is easy to spend hundreds of billions in response to terrorist threats but the reality is that the resources needed to disrupt a modern economy are small, and a Department of Homeland Security, however heavily funded, provides only minimal protection from suicidal terrorists. The challenge isn't just to provide a high-tech military response to terrorism but to build a global society that is environmentally sustainable, socially equitable, and democratically based—one where there is hope for everyone. Such an effort would more effectively undermine the spread of terrorism than a doubling of military expenditures.

We can build an economy that doesn't destroy its natural support systems, a global community where the basic needs of all the Earth's people are satisfied, and a world that will allow us to think of ourselves as civilized. This is entirely doable. To paraphrase former President Franklin Roosevelt at another of those hinge points in history, let no one say it cannot be done.

The choice is ours—yours and mine. We can stay with business as usual and preside over a global bubble economy that keeps expanding until it bursts, leading to economic decline. Or we can adopt Plan B and be the generation that stabilizes population, eradicates poverty, and stbilizes climate. Historians will record the choice—but it is ours to make.

Lester R. Brown is president of the Earth Policy Institute. This article is adapted from his recently released book *Plan B: Rescuing a Planet Under Stress and a Civilization in Trouble*, which is available for free downloading at **www.earth-policy.org**

Globalizing Greenwash

After a string of environmental disasters associated with its projects and years of criticism, the World Bank promised in the early 1990s to clean up its act and become a lean green funding machine. Pamela Foster inspects the Bank's track record over the last 10 years and isn't convinced.

Pamela Foster

FOR 15 days in 1991, Medha Patkar, social scientist and community leader, was effectively left to starve by those backing the construction of the giant Sardar Sarovar dam project in the Narmada valley in India. This slight woman was on a hunger strike until her protests, along with those of thousands of others opposing the project, were heard by the World Bank, one of the dam's main backers. In an unprecedented move, the World Bank agreed to an independent review of the project.

Completed in 1992, the review substantiated the people's claims about the negative environmental and social impacts of the dam. It found that the Indian Government's calculations of the amount of energy that would be produced and the number of people affected were completely inaccurate. It also found the rehabilitation and relocation efforts of the Government to be insufficient and in violation of human rights.

Not long after this damning report, criticism again rained on the Bank—an internal review of the Bank's loan portfolio was leaked to the public. It revealed that a large number of projects were unsatisfactory using even the Bank's own narrow set of criteria. The report linked the decline in project quality to a 'pervasive culture of approval' for loans, whereby pressure to lend overwhelms all other considerations.

By 1993, it appeared that momentum might have been created for change. With clear signals coming from the Bank, India did not submit applications for further financing of the Sardar Sarovar project. The World Bank also created an independent Inspection Panel to address complaints related to Bank projects and the Bank's failure to follow its own rules. This kind of accountability system was unique among financial institutions.

In 1995, following the first-ever Inspection Panel investigation, the World Bank dropped its commitment to funding the Arun-III dam in Nepal, another mega-project which would have had disastrous consequences for one of the world's poorest countries. An editorial on the Arun-III dam in the Ottawa Citizen of 22 August 1995 said: 'The World Bank has justly been criticized in the past for acting more like a bank than an aid agency. It has promoted mega-projects such as dams, airports and large-scale irrigation systems that wreaked environmental havoc and led countries into a spiralling debt while doing little to help the poor. Now the World Bank seems to have reached a turning point … As the Bank marks its 50th Anniversary, however, there appears to be a real shift in its attitude towards development.' But nearly nine years on, has there been a real shift in attitude at the Bank?

Reigniting wars

Few are celebrating the 60th birthday of the World Bank. In 2003, it invested in an oilfield and pipeline development stretching across Central Asia. Regional and international non-governmental organizations (NGOs) investigating the Baku-Tbilisi-Ceyhan project have called attention to the fact that it would cause serious human rights abuses, could spark or reignite regional wars, would rob local people of their land and livelihoods, and would deliver yet more oil to already saturated Western markets, further contributing to climate change. An analysis by a coalition of organizations monitoring the project pointed out no less than 170 partial or full violations of the World Bank's own policies. Yet the Bank approved it anyway, arguing that its involvement raises the environmental and social standards of the project.

The Bank used similar arguments to justify its financing of Exxon and Chevron for the Chad-Cameroon Oil and Pipeline project in 2000. On 10 October 2003, the date of the official inauguration of the Chad-Cameroon Oil Pipeline, a coalition of human rights associations in Chad called for a national day of mourning, arguing that the oil revenues 'will only be another weapon in the hands of a plundering oligarchy to oppress the Chadian people.'

Since refusing to disburse further funds for the Sardar Sarovar dam in 1993, the World Bank had not funded any large dam projects in India. Yet, late in 2003, a senior Bank official told the Indian newspaper *Economic Times* that the process of funding hydro-power projects in India was already under way. This news is consistent with the 'High Risk/High Reward' strategy that the Bank adopted earlier in 2003. It renews the Bank's commitment to mega-projects like large hydro dams with no explanation as to how the Bank will address their known negative impacts.

This development is particularly worrying because the Bank's own policies intended to protect the environment and vulnerable social groups have been steadily diluted from day one. Introduced in the early 1980s, these so-called Safeguard Policies cover issues such as environmental assessment, forests, indigenous peoples and natural habitat. The Bank has been 'reformatting' them since 1996. Many NGOs fear that policies are being made so flexible that staff or borrowers can never be accused of having violated them. In 2002, the Bank changed its forest policy to allow it to support logging in tropical forests.

A report from the Operations and Evaluation Department of the World Bank that same year summarized its performance in the area of the environment. 'Environmental sustainability was not integrated into the Bank's core objectives ... there has been a lack of consistent management commitment to the environment, coupled with a lack of consistent management accountability.'

Disarming through dialogue

What the Bank has done is insist on privatization, liberalization and exploitation of natural resources, despite overwhelming evidence that this approach results in poverty, environmental degradation and increasing inequity.

Whereas the Bank used to dismiss its many opponents, it now attempts to disarm through dialogue. Consultation processes initiated by the Bank are proliferating, but with no corresponding changes in policy or practice.

In 1998, for example, at the urging of communities around the world, the Bank appointed an independent World Commission on Dams (WCD). The WCD comprised 12 representatives of industry, government and dam-affected peoples including community activists like Medha Patkar. After 30 months of intensive study and consultation, the WCD released a report that developed new standards and policies for guiding future projects. The Bank has yet to adopt its recommendations.

Instead, the Bank adopted a Water Resources Sector Strategy in 2003 that embraces high-risk dam projects and the privatization of water services, a strategy that has been shown to lead to further impoverishment of the poor as the private sector prices essential services out of their grasp. The World Bank also continues to justify investments in fossil fuels as 'development' projects. From mid-1992 to June 2002, the World Bank supported 226 fossil fuel projects with

over \$22 billion in financing despite its self-affirmed commitment to combat climate change.

In the name of poverty alleviation, the Bank has entered into partnership with some of the most notorious producers of hazardous pesticides, again undermining its stated policy commitments to the environment. Personnel exchanges routinely occur between the World Bank and major pesticide companies—companies whose misdeeds have included illegal toxic shipments, chemical dumping and accidents, exposing humans to high levels of toxins and false advertising.

When presented with this evidence, the World Bank relies on its own External Affairs and Publications departments to manage criticism. Since the Earth Summit in 1992, the Bank's External Affairs department budget has increased 52 per cent.

While the World Bank's External Affairs department globalizes greenwash, its Publications department greenwashes globalization. The flagship document of the World Bank is the annual World Development Report, which the Bank uses to provide an 'intellectual framework' for describing its work as development. Positioned as 'secular bibles' on current development thinking, its two World Development Reports on Environment were published in 1992 and 2003 in order to influence outcomes of the World Summits on Sustainable Development. Defining development largely as growth, the reports avoid addressing contradictions between particular approaches to growth and poverty eradication, environmental protection and reductions in inequality—in other words, justifying business as usual.

In 2002 the evaluations department of the Bank again warned: 'Unless and until ... the environment becomes part of the Bank's core objectives ... the tension between the Bank and its stakeholders that has characterized the past decade will continue and probably intensify.'

Probably? The Bank has shown itself consistently to be unwilling to or perhaps incapable of change.

Pamela Foster is co-ordinator for the Halifax Initiative, a Canadian coalition

IMF and the Environment

'Only sustainable economic growth—a central aim of the IMF's policy advice—can generate the additional resources needed to address environmental problems. Ideally, the environment benefits from virtuous circles in which sustainable economic growth reduces poverty, increases resources available to improve the environment, and is itself reinforced by these trends.'— IMF Factsheet on the Environment 2003

The IMF claims its goal is to increase financial resources available to governments. Laudable as this is, IMF loans and policy advice exploit natural and human resources to increase the pool of money available for, among other things, repaying debts countries owe it, the World Bank and their largest shareholders. The outcome is that there remains a net flow of financial, natural and human resources from the South to the North.

The IMF approach to growth involves pressuring countries to:

- reduce government spending
- promote export-led growth
- increase foreign investment

Each of these planks has an enormous negative environmental impact and, at best, mixed economic impacts.

Reductions in government spending reduce financial resources needed to enforce environmental regulations. Budget-cutting for an IMF loan in Russia led to a 40-per-cent reduction in funding for protected areas, increasing poaching and illegal logging. An IMF loan after the 1999 financial crisis in Brazil led to a 90-per-cent cut in the largest official programme for the protection of the Amazon.

To promote export-led growth in the mid-1990s, the IMF prevailed upon Cameroon to devalue its currency and cut export taxes on forest products. This made logging more profitable and by 2000, over 75 per cent of the country's forest cover had been or was scheduled to be logged.

Restrictions on foreign investment are seen as barriers by the IMF, rather than legitimate policy tools of national governments. In the Philippines, the IMF advised the Government to put in place new laws to facilitate foreign ownership in the country's mining sector, resulting in few benefits for local communities and increased social and environmental stress.

The IMF claims that it is too difficult to link a particular outcome to a particular policy. It has a small team in its Fiscal Affairs Department with special responsibility for following environmental issues relevant to its work. It also relies on the World Bank, 'given the Bank's substantial expertise in the area of the environment.' A vicious not virtuous circle, indeed.

of development, environment, labour, human rights and faith groups deeply concerned about the international financial system and its institutions. info@halifaxinitiative.org

From *New Internationalist*, March 2004, pp. 20-21. Copyright © 2004 by New Internationalist. Reprinted by permission.

UNIT 2
Population, Policy, and Economy

Unit Selections

7. **Population and Consumption: What We Know, What We Need to Know**, Robert W. Kates
8. **An Economy for the Earth**, Lester R. Brown
9. **Factory Farming in the Developing World**, Danielle Nierenberg
10. **Why Race Matters in the Fight for a Healthy Planet**, Jennifer Hattam
11. **Will Frankenfood Save the Planet?**, Jonathan Rauch

Key Points to Consider

• Why should policy makers in the more developed countries of the world become more aware of the true dimensions of the world's food problem? How can increased awareness of food scarcity and misallocation lead to solutions for both food production and environmental protection?

• How are present economic theories insufficient in addressing environmental issues? How might an "eco-economy" take on different characteristics than the traditional economies that exist in the world's developing countries?

• How does "factory farming," which may increase food production, also increase problems of environmental contamination? What is the relationship between regulations developed to control factory farming and the migration of this agro-business system to new areas?

• How serious is the argument that racial or economic status can be related to environmental quality? Do poor people and non-white populations suffer from a lack of "environmental justice"?

• Assess the potential benefits and damages of genetic engineering of agricultural products. Is there or should there be an alliance of interests between scientists involved in biotechnology and environmental organizations?

 **Links: www.dushkin.com/online/**
These sites are annotated in the World Wide Web pages.

The Hunger Project
   http://www.thp.org
Poverty Mapping
   http://www.povertymap.net
World Health Organization
   http://www.who.int
World Population and Demographic Data
   http://geography.about.com/cs/worldpopulation/
WWW Virtual Library: Demography & Population Studies
   http://demography.anu.edu.au/VirtualLibrary/

One of the greatest setbacks on the road to the development of more stable and sensible population policies came about as a result of inaccurate population growth projections made in the late 1960s and early 1970s. The world was in for a population explosion, the experts told us back then. But shortly after the publication of the heralded works *The Population Bomb* (Paul Ehrlich, 1975) and *Limits to Growth* (D. H. Meadows et al., 1974), the growth rate of the world's population began to decline slightly. There was no cause and effect relationship at work here. The decline in growth was simply demographic transition at work, a process in which declining population growth tends to accompany increasing levels of economic development. Since the alarming predictions did not come to pass, the world began to relax a little. However, two facts still remain: population growth in biological systems must be limited by available resources and the availability of Earth's resources is finite.

That population growth cannot continue indefinitely is a mathematical certainty. But it is also a certainty that contemporary notions of a continually expanding economy must give way before the realities of a finite resource base. Consider the following: In developing countries, high and growing rural population densities have forced the use of increasingly marginal farmland once considered to be too steep, too dry, too wet, too sterile, or too far from markets for efficient agricultural use. Farming this land damages soil and watershed systems, creates deforestation problems, and adds relatively little to total food production. In the more developed world, farmers also have been driven—usually by market forces—to farm more marginal lands and to rely more on environmentally harmful farming methods utilizing high levels

of agricultural chemicals (such as pesticides and artificial fertilizers). These chemicals create hazards for all life and rob the soil of its natural ability to renew itself. The increased demand for economic expansion has also created an increase in the use of precious groundwater reserves for irrigation purposes, depleting those reserves beyond their natural capacity to recharge and creating the potential for once-fertile farmland and grazing land to be transformed into desert. The continued demand for higher production levels also contributes to a soil erosion problem that has reached alarming proportions in all agricultural areas of the world, whether high or low on the scale of economic development. The need to increase the food supply and its consequent effects on the agricultural environment are not the only results of continued population growth. For industrialists, the larger market creates an almost irresistible temptation to accelerate production, requiring the use of more marginal resources and resulting in the destruction of more fragile ecological systems, particularly in the tropics. For consumers, the increased demand for products means increased competition for scarce resources, driving up the cost of those resources until only the wealthiest can afford what our grandfathers would have viewed as an adequate standard of living.

The articles selected for this second unit all relate, in one way or another, to the theory and reality of population growth and its relationship to public policy and economic growth. In the first selection, "Population and Consumption: What We Know, What We Need to Know," geographer and MacArthur Fellow Robert Kates argues that the present set of environmental problems are tied to both the expanding human population in strict numerical terms and the tendency of that growing population to demand more per capita shares of the world's dwindling resources. In the following selection— "An Economy for the Earth"—Lester Brown of the Earth Policy Institute takes a theoretical approach in contending that economic theory does not explain why so many environmental problems exist, whereas ecological theory does. What is required is a merger of the two sets of theories.

The next articles in this section move from the theoretical to the practical and from global to local scales in addressing issues of agriculture and environment. In "Factory Farming in the Developing World," Danielle Nierenberg of the World Watch Institute notes that, while factory farming has allowed meat to become more common in the diet of developing countries, the farming system itself produces significant damage at both local and global scales. She suggests that part of the problem is in thinking about factory farms—or the presence of meat in the diet—as symbols of wealth. Wealth also defines class and race and the next article in the section is devoted to a discussion of "environmental justice" or the notion that race becomes an issue when dealing with environmental topics. In "Why Race Matters In the Fight for a Healthy Planet," Jennifer Hattam of the Sierra Club notes that environmental problems are colored by race and

class issues that must be overcome if the environmental problems themselves are to be overcome. The third article in this group, by journalist Jonathan Rauch, addresses the specific problem of genetic engineering, particularly of agricultural products. Rauch contends that while genetic engineering may be the most environmentally-beneficial technology in decades, its benefits are limited by environmental activists who are—with quite good reason—suspicious of claims by business and industry that they have "solved" environmental problems with technology. If the new biotechnology is as environmentally-friendly as Rauch suggests, then the natural allies of genetic engineers are environmentalists.

All the authors of selections in this unit make it clear that the global environment is being stressed by population growth as well as environmental and economic policies that result in more environmental pressure and degradation. While it should be evident that we can no longer afford to permit the unplanned and unchecked growth of the planet's dominant species, it should also be apparent that the unchecked growth of economic systems without some kind of environmental accounting systems is just as dangerous.

Population and Consumption

What We Know, What We Need to Know

by Robert W. Kates

Thirty years ago, as Earth Day dawned, three wise men recognized three proximate causes of environmental degradation yet spent half a decade or more arguing their relative importance. In this classic environmentalist feud between Barry Commoner on one side and Paul Ehrlich and John Holdren on the other, all three recognized that growth in population, affluence, and technology were jointly responsible for environmental problems, but they strongly differed about their relative importance. Commoner asserted that technology and the economic system that produced it were primarily responsible.[1] Ehrlich and Holdren asserted the importance of all three drivers: population, affluence, and technology. But given Ehrlich's writings on population,[2] the differences were often, albeit incorrectly, described as an argument over whether population or technology was responsible for the environmental crisis.

Now, 30 years later, a general consensus among scientists posits that growth in population, affluence, and technology are jointly responsible for environmental problems. This has become enshrined in a useful, albeit overly simplified, identity known as IPAT, first published by Ehrlich and Holdren in *Environment* in 1972[3] in response to the more limited version by Commoner that had appeared earlier in *Environment* and in his famous book *The Closing Circle*.[4] In this identity, various forms of environmental or resource impacts (I) equals population (P) times affluence (A) (usually income per capita) times the impacts per unit of income as determined by technology (T) and the institutions that use it. Academic debate has now shifted from the greater or lesser importance of each of these driving forces of environmental degradation or resource depletion to debate about their interaction and the ultimate forces that drive them.

However, in the wider global realm, the debate about who or what is responsible for environmental degradation lives on. Today, many Earth Days later, international debates over such major concerns as biodiversity, climate change, or sustainable development address the population and the affluence terms of Holdrens' and Ehrlich's identity, specifically focusing on the character of consumption that affluence permits. The concern with technology is more complicated because it is now widely recognized that while technology can be a problem, it can be a

solution as well. The development and use of more environmentally benign and friendly technologies in industrialized countries have slowed the growth of many of the most pernicious forms of pollution that originally drew Commoner's attention and still dominate Earth Day concerns.

A recent report from the National Research Council captures one view of the current public debate, and it begins as follows:

> *For over two decades, the same frustrating exchange has been repeated countless times in international policy circles. A government official or scientist from a wealthy country would make the following argument: The world is threatened with environmental disaster because of the depletion of natural resources (or climate change or the loss of biodiversity), and it cannot continue for long to support its rapidly growing population. To preserve the environment for future generations, we need to move quickly to control global population growth, and we must concentrate the effort on the world's poorer countries, where the vast majority of population growth is occurring.*

Government officials and scientists from low-income countries would typically respond:

> *If the world is facing environmental disaster, it is not the fault of the poor, who use few resources. The fault must lie with the world's wealthy countries, where people consume the great bulk of the world's natural resources and energy and cause the great bulk of its environmental degradation. We need to curtail overconsumption in the rich countries which use far more than their fair share, both to preserve the environment and to allow the poorest people on earth to achieve an acceptable standard of living.*[5]

It would be helpful, as in all such classic disputes, to begin by laying out what is known about the relative responsibilities of both population and consumption for the environmental crisis, and what might need to be known to address them. However, there is a profound asymmetry that must fuel the frustration of the developing countries' politicians and scientists: namely, how much people know about population and how little they know about consumption. Thus, this article begins by examining these differences in knowledge and action and concludes with the alternative actions needed to go from more to enough in both population and consumption.[6]

Population

What population is and how it grows is well understood even if all the forces driving it are not. Population begins with people and their key events of birth, death, and location. At the margins, there is some debate over when life begins and ends or whether residence is temporary or permanent, but little debate in between. Thus, change in the world's population or any place is the simple arithmetic of adding births, subtracting deaths, adding immigrants, and subtracting outmigrants. While whole subfields of demography are devoted to the arcane details of these additions and subtractions, the error in estimates of population for almost all places is probably within 20 percent and for countries with modern statistical services, under 3 percent—better estimates than for any other living things and for most other environmental concerns.

Current world population is more than six billion people, growing at a rate of 1.3 percent per year. The peak annual growth rate in all history—about 2.1 percent—occurred in the early 1960s, and the peak population increase of around 87 million per year occurred in the late 1980s. About 80 percent or 4.8 billion people live in the less developed areas of the world, with 1.2 billion living in industrialized countries. Population is now projected by the United Nations (UN) to be 8.9 billion in 2050, according to its medium fertility assumption, the one usually considered most likely, or as high as 10.6 billion or as low as 7.3 billion.[7]

A general description of how birth rates and death rates are changing over time is a process called the demographic transition.[8] It was first studied in the context of Europe, where in the space of two centuries, societies went from a condition of high births and high deaths to the current situation of low births and low deaths. In such a transition, deaths decline more rapidly than births, and in that gap, population grows rapidly but eventually stabilizes as the birth decline matches or even exceeds the death decline. Although the general description of the transition is widely accepted, much is debated about its cause and details.

The world is now in the midst of a global transition that, unlike the European transition, is much more rapid. Both births and deaths have dropped faster than experts expected and history foreshadowed. It took 100 years for deaths to drop in Europe compared to the drop in 30 years in the developing world. Three is the current global average births per woman of reproductive age. This number is more than halfway between the average of five children born to each woman at the post World War II peak of population growth and the average of 2.1 births required to achieve eventual zero population growth.[9] The death transition is more advanced, with life expectancy currently at 64 years. This represents three-quarters of the transition between a life expectancy of 40 years to one of 75 years. The current rates of decline in births outpace the estimates of the demographers, the UN having reduced its latest medium ex-

pectation of global population in 2050 to 8.9 billion, a reduction of almost 10 percent from its projection in 1994.

Demographers debate the causes of this rapid birth decline. But even with such differences, it is possible to break down the projected growth of the next century and to identify policies that would reduce projected populations even further. John Bongaarts of the Population Council has decomposed the projected developing country growth into three parts and, with his colleague Judith Bruce, has envisioned policies that would encourage further and more rapid decline.[10] The first part is unwanted fertility, making available the methods and materials for contraception to the 120 million married women (and the many more unmarried women) in developing countries who in survey research say they either want fewer children or want to space them better. A basic strategy for doing so links voluntary family planning with other reproductive and child health services.

Yet in many parts of the world, the desired number of children is too high for a stabilized population. Bongaarts would reduce this desire for large families by changing the costs and benefits of childrearing so that more parents would recognize the value of smaller families while simultaneously increasing their investment in children. A basic strategy for doing so accelerates three trends that have been shown to lead to lower desired family size: the survival of children, their education, and improvement in the economic, social, and legal status for girls and women.

However, even if fertility could immediately be brought down to the replacement level of two surviving children per woman, population growth would continue for many years in most developing countries because so many more young people of reproductive age exist. So Bongaarts would slow this momentum of population growth by increasing the age of childbearing, primarily by improving secondary education opportunity for girls and by addressing such neglected issues as adolescent sexuality and reproductive behavior.

How much further could population be reduced? Bongaarts provides the outer limits. The population of the developing world (using older projections) was expected to reach 10.2 billion by 2100. In theory, Bongaarts found that meeting the unmet need for contraception could reduce this total by about 2 billion. Bringing down desired family size to replacement fertility would reduce the population a billion more, with the remaining growth—from 4.5 billion today to 7.3 billion in 2100—due to population momentum. In practice, however, a recent U.S. National Academy of Sciences report concluded that a 10 percent reduction is both realistic and attainable and could lead to a lessening in projected population numbers by 2050 of upwards of a billion fewer people.[11]

Consumption

In contrast to population, where people and their births and deaths are relatively well-defined biological events, there is no consensus as to what consumption includes. Paul Stern of the National Research Council has described the different ways physics, economics, ecology, and sociology view consumption.[12] For physicists, matter and energy cannot be consumed, so consumption is conceived as transformations of matter and energy with increased entropy. For economists, consumption is spending on consumer goods and services and thus distinguished from their production and distribution. For ecologists, consumption is obtaining energy and nutrients by eating something else, mostly green plants or other consumers of green plants. And for some sociologists, consumption is a status symbol—keeping up with the Joneses—when individuals and households use their incomes to increase their social status through certain kinds of purchases. These differences are summarized in the box below.

In 1977, the councils of the Royal Society of London and the U.S. National Academy of Sciences issued a joint statement on consumption, having previously done so on population. They chose a variant of the physicist's definition:

> *Consumption is the human transformation of materials and energy. Consumption is of concern to the extent that it makes the transformed materials or energy less available for future use, or negatively impacts biophysical systems in such a way as to threaten human health, welfare, or other things people value.*[13]

On the one hand, this society/academy view is more holistic and fundamental than the other definitions; on the other hand, it is more focused, turning attention to the environmentally damaging. This article uses it as a working definition with one modification, the addition of information to energy and matter, thus completing the triad of the biophysical and ecological basics that support life.

In contrast to population, only limited data and concepts on the transformation of energy, materials, and information exist.[14] There is relatively good global knowledge of energy transformations due in part to the common units of conversion between different technologies. Between 1950 and today, global energy production and use increased more than fourfold.[15] For material transformations, there are no aggregate data in common units on a global basis, only for some specific classes of materials including materials for energy production, construction, industrial minerals and metals, agricultural crops, and water.[16] Calculations of material use by volume, mass, or value lead to different trends.

Trend data for per capita use of physical structure materials (construction and industrial minerals, metals, and forestry products) in the United States are relatively complete. They show an inverted S shaped (logistic) growth pattern: modest doubling between 1900 and the depression of the 1930s (from two to four metric tons), followed by a steep quintupling with economic recovery until the early 1970s (from two to eleven tons), followed by a leveling off since then with fluctuations related to economic downturns (see Figure 1).[17] An aggregate analysis of all current material production and consumption in the United States averages more than 60 kilos per person per day (excluding water). Most of this material flow is split between energy and related products (38 percent) and minerals for construction (37 percent), with the remainder as industrial minerals (5 percent), metals (2 percent), products of fields (12 percent), and forest (5 percent).[18]

A massive effort is under way to catalog biological (genetic) information and to sequence the genomes of microbes, worms, plants, mice, and people. In contrast to the molecular detail, the number and diversity of organisms is unknown, but a conservative estimate places the number of species on the order of 10 million, of which only one-tenth have been described.[19] Although there is much interest and many anecdotes, neither concepts nor data are available on most cultural information. For example, the number of languages in the world continues to decline while the number of messages expands exponentially.

What Is Consumption?

Physicist: "What happens when you transform matter/energy"

Ecologist: "What big fish do to little fish"

Economist: "What consumers do with their money"

Sociologist: "What you do to keep up with the Joneses"

Trends and projections in agriculture, energy, and economy can serve as surrogates for more detailed data on energy and material transformation.[20] From 1950 to the early 1990s, world population more than doubled (2.2 times), food as measured by grain production almost tripled (2.7 times), energy more than quadrupled (4.4 times), and the economy quintupled (5.1 times). This 43-year record is similar to a current 55-year projection (1995–2050) that assumes the continuation of current trends or, as some would note, "business as usual." In this 55-year projection, growth in half again of population (1.6 times) finds almost a doubling of agriculture (1.8 times), more than twice as much energy used (2.4 times), and a quadrupling of the economy (4.3 times).[21]

Thus, both history and future scenarios predict growth rates of consumption well beyond population. An attractive similarity exists between a demographic transition that moves over time from high births and high deaths to low births and low deaths with an energy, materials, and information transition. In this transition, societies will use increasing amounts of energy and materials as consumption increases, but over time the energy and materials input per unit of consumption decrease and information substitutes for more material and energy inputs.

Some encouraging signs surface for such a transition in both energy and materials, and these have been variously labeled as decarbonization and dematerialization.[22] For more than a century, the amount of carbon per unit of energy produced has been decreasing. Over a shorter period, the amount of energy used to produce a unit of production has also steadily declined. There is also evidence for dematerialization, using fewer materials for a unit of production, but only for industrialized countries and for some specific materials. Overall, improvements in technology

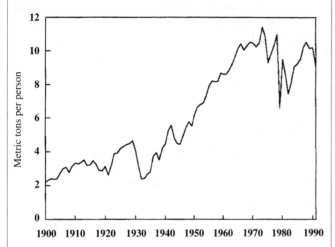

Figure 1. Consumption of physical structure materials in the United States, 1900-1991

SOURCE: I. Wernick, "Consuming Materials: The American Way," *Technological Forecasting and Social Change*, 53 (1996): 114.

and substitution of information for energy and materials will continue to increase energy efficiency (including decarbonization) and dematerialization per unit of product or service. Thus, over time, less energy and materials will be needed to make specific things. At the same time, the demand for products and services continues to increase, and the overall consumption of energy and most materials more than offsets these efficiency and productivity gains.

What to Do about Consumption

While quantitative analysis of consumption is just beginning, three questions suggest a direction for reducing environmentally damaging and resource-depleting consumption. The first asks: *When is more too much for the life-support systems of the natural world and the social infrastructure of human society?* Not all the projected growth in consumption may be resource-depleting— "less available for future use"—or environmentally damaging in a way that "negatively impacts biophysical systems to threaten human health, welfare, or other things people value."[23] Yet almost any human-induced transformations turn out to be either or both resource-depleting or damaging to some valued environmental component. For example, a few years ago, a series of eight energy controversies in Maine were related to coal, nuclear, natural gas, hydroelectric, biomass, and wind generating sources, as well as to various energy policies. In all the controversies, competing sides, often more than two, emphasized environmental benefits to support their choice and attributed environmental damage to the other alternatives.

Despite this complexity, it is possible to rank energy sources by the varied and multiple risks they pose and, for those concerned, to choose which risks they wish to minimize and which they are more willing to accept. There is now almost 30 years of experience with the theory and methods of risk assessment and 10 years of experience with the identification and setting of en-

vironmental priorities. While there is still no readily accepted methodology for separating resource-depleting or environmentally damaging consumption from general consumption or for identifying harmful transformations from those that are benign, one can separate consumption into more or less damaging and depleting classes and *shift* consumption to the less harmful class. It is possible to *substitute* less damaging and depleting energy and materials for more damaging ones. There is growing experience with encouraging substitution and its difficulties: renewables for nonrenewables, toxics with fewer toxics, ozone-depleting chemicals for more benign substitutes, natural gas for coal, and so forth.

The second question, *Can we do more with less?*, addresses the supply side of consumption. Beyond substitution, shrinking the energy and material transformations required per unit of consumption is probably the most effective current means for reducing environmentally damaging consumption. In the 1997 book, *Stuff: The Secret Lives of Everyday Things*, John Ryan and Alan Durning of Northwest Environment Watch trace the complex origins, materials, production, and transport of such everyday things as coffee, newspapers, cars, and computers and highlight the complexity of reengineering such products and reorganizing their production and distribution.[24]

Yet there is growing experience with the three Rs of consumption shrinkage: reduce, recycle, reuse. These have now been strengthened by a growing science, technology, and practice of industrial ecology that seeks to learn from nature's ecology to reuse everything. These efforts will only increase the existing favorable trends in the efficiency of energy and material usage. Such a potential led the Intergovernmental Panel on Climate Change to conclude that it was possible, using current best practice technology, to reduce energy use by 30 percent in the short run and 50–60 percent in the long run.[25] Perhaps most important in the long run, but possibly least studied, is the potential for and value of substituting information for energy and materials. Energy and materials per unit of consumption are going down, in part because more and more consumption consists of information.

The third question addresses the demand side of consumption—*When is more enough?*[26] Is it possible to reduce consumption by more satisfaction with what people already have, by *satiation*, no more needing more because there is enough, and by *sublimation*, having more satisfaction with less to achieve some greater good? This is the least explored area of consumption and the most difficult. There are, of course, many signs of *satiation* for some goods. For example, people in the industrialized world no longer buy additional refrigerators (except in newly formed households) but only replace them. Moreover, the quality of refrigerators has so improved that a 20-year or more life span is commonplace. The financial pages include frequent stories of the plight of this industry or corporation whose markets are saturated and whose products no longer show the annual growth equated with profits and progress. Such enterprises are frequently viewed as failures of marketing or entrepreneurship rather than successes in meeting human needs sufficiently and efficiently. Is it possible to reverse such views, to create a standard of satiation, a satisfaction in a need well met?

Can people have more satisfaction with what they already have by using it more intensely and having the time to do so? Economist Juliet Schor tells of some overworked Americans who would willingly exchange time for money, time to spend with family and using what they already have, but who are constrained by an uncooperative employment structure.[27] Proposed U.S. legislation would permit the trading of overtime for such compensatory time off, a step in this direction. *Sublimation*, according to the dictionary, is the diversion of energy from an immediate goal to a higher social, moral, or aesthetic purpose. Can people be more satisfied with less satisfaction derived from the diversion of immediate consumption for the satisfaction of a smaller ecological footprint?[28] An emergent research field grapples with how to encourage consumer behavior that will lead to change in environmentally damaging consumption.[29]

A small but growing "simplicity" movement tries to fashion new images of "living the good life."[30] Such movements may never much reduce the burdens of consumption, but they facilitate by example and experiment other less-demanding alternatives. Peter Menzel's remarkable photo essay of the material goods of some 30 households from around the world is powerful testimony to the great variety and inequality of possessions amidst the existence of alternative life styles.[31] Can a standard of "more is enough" be linked to an ethic of "enough for all"? One of the great discoveries of childhood is that eating lunch does not feed the starving children of some far-off place. But increasingly, in sharing the global commons, people flirt with mechanisms that hint at such—a rationing system for the remaining chlorofluorocarbons, trading systems for reducing emissions, rewards for preserving species, or allowances for using available resources.

A recent compilation of essays, *Consuming Desires: Consumption, Culture, and the Pursuit of Happiness*,[32] explores many of these essential issues. These elegant essays by 14 well-known writers and academics ask the fundamental question of why more never seems to be enough and why satiation and sublimation are so difficult in a culture of consumption. Indeed, how is the culture of consumption different for mainstream America, women, inner-city children, South Asian immigrants, or newly industrializing countries?

Why We Know and Don't Know

In an imagined dialog between rich and poor countries, with each side listening carefully to the other, they might ask themselves just what they actually know about population and consumption. Struck with the asymmetry described above, they might then ask: "Why do we know so much more about population than consumption?"

The answer would be that population is simpler, easier to study, and a consensus exists about terms, trends, even policies. Consumption is harder, with no consensus as to what it is, and with few studies except in the fields of marketing and advertising. But the consensus that exists about population comes from substantial research and study, much of it funded by governments and groups in rich countries, whose asymmetric concern readily identifies the troubling fertility behavior of others and only reluctantly considers their own consumption behavior. So while consumption is harder, it is surely studied less (see Table 1).

The asymmetry of concern is not very flattering to people in developing countries. Anglo-Saxon tradition has a long history of dominant thought holding the poor responsible for their con-

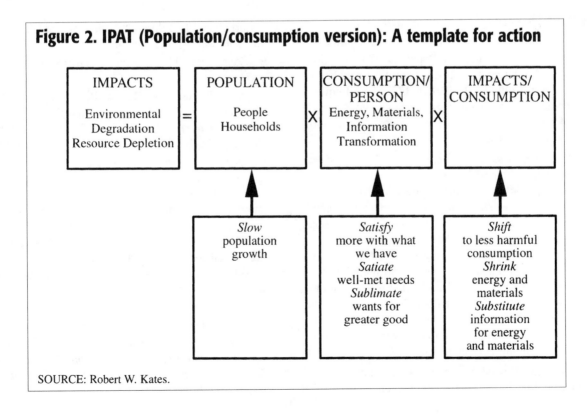

Figure 2. IPAT (Population/consumption version): A template for action

IMPACTS		POPULATION		CONSUMPTION/ PERSON		IMPACTS/ CONSUMPTION
Environmental Degradation Resource Depletion	=	People Households	X	Energy, Materials, Information Transformation	X	

	Slow population growth	*Satisfy* more with what we have *Satiate* well-met needs *Sublimate* wants for greater good	*Shift* to less harmful consumption *Shrink* energy and materials *Substitute* information for energy and materials

SOURCE: Robert W. Kates.

dition—they have too many children—and an even longer tradition of urban civilization feeling besieged by the barbarians at their gates. But whatever the origins of the asymmetry, its persistence does no one a service. Indeed, the stylized debate of population versus consumption reflects neither popular understanding nor scientific insight. Yet lurking somewhere beneath the surface concerns lies a deeper fear.

Table 1. A comparison of population and consumption

Population	Consumption
Simpler, easier to study	More complex
Well-funded research	Unfunded, except marketing
Consensus terms, trends	Uncertain terms, trends
Consensus policies	Threatening policies

SOURCE: Robert W. Kates.

Consumption is more threatening, and despite the North–South rhetoric, it is threatening to all. In both rich and poor countries alike, making and selling things to each other, including unnecessary things, is the essence of the economic system. No longer challenged by socialism, global capitalism seems inherently based on growth—growth of both consumers and their consumption. To study consumption in this light is to risk concluding that a transition to sustainability might require profound changes in the making and selling of things and in the opportunities that this provides. To draw such conclusions, in the absence of convincing alternative visions, is fearful and to be avoided.

What We Need to Know and Do

In conclusion, returning to the 30-year-old IPAT identity—a variant of which might be called the Population/Consumption (PC) version—and restating that identity in terms of population and consumption, it would be: $I = P*C/P*I/C$, where I equals environmental degradation and/or resource depletion; P equals the number of people or households; and C equals the transformation of energy, materials, and information (see Figure 2).

With such an identity as a template, and with the goal of reducing environmentally degrading and resource-depleting influences, there are at least seven major directions for research and policy. To reduce the level of impacts per unit of consumption, it is necessary to separate out more damaging consumption and shift to less harmful forms, *shrink* the amounts of environmentally damaging energy and materials per unit of consumption, and *substitute* information for energy and materials. To reduce consumption per person or household, it is necessary to *satisfy* more with what is already had, *satiate* well-met consumption needs, and *sublimate* wants for a greater good. Finally, it is possible to *slow* population growth and then to *stabilize* population numbers as indicated above.

However, as with all versions of the IPAT identity, population and consumption in the PC version are only proximate

driving forces, and the ultimate forces that drive consumption, the consuming desires, are poorly understood, as are many of the major interventions needed to reduce these proximate driving forces. People know most about slowing population growth, more about shrinking and substituting environmentally damaging consumption, much about shifting to less damaging consumption, and least about satisfaction, satiation, and sublimation. Thus the determinants of consumption and its alternative patterns have been identified as a key understudied topic for an emerging sustainability science by the recent U.S. National Academy of Science study.[33]

But people and society do not need to know more in order to act. They can readily begin to separate out the most serious problems of consumption, shrink its energy and material throughputs, substitute information for energy and materials, create a standard for satiation, sublimate the possession of things for that of the global commons, as well as slow and stabilize population. To go from more to enough is more than enough to do for 30 more Earth Days.

Robert W. Kates is an independent scholar in Trenton, Maine; a geographer; university professor emeritus at Brown University; and an executive editor of *Environment*. The research for "Population and Consumption: What We Know, What We Need to Know" was undertaken as a contribution to the recent National Academies/National Research Council report, *Our Common Journey: A Transition Toward Sustainability*. The author retains the copyright to this article. Kates can be reached at RR1, Box 169B, Trenton, ME 04605.

NOTES

1. B. Commoner, M. Corr, and P. Stamler, "The Causes of Pollution," *Environment*, April 1971, 2–19.
2. P. Ehrlich, *The Population Bomb* (New York: Ballantine, 1966).
3. P. Ehrlich and J. Holdren, "Review of The Closing Circle," *Environment*, April 1972, 24–39.
4. B. Commoner, *The Closing Circle* (New York: Knopf, 1971).
5. P. Stern, T. Dietz, V. Ruttan, R. H. Socolow, and J. L. Sweeney, eds., *Environmentally Significant Consumption: Research Direction* (Washington, D.C.: National Academy Press, 1997), 1.
6. This article draws in part upon a presentation for the 1997 De Lange-Woodlands Conference, an expanded version of which will appear as: R. W. Kates, "Population and Consumption: From More to Enough," in *In Sustainable Development: The Challenge of Transition*, J. Schmandt and C. H. Wards, eds. (Cambridge, U.K.: Cambridge University Press, forthcoming), 79–99.
7. United Nations, Population Division, *World Population Prospects: The 1998 Revision* (New York: United Nations, 1999).
8. K. Davis, "Population and Resources: Fact and Interpretation," K. Davis and M. S. Bernstam, eds., in *Resources, Environment and Population: Present Knowledge, Future Options*, supplement to *Population and Development Review*, 1990: 1–21.

9. Population Reference Bureau, *1997 World Population Data Sheet of the Population Reference Bureau* (Washington, D.C.: Population Reference Bureau, 1997).

10. J. Bongaarts, "Population Policy Options in the Developing World," *Science*, 263: (1994), 771–776; and J. Bongaarts and J. Bruce, "What Can Be Done to Address Population Growth?" (unpublished background paper for The Rockefeller Foundation, 1997).

11. National Research Council, Board on Sustainable Development, *Our Common Journey: A Transition Toward Sustainability* (Washington, D.C.: National Academy Press, 1999).

12. See Stern, et al., note 5 above.

13. Royal Society of London and the U.S. National Academy of Sciences, "Towards Sustainable Consumption," reprinted in *Population and Development Review*, 1977, 23 (3): 683–686.

14. For the available data and concepts, I have drawn heavily from J. H. Ausubel and H. D. Langford, eds., *Technological Trajectories and the Human Environment.* (Washington, D.C.: National Academy Press, 1997).

15. L. R. Brown, H. Kane, and D. M. Roodman, *Vital Signs 1994: The Trends That Are Shaping Our Future* (New York: W. W. Norton and Co., 1994).

16. World Resources Institute, United Nations Environment Programme, United Nations Development Programme, World Bank, *World Resources, 1996–97* (New York: Oxford University Press, 1996); and A. Gruebler, *Technology and Global Change* (Cambridge, Mass.: Cambridge University Press, 1998).

17. I. Wernick, "Consuming Materials: The American Way," *Technological Forecasting and Social Change*, 53 (1996): 111–122.

18. I. Wernick and J. H. Ausubel, "National Materials Flow and the Environment," *Annual Review of Energy and Environment*, 20 (1995): 463–492.

19. S. Pimm, G. Russell, J. Gittelman, and T. Brooks, "The Future of Biodiversity," *Science*, 269 (1995): 347–350.

20. Historic data from L. R. Brown, H. Kane, and D. M. Roodman, note 15 above.

21. One of several projections from P. Raskin, G. Gallopin, P. Gutman, A. Hammond, and R. Swart, *Bending the Curve: Toward Global Sustainability*, a report of the Global Scenario Group, Polestar Series, report no. 8 (Boston: Stockholm Environmental Institute, 1995).

22. N. Nakicénovíc, "Freeing Energy from Carbon," in *Technological Trajectories and the Human Environment*, eds., J. H. Ausubel and H. D. Langford. (Washington, D.C.: National Academy Press, 1997); I. Wernick, R. Herman, S. Govind, and J. H. Ausubel, "Materialization and Dematerialization: Measures and Trends," in J. H. Ausubel and H. D.

Langford, eds., *Technological Trajectories and the Human Environment* (Washington, D.C.: National Academy Press, 1997), 135–156; and see A. Gruebler, note 16 above.

23. Royal Society of London and the U.S. National Academy of Science, note 13 above.

24. J. Ryan and A. Durning, *Stuff: The Secret Lives of Everyday Things* (Seattle, Wash.: Northwest Environment Watch, 1997).

25. R. T. Watson, M. C. Zinyowera, and R. H. Moss, eds., *Climate Change 1995: Impacts, Adaptations, and Mitigation of Climate Change—Scientific-Technical Analyses* (Cambridge, U.K.: Cambridge University Press, 1996).

26. A sampling of similar queries includes: A. Durning, *How Much Is Enough?* (New York: W. W. Norton and Co., 1992); Center for a New American Dream, *Enough!: A Quarterly Report on Consumption, Quality of Life and the Environment* (Burlington, Vt.: The Center for a New American Dream, 1997); and N. Myers, "Consumption in Relation to Population, Environment, and Development," *The Environmentalist*, 17 (1997): 33–44.

27. J. Schor, *The Overworked American* (New York: Basic Books, 1991).

28. A. Durning, *How Much Is Enough?: The Consumer Society and the Future of the Earth* (New York: W. W. Norton and Co., 1992); Center for a New American Dream, note 26 above; and M. Wackernagel and W. Ress, *Our Ecological Footprint: Reducing Human Impact on the Earth* (Philadelphia. Pa.: New Society Publishers, 1996).

29. W. Jager, M. van Asselt, J. Rotmans, C. Vlek, and P. Costerman Boodt, *Consumer Behavior: A Modeling Perspective in the Contest of Integrated Assessment of Global Change*, RIVM report no. 461502017 (Bilthoven, the Netherlands: National Institute for Public Health and the Environment, 1997); and P. Vellinga, S. de Bryn, R. Heintz, and P. Molder, eds., *Industrial Transformation: An Inventory of Research*. IHDP-IT no. 8 (Amsterdam, the Netherlands: Institute for Environmental Studies, 1997).

30. H. Nearing and S. Nearing. *The Good Life: Helen and Scott Nearing's Sixty Years of Self-Sufficient Living* (New York: Schocken, 1990); and D. Elgin, *Voluntary Simplicity: Toward a Way of Life That Is Outwardly Simple, Inwardly Rich* (New York: William Morrow, 1993).

31. P. Menzel, *Material World: A Global Family Portrait* (San Francisco: Sierra Club Books, 1994).

32. R. Rosenblatt, ed., *Consuming Desires: Consumption, Culture, and the Pursuit of Happiness* (Washington, D.C.: Island Press, 1999).

33. National Research Council, Board on Sustainable Development, *Our Common Journey: A Transition Toward Sustainability* (Washington, D.C.: National Academy Press, 1999).

From *Environment*, April 2000, pp. 10–19. Reprinted with permission of the Helen Dwight Reid Educational Foundation. Published by Heldref Publications, 1319 Eighteenth St., NW, Washington, DC 20036-1802. © 2000.

AN
Economy
FOR THE
Earth

by LESTER R. BROWN

In 1543, Polish astronomer Nicolaus Copernicus published "On the Revolutions of the Celestial Spheres," in which he challenged the view that the sun revolves around the Earth, arguing instead that the Earth revolves around the sun. With his new model of the solar system, he began a wide-ranging debate among scientists, theologians, and others. His alternative to the earlier Ptolemaic model, which had the Earth at the center of the universe, led to a revolution in thinking, to a new worldview.

Today we need a similar shift in our worldview in how we think about the relationship between the Earth and the economy. The issue now is not which celestial sphere revolves around the other but whether the environment is part of the economy or the economy is part of the environment. Economists see the environment as a subset of the economy. Ecologists, on the other hand, see the economy as a subset of the environment.

Like Ptolemy's view of the solar system, the economists' view is confusing efforts to understand our modern world. This has resulted in an economy that is out of sync with the ecosystem on which it depends.

Economic theory and economic indicators don't explain how the economy is disrupting and destroying the Earth's natural systems. Economic theory doesn't explain why Arctic Sea ice is melting, why grasslands are turning into desert in northwestern China, why coral reefs are dying in the South Pacific, or why the Newfoundland cod fishery collapsed. Nor does it explain why we are in the early stages of the greatest extinction of plants and animals since the dinosaurs disappeared 65 million years ago. Yet economics is essential to measuring the cost to society of these excesses.

Evidence that the economy is in conflict with the Earth's natural systems can be seen in the daily news reports of shrinking forests, eroding soils, deteriorating rangelands, expanding deserts, rising carbon dioxide levels, falling water tables, rising temperatures, more destructive storms, melting glaciers, rising sea level, dying coral reefs, collapsing fisheries, and disappearing species. These trends, which mark an increasingly stressed relationship between the economy and the ecosystem, are taking a growing economic toll. At some point this could overwhelm the worldwide forces of progress, leading to economic decline. The challenge for our generation is to reverse these trends before environmental deterioration leads to long-term economic decline—as it did for so many earlier civilizations.

These increasingly visible trends indicate that, if the operation of the subsystem—the economy—is not compatible with the behavior of the larger system—the ecosystem—both will eventually suffer. The larger the economy becomes relative to the ecosystem, and the more it presses against the Earth's natural limits, the more destructive this incompatibility will be.

An environmentally sustainable economy—an *eco-economy*—requires that the principles of ecology establish the framework for the formulation of economic policy and that economists and ecologists work together to fashion the new economy. Ecologists understand that all economic activity, indeed all life, depends on the Earth's ecosystem—the complex of individual species living together, interacting with each other and their physical habitat. These millions of species exist in an intricate balance, woven together by food chains, nutrient cycles, the hydrological cycle, and the climate system. Economists know how to translate goals into policy. Ecologists and economists working together can design and build an eco-economy that can sustain progress.

Just as recognition that the Earth is not the center of the solar system set the stage for advances in astronomy, physics, and related sciences, so will recognition that the economy isn't the center of our world create the conditions to sustain economic progress and improve the human condition.

Converting the world economy into an eco-economy, however, is a monumental undertaking, as the current gap between economists and ecologists in their perception of the world could not be wider. There is no precedent for transforming an economy shaped largely by market forces into one shaped by the principles of ecology.

The scale of projected economic growth outlines the dimensions of the challenge. The growth in world output of goods and services from $6 trillion in 1950 to $43 trillion in 2000 has caused environmental devastation on a scale that we could not easily have imagined a half-century ago. If the world economy continues to expand at 3 percent annually, the output of goods and services will increase fourfold over the next half-century, reaching $172 trillion.

Building an eco-economy in the time available requires rapid systemic change. We won't succeed with a project here and a project there. We are winning occasional battles, but we are losing the war because we don't have a strategy for the systemic economic change that will put the world on a development path that is environmentally sustainable.

Although the concept of environmentally sustainable development evolved a quarter-century ago, not one country has a strategy to build an eco-economy—to restore balances, to stabilize population and water tables, and to conserve forests, soils, and diversity of plant and animal life. We can find individual countries that are succeeding with one or more elements of restructuring but not one that is progressing satisfactorily on all fronts.

Nevertheless, glimpses of the eco-economy are clearly visible in some countries. For example, thirty-one nations in Europe, as well as Japan, have stabilized their population size, satisfying one of the most basic conditions of an eco-economy. Europe has stabilized its population within its food-producing capacity, leaving it with an exportable surplus of grain to help meet the deficits in developing countries. Furthermore, China—the world's most populous country—now has lower fertility than the United States and is moving toward population stability.

Currently Denmark is the eco-economy leader. It has stabilized its population, banned the construction of coal-fired power plants, banned the use of nonrefillable beverage containers, and is now getting 15 percent of its electricity from wind. In addition, it has restructured its urban transport network; now 32 percent of all trips in Copenhagen are on bicycle. Denmark is still not close to balancing carbon emissions and fixation, but it is moving in that direction.

Other countries have also achieved specific goals. A reforestation program in South Korea, begun more than a generation ago, has blanketed that country's hills and mountains with trees. Costa Rica has a plan to shift entirely to renewable energy by 2025. Iceland plans to be the world's first hydrogen-powered economy.

Building an eco-economy will affect every facet of our lives. It will alter how we light our homes, what we eat, where we live, how we use our leisure time, and how many children we have. It will give us a world in which we are a part of nature instead of estranged from it.

According to Seth Dunn in the November/December 2000 issue of *World Watch* magazine, a consortium of corporations led by Shell Hydrogen and DaimlerChrysler reached an agreement in 1999 with the government of Iceland to establish this new economy. Shell is interested because it wants to begin developing its hydrogen production and distribution capacity, and DaimlerChrysler expects to have the first fuel cell-powered automobile on the market. Shell plans to open its first chain of hydrogen stations in Iceland.

So we can see pieces of the eco-economy emerging, but systemic change requires a fundamental shift in market signals—signals that respect the principles of ecological sustainability. Unless we are prepared to shift taxes from income to environmentally destructive activities—such as carbon emissions and the wasteful use of water—we won't succeed in building an eco-economy.

It is a huge undertaking to restore the balances of nature. For energy, this will depend on shifting from a carbon-based economy to a hydrogen-based one. Even the most progressive oil companies, such as British Petroleum and Royal Dutch Shell, which are all talking extensively about building a solar/hydrogen energy economy, are still investing overwhelmingly in oil, with funds going into climate-benign sources accounting for a minute share of their investment.

Reducing soil erosion to the level of new soil formation will require changes in farming practices. In some situations, it will mean shifting from intense tillage to minimum tillage or no tillage. Agroforestry will loom large in an eco-economy.

Restoring forests that recycle rainfall inland and control flooding is itself a huge undertaking. It means reversing decades of tree cutting and land clearing with forest restoration—an activity that will require millions of people planting billions of trees.

Building an eco-economy will affect every facet of our lives. It will alter how we light our homes, what we eat, where we live, how we use our leisure time, and how many children we have. It will give us a world in which we are *a part of* nature instead of estranged from it.

In May 2001, the Bush White House released with great fanfare a twenty-year plan for the United States' energy economy. It disappointed many people because it largely overlooked the enormous potential for raising energy efficiency. It also overlooked the huge potential of wind power, which is likely to add more to U.S. generating capacity over the next twenty years than coal does. The plan was indicative of the problems some governments are having in fashioning an energy economy that is compatible with the Earth's ecosystem.

Prepared under the direction of Vice-President Dick Cheney, the administration's plan centers on expanding production of fossil fuels—something more appropriate for the early twentieth century than the early twenty-first. It emphasizes the role of coal, but its architects were apparently unaware that world coal use peaked in 1996 and has declined some 7 percent since then as other countries have turned away from this fuel. Even China, which rivals the United States as a coal-burning country, has reduced its coal use by an estimated 14 percent since 1996.

Bush's energy plan notes that the 2 percent of U.S. electricity generation that today comes from renewable sources, excluding hydropower, would increase to 2.8 percent in 2020. But months before the Bush plan was released, the American Wind Energy Association was projecting a staggering 60 percent growth in U.S. wind-generating capacity in 2001. Worldwide, use of wind power alone has multiplied nearly fourfold over the last five years—a growth rate matched only by the computer industry.

The solar cell is a relatively new source of alternative energy and, after wind power, is already the second fastest growing source. In 1952, three scientists at Bell Labs in Princeton, New Jersey, discovered that sunlight striking a silicon-based material produces electricity. The discovery of this photovoltaic, or solar cell, opened up a vast new potential for power generation. Initially very costly, solar cells were used mostly for high-value purposes such as providing the electricity to operate satellites. As the solar cell became economical, it opened the potential for providing electricity to remote sites not yet linked to an electrical grid. It is already more economical in remote areas to install solar cells than to build a power plant and connect villages by grid. By the end of 2000, about a million homes worldwide were getting their electricity from solar cell installations. An estimated 700,000 of these were in villages in developing countries.

Today, as the cost of solar cells continues to decline, this technology is becoming competitive with large, centralized power sources. For many of the two billion people in the world who don't have access to conventional electricity sources, small solar cell arrays provide an affordable shortcut. In the developing world, in some communities not serviced by a centralized power system, local entrepreneurs are investing in solar cell generating facilities and selling the energy to village families.

Perhaps the most exciting technological advance has been the development of a photovoltaic roofing material in Japan. A joint effort involving the construction industry, the solar cell manufacturing industry, and the Japanese government plans to have 4,600 megawatts of electrical generating capacity in place by 2010—enough to satisfy all of the electricity needs of a country like Estonia. With photovoltaic roofing material, the roof of a building becomes the power plant. In some countries, including Germany and Japan, buildings now have a two-way meter—selling electricity to the local utility when they have an excess and buying it when they don't have enough.

In contrast to these other sources of renewable energy, geothermal energy comes from within the Earth itself. Produced radioactively and by the pressures of gravity, it is a vast resource, most of which is deep within the planet. Geothermal energy can be economically tapped when it is relatively close to the surface, as evidenced by hot springs, geysers, and volcanic activity. It is used directly both to supply heat and generate electricity.

Geothermal energy is much more abundant in some parts of the world than in others. The richest region is the vast Pacific Rim: along the western coastal regions of Latin America, Central America, and North America; and widely distributed in eastern Russia, Japan, the Korean Peninsula, China, and island nations like the Philippines, Indonesia, New Guinea, Australia, and New Zealand.

This energy source is essentially inexhaustible. Hot baths, for example, have been used for millennia. It is possible to extract heat faster than it is generated at any local site, but this is a matter of adjusting the extraction of heat to the amount generated. In contrast to oil fields, which are eventually depleted, properly managed geothermal fields keep producing indefinitely. In a time of mounting concern about climate change, many governments are beginning to exploit the geothermal potential—as in Iceland, where it heats some 85 percent of buildings; as in Japan, for hot baths when springs bring geothermal energy to the surface; as in the United States for generating electricity. In fact, the U.S. Department of Energy announced in 2000 that it was launching a program to develop the rich geothermal energy resources in the western United States. The goal is to have 10 percent of the electricity in the West coming from geothermal energy by 2020.

Although the Bush energy plan doesn't reflect it, the world energy economy is on the verge of a major transformation. Historically, the twentieth century was the century of fossil fuels: first coal, then oil, and finally natural gas were the workhorses of the world economy. But with the advent of the twenty-first century, the sun is setting on the fossil fuel era. The last several decades have shown a steady shift from the most polluting and climate-disrupting fuels toward clean, climate-benign energy sources.

Even the oil companies are now beginning to recognize that the time has come for an energy transition. After years of denying any link between fossil fuel burning and climate change, John Browne, the chief executive officer of BP, announced his new position in a historic speech at Stanford University in May 1997:

> My colleagues and I now take the threat of global warming seriously. The time to consider the policy dimensions of climate change is not when the link between greenhouse gases and climate change is conclusively proven but when the possibility cannot be discounted and is taken seriously by the society of which we are a part. We in BP have reached that point.

At an energy conference in Houston, Texas, in February 1999, Michael Bowlin, CEO of ARCO, said that the beginning of the end of the age of oil was in sight. He went on to discuss the need to shift from a carbon-based to a hydrogen-based energy economy.

The signs of restructuring the global energy economy are unmistakable. Events are moving far faster than would have been expected even a few years ago, driven in part by the mounting evidence that the Earth is indeed warming up and that the burning of fossil fuels is responsible. But can we do what needs to be done fast enough?

We know that social change often takes time. In Eastern Europe, it was fully four decades from the imposition of communism until its demise. Thirty-four years passed between the first U.S. Surgeon General's report on smoking and health and the landmark agreement between the tobacco industry and state governments. Thirty-eight years have passed since biologist Rachel Carson published *Silent Spring*, the wakeup call that gave rise to the modern environmental movement.

Sometimes things move much faster, especially when the magnitude of the threat is understood and the nature of the response is obvious, such as the U.S. response to the attack on Pearl Harbor. Within one year, the U.S. economy had largely been reconstructed. In less than four years, the war was over.

Accelerating the transition to a sustainable future means overcoming the inertia of both individuals and institutions. In some ways, inertia is our worst enemy. As individuals we often resist change. When we are gathered into large organizations, we resist it even more.

At the institutional level, we are looking for massive changes in industry, especially in energy. We are looking for changes in the material economy, shifting from a throwaway mentality to a closed loop/recycle mindset. If future food needs are to be satisfied adequately, we need a worldwide effort to reforest the land, conserve soil, and raise water productivity. Stabilizing population growth means quite literally a revolution in human reproductive behavior—one that recognizes a sustainable future is possible only if we average two children per couple. This isn't a debatable point. It is a mathematical reality.

The big remaining challenge is on the educational front: how can we help literally billions of people in the world understand not only the need for change but how that change can bring a life far better than they have today?

In this connection, I am frequently asked if it is too late. My response is: "Too late for what?" Is it too late to save the Aral Sea? Yes, the Aral Sea is dead; its fish have died, and its fisheries have collapsed. Is it too late to save the glaciers in Glacier National Park in the United States? Most likely. They are already half gone, and it would be virtually impossible now to reverse the rise in temperature in time to save them. Is it too late to avoid a rise in temperature from the buildup of greenhouse gases? Yes. A greenhouse gas-induced rise in temperature is apparently already underway. But is it too late to avoid runaway climate change? Perhaps not, if we quickly restructure the energy economy.

For many specifics, the answer is, yes, it is too late. But there is a broader, more fundamental question: is it too late to reverse the trends that will eventually lead to economic decline? Here I think the answer is no—not if we act quickly.

Perhaps the biggest challenge we face is shifting from a carbon-based to a hydrogen-based energy economy—basically moving from fossil fuels to renewable sources of energy, such as solar, wind, and geothermal. How fast can we make this change? Can it be done before we trigger irreversible damage, such as a disastrous rise in sea level? As I indicated, we know from the United States' response to the attack on Pearl Harbor that economic restructuring can occur at an incredible pace if a society is convinced of the need for it.

We study the archaeological sites of civilizations that moved onto economic paths that were environmentally destructive and could not make the needed course corrections in time. We face the same risk.

There is no middle path. Do we join together to build an economy that is sustainable, or do we stay with our environmentally unsustainable economy until it declines? It isn't a goal that can be compromised. One way or another, the choice will be made by our generation. But what we choose will affect life on Earth for all generations to come.

Lester R. Brown is president and founder of the Earth Policy Institute; the founder and former president of the Worldwatch Institute; a MacArthur Fellow; and the recipient of twenty-two honorary degrees and many awards, including the 1987 UN Environment Prize, the 1989 World Wide Fund for Nature Gold Medal, the 1991 Humanist of the Year Award, and the 1994 Blue Planet Prize for his "exceptional contributions to solving global environmental problems." He has authored or coauthored forty-seven books, nineteen monographs, and countless articles. This article is adapted from Eco-Economy: Building an Economy for the Earth, *which is available online at www.earthpolicy.org.*

Originally appeared in *The Humanist*, May/June 2002, pp. 32-34. Excerpted from Chapters 1 & 5 in Lester R. Brown's, *Eco-Economy: Building an Economy for the Earth*, W. W. Norton & Company, NY: 2001. © 2002 by Earth Policy Institute. Reprinted by permission.

Factory Farming in the Developing World

In some critical respects, this is not progress at all.

by Danielle Nierenberg

Walking through Bobby Inocencio's farm in the hills of Rizal province in the Philippines is like taking a step back to a simpler time. Hundreds of chickens (a cross between native Filipino chickens and a French breed) roam around freely in large, fenced pens. They peck at various indigenous plants, they eat bugs, and they fertilize the soil, just as domesticated chickens have for ages.

The scene may be old, but Inocencio's farm is anything but simple. What he has recreated is a complex and successful system of raising chickens that benefits small producers, the environment, and even the chickens. Once a "factory farmer," Inocencio used to raise white chickens for Pure Foods, one of the biggest companies in the Philippines.

Thousands of birds were housed in long, enclosed metal sheds that covered his property. Along with the breed stock and feeds he had to import, Inocencio also found himself dealing with a lot of imported diseases and was forced to buy expensive antibiotics to keep the chickens alive long enough to take them to market. Another trick of the trade Inocencio learned was the use of growth promotants that decrease the time it takes for chickens to mature.

Eventually he noticed that fewer and fewer of his neighbors were raising chickens, which threatened the community's food security by reducing the locally available supply of chickens and eggs. As the community dissolved and farms (and farming methods) that had been around for generations went virtually extinct, Inocencio became convinced that there had to be a different way to raise chickens and still compete in a rapidly globalizing marketplace. "The business of the white chicken," he says, "is controlled by the big guys." Not only do small farmers have to compete with the three big companies that control white chickens in the Philippines, but they must also contend

with pressure from the World Trade Organization (WTO) to open up trade. In the last two decades the Filipino poultry production system has transitioned from mainly backyard farms to a huge industry. In the 1980s the country produced 50 million birds annually. Today that figure has increased some ten-fold. The large poultry producers have benefited from this population explosion, but average farmers have not. So Inocencio decided to go forward by going back and reviving village-level poultry enterprises that supported traditional family farms and rural communities.

Inocencio's farm and others like it show that the Philippines can support indigenous livestock production and stand up to the threat of the factory farming methods now spreading around the world. Since 1997, his Teresa Farms has been raising free range chickens and teaching other farmers how to do the same. He says that the way he used to raise chickens, by concentrating so many of them in a small space, is dangerous. Diseases such as avian flu, leukosis J (avian leukemia), and Newcastle disease are spread from white chickens to the Filipino native chicken populations, in some cases infecting eggs before the chicks are even born. "The white chicken," says Inocencio, "is weak, making the system weak. And if these chickens are weak, why should we be raising them? Limiting their genetic base and using breeds that are not adapted to conditions in the Philippines is like setting up the potential for a potato blight on a global scale." Now Teresa Farms chickens are no longer kept in long, enclosed sheds, but roam freely in large tree-covered areas of his farm that he encloses with recycled fishing nets.

Inocencio's chickens also don't do drugs. Antibiotics, he says, are not only expensive but encourage disease. He found the answer to the problem of preventing diseases in chickens literally in his own back yard. His chickens eat spices and native plants with antibacterial and other medicinal properties. Chili, for instance, is mixed in grain to treat respiratory problems, stimulate appetite during heat stress, de-worm the birds, and to treat Newcastle disease. Native plants growing on the farm, including *ipil-ipil* and *damong maria*, are also used as low-cost alternatives to antibiotics and other drugs.

There was a time when most farms in the Philippines, the United States, and everywhere else functioned much like Bobby Inocencio's. But today the factory model of raising animals in intensive conditions is spreading around the globe.

A New Jungle

Meat once occupied a very different dietary place in most of the world. Beef, pork, and chicken were considered luxuries, and were eaten on special occasions or to enhance the flavor of other foods. But as agriculture became more mechanized, so did animal production. In the United States, livestock raised in the West was herded or transported east to slaughterhouses and packing mills. Upton Sinclair's *The Jungle*, written almost a century ago when the United States lacked many food-safety and labor regulations, described the appalling conditions of slaughterhouses in Chicago in the early 20th century and was a shocking expose of meat production and the conditions inflicted on both animals and humans by the industry. Workers were treated much like animals themselves, forced to labor long hours for very little pay under dangerous conditions, and with no job security.

If *The Jungle* were written today, however, it might not be set in the American Midwest. Today, developing nations like the Philippines are becoming the centers of large-scale livestock production and processing to feed the world's growing appetite for cheap meat and other animal products. But the problems Sinclair pointed to a century ago, including hazardous working conditions, unsanitary processing methods, and environmental contamination, still exist. Many have become even worse. And as environmental regulations in the European Union and the United States become stronger, large agribusinesses are moving their animal production operations to nations with less stringent enforcement of environmental laws.

These intensive and environmentally destructive production methods are spreading all over the globe, to Mexico, India, the former Soviet Union, and most rapidly throughout Asia. Wherever they crop up, they create a web of related food safety, animal welfare, and environmental problems. Philip Lymbery, campaign director of the World Society for the Protection of Animals, describes the growth of industrial animal production this way: Imagine traditional livestock production as a beach and factory farms as a tide. In the United States, the tide has completely covered the beach, swallowing up small farms and concentrating production in the hands of a few large companies. In Taiwan, it is almost as high. In the Philippines, however, the tide is just hitting the beach. The industrial, factory-farm methods of raising and slaughtering animals—methods that were conceived and developed in the United States and Western Europe—have not yet swept over the Philippines, but they are coming fast.

An Appetite for Destruction

Global meat production has increased more than fivefold since 1950, and factory farming is the fastest growing method of animal production worldwide. Feedlots are responsible for 43 percent of the world's beef, and more than half of the world's pork and poultry are raised in factory farms. Industrialized countries dominate production, but developing countries are rapidly expanding and intensifying their production systems. According to the United Nations Food and Agriculture Organization (FAO), Asia (including the Philippines) has the fastest developing livestock sector. On the islands that make up the Philippines, 500 million chickens and 20 million hogs are slaughtered each year.

Despite the fact that many health-conscious people in developed nations are choosing to eat less meat, worldwide meat consumption continues to rise. Consumption is growing fastest in the developing countries. Two-thirds of the gains in meat consumption in 2002 were in the developing world, where urbanization, rising incomes, and globalized trade are changing diets and fueling appetites for meat and animal products. Because eating meat has been perceived as a measure of economic and social development, the Philippines and other poor nations are eager to climb up the animal-protein ladder. People in the Philippines still eat relatively little meat, but their consumption is growing. As recently as 1995, the average Filipino ate 21 kilograms of meat per year. Since then, average consumption has soared to almost 30 kilograms per year, although that is still less than half the amount in Western countries, where per-capita consumption is 80 kilograms per year.

This push to increase both production and consumption in the Philippines and other developing nations is coming from a number of different directions. Since the end of World War II, agricultural development has been considered a part of the foreign aid and assistance given to developing nations. The United States and international development agencies have been leaders in promoting the use of pesticides, artificial fertilizers, and other chemicals to boost agricultural production in these countries, often at the expense of the environment. American corporations like Purina Mills and Tyson Foods are also opening up feed mills and farms so they can expand business in the Philippines.

But Filipinos are also part of the push to industrialize agriculture. "This is not an idea only coming from the West," says Dr. Abe Agulto, president of the Philippine Society for the Protection of Animals, "but also coming from us." Meat equals wealth in much of the world and many Filipino businesspeople have taken up largescale livestock production to supply the growing demand for meat. But small farmers don't get much financial support in the Philippines. It's not farms like Bobby Inocencio's that are likely to get government assistance, but the big production facilities that can crank out thousands of eggs, chicks, or piglets a year.

The world's growing appetite for meat is not without its consequences, however. One of the first indications that meat production can be hazardous arises long before animals ever reach the slaughterhouse. Mountains of smelly and toxic manure are created by the billions of animals raised for human consumption in the world each year. In the United States, people in North Carolina know all too well the effects of this liquid and solid waste. Hog production there has increased faster than anywhere else in the nation, from 2 million hogs per year in 1987 to 10 million hogs per year today. Those hogs produce more than 19 million tons of manure each year and most of it gets stored in lagoons, or large uncovered containment pits. Many of those lagoons flooded and burst when Hurricane Floyd swept through the region in 1999. Hundreds of acres of land and miles of waterway were flooded with excrement, resulting in massive fish kills and millions of dollars in cleanup costs. The lagoons' contents are also known to leak out and seep into groundwater.

Some of the same effects can now be seen in the Philippines. Not far from Teresa Farms sits another, very different, farm that produces the most frequently eaten meat product in the world. Foremost Farms is the largest piggery, or pig farm, in all of Asia. An estimated 100,000 pigs are produced there every year.

High walls surround Foremost and prevent people in the community from getting in or seeing what goes on inside. What they do get a whiff of is the waste. Not only do the neighbors smell the manure created by the 20,000 hogs kept at Foremost or the 10,000 hogs kept at nearby Holly Farms, but their water supply has also been polluted by it. In fact, they've named the river where many of them bathe and get drinking water the River Stink. Apart from the stench, some residents have complained of skin rashes, infections, and other health problems from the water. And instead of keeping the water clean and installing effective waste treatment, the firms are just digging deeper drinking wells and giving residents free access to them. Many in the community are reluctant to complain about the smell because they fear losing their water supply. Even the mayor of Bulacan, the nearby village, has said "we give these farms leeway as much as possible because they provide so much economically."

It would be easy to assume that some exploitive foreign corporation owns Foremost, but in fact the owner is Lucia Tan, a Filipino. Tan is not your average Filipino, however, but the richest man in the Philippines. In addition to Foremost Farms, he owns San Miguel beer and Philippine Airlines. Tan might be increasing his personal wealth, but his farm and others like it are gradually destroying traditional farming methods and threatening indigenous livestock breeds in the Philippines. As a result, many small farmers can no longer afford to produce hogs for sale or for their own consumption, which forces them to become consumers of Tan's pork. Most of the nation's 11 million hogs are still kept in back yards, but because of farms like Foremost, factory farming is growing. Almost one-quarter of the breeding herd is now factory farmed. More than 1 million pigs are raised in factory farms every year in Bulacan alone.

Chicken farms in the Philippines are also becoming more intensive. The history of intensive poultry production in the Philippines is not long. Forty years ago, the nation's entire population was fed on native eggs and chickens produced by family farmers. Now, most of those farmers are out of business. They have lost not only their farms, but livestock diversity and a way of life as well.

The loss of this way of life to the industrialized farm-to-abattoir system has made the process more callous at every stage. Adopting factory farming methods works to diminish farmers' concern for the welfare of their livestock. Chickens often can't walk properly because they have been pumped full of growth-promoting antibiotics to gain weight as quickly as possible. Pigs are confined to gestation crates where they can't turn around. Cattle are crowded together in feedlots that are seas of manure.

Most of the chickens in the country are from imported breed stock and the native Filipino chicken has practically disappeared because of viral diseases spread by foreign breeds. Almost all of the hens farmed commercially for their eggs are confined in wire battery cages that cram three or four hens together, giving each bird an area less than the size of this page to stand on.

Unlike laying hens, chickens raised for meat in the Philippines are not housed in cages. But they're not pecking around in back yards, either. Over 90 percent of the meat chickens raised in the Philippines live in long sheds that house thousands of birds. At this time, most Filipino producers allow fowl to have natural ventilation and lighting and some roaming room, but they are under pressure to adopt more "modern" factory-farm standards to increase production.

The problems of a system that produces a lot of animals in crowded and unsanitary conditions can also be seen off the farm. The *baranguay* (neighborhood) of Tondo in Manila is best known for the infamous "Smoky Mountain" garbage dump that collapsed on scavengers in 2000, killing at least 200 people. But another hazard also sits in the heart of Tondo. Surrounded by tin houses, stores, and bars, the largest government-owned slaughtering facility in the country processes more than 3,000 swine, cattie, and *caraboa* (water buffalo) per day, all brought from farms just outside the city limits. The slaughterhouse does have a waste treatment system where the blood and other waste is supposed to be treated before it is released into the city's sewer system and nearby Manila Bay. Unfortunately, that's not what's going on. Instead, what can't be cut up and sold for human consumption is dumped into the sewer.

Some 60 men are employed at the plant. They stun, bludgeon, and slaughter animals by hand and at a breakneck pace. They wear little protective gear as they slide around on floors slippery with blood, which makes it hard to stun animals on the first try, or sometimes even the second, or to butcher meat without injuring themselves.

The effects of producing meat this way also show up in rising cases of food-borne illness, emerging animal diseases that can spread to humans, and in an increasingly overweight Filipino population that doesn't remember where meat comes from.

There are few data on the incidence of food-borne illness in the Philippines or most other developing nations, and even fewer about how much of it might be related to eating unsafe meat. What food safety experts do know is that food-borne illness is one of the most widespread health problems worldwide.

And it could be an astounding 300–350 times more frequent than reported, according to the World Health Organization. Developing nations bear the greatest burden because of the presence of a wide range of parasites, toxins, and biological hazards and the lack of surveillance, prevention, and treatment measures—all of which ensnarl the poor in a chronic cycle of infection. According to the FAO, the trend toward increased commercialization and intensification of livestock production is leading to a variety of food safety problems. Crowded, unsanitary conditions and poor waste treatment in factory farms exacerbate the rapid movement of animal diseases and food-borne infections. E. coli 0157:H7, for instance, is spread from animals to humans when people eat food contaminated by manure. Animals raised in intensive conditions often arrive at slaughterhouses covered in feces, thus increasing the chance of contamination during slaughtering and processing.

Cecilia Ambos is one of the meat inspectors at the Tondo slaughterhouse. Cecilia or another inspector is required to be on site at all times, but she says she rarely has to go to the killing floor. Inspections of carcasses only occur, she said, if one of the workers alerts the inspector. That doesn't happen very often, and not because the animals are all perfectly healthy. Consider that the men employed at the plant are paid about $5 per day, which is less than half of the cost of living—and are working as fast as they can to slaughter a thousand animals per shift. It's unlikely that they have the time or the knowledge to notice problems with the meat.

Since the 1960s, farm-animal health in the United States has depended not on humane farming practices but on the use of antibiotics. Many of the same drugs used to treat human illnesses are also used in animal production, thus reducing the arsenal of drugs available to fight food-borne illnesses and other health problems. Because antibiotics are given to livestock to prevent disease from spreading in crowded conditions and to increase growth, antibiotic resistance has become a global threat. In the Philippines, chicken, egg, and hog producers use antibiotics not because their birds or hogs are sick, but because drug companies and agricultural extension agents have convinced them that these antibiotics will ensure the health of their birds or pigs and increase their weight.

Livestock raised intensively can also spread diseases to humans. Outbreaks of avian flu in Hong Kong during the past five years have led to massive culls of thousands of chickens. When the disease jumped the species barrier for the first time in 1997, six of the eighteen people infected died. Avian flu spread to people living in Hong Kong again this February, killing two. Dr. Gary Smith, of the University of Pennsylvania School of Veterinary Medicine, also warns that "it is not high densities [of animals] that matter, but the increased potential for transmission between firms that we should be concerned about. The nature of the farming nowadays is such that there is much more movement of animals between farms than there used to be, and much more transport of associated materials between farms taking place rapidly. The problem is that the livestock industry is operating on a global, national, and county level." The foot-and-mouth disease epidemic in the United Kingdom is a perfect example of how just a few cows can spread a disease across an entire nation.

Modern Methods, Modern Policies?

The expansion of factory farming methods in the Philippines is raising the probability that it will become another fast food nation. Factory farms are supplying much of the pork and chicken preferred by fast food restaurants there. American-style fast food was unknown in the Philippines until the 1970s, when Jollibee, the Filipino version of McDonald's, opened its doors. Now, thanks to fast food giants like McDonald's, Kentucky Fried Chicken, Burger King, and others, the traditional diet of rice, vegetables, and a little meat or fish is changing—and so are rates of heart disease, diabetes, and stroke, which have risen to numbers similar to those in the United States and other western nations.

The Filipino government doesn't see factory farming as a threat. To the contrary, many officials hope it will be a solution to their country's economic woes, and they're making it easier for large farms to dominate livestock production. For instance, the Department of Agriculture appears to have turned a blind eye when many farms have violated environmental and animal welfare regulations. The government has also encouraged big farms to expand by giving them loans. But as the farms get bigger and produce more, domestic prices for chicken and pork fail, forcing more farmers to scale up their production methods. And because the Philippines (and many other nations) are prevented by the Global Agreement on Tariffs and Trade and the WTO from imposing tariffs on imported products, the Philippines is forced to allow cheap, factory-farmed American pork and poultry into the country. These products are then sold at lower prices than domestic meat.

Rafael Mariano, a leader in the Peasant Movement for the Philippines (KMP), has not turned away from the problems caused by factory farming in the Philippines. He and the 800,000 farmers he works with believe that "factory farming is not acceptable, we have our own farming." But farmers, he says, are told by big agribusiness companies that their methods are old fashioned, and that to compete in the global market they must forget what they have learned from generations of farming. Rafael and KMP are working to promote traditional methods of livestock production that benefit small farmers and increase local food security. This means doing what farmers used to do: raising both crops and animals. In mixed crop–livestock farms, animals and crops are parts of a self-sustaining system. Some farmers in the Philippines raise hogs, chickens, tilapia, and rice on the same farm. The manure from the hogs and chickens is used to fertilize the algae in ponds needed for both tilapia and rice to grow. These farms produce little waste, provide a variety of food for the farm, and give farmers social security when prices for poultry, pork, and rice go down.

The Philippines is not the only country at risk from the spread of factory farms. Argentina, Brazil, Canada, China, India, Mexico, Pakistan, South Africa, Taiwan, and Thailand are all seeing growth in industrial animal production. As regu-

The Return of the Native... Chicken

Bobby Inocencio believes that the happier his chickens are, the healthier they will be, both on the farm and at the table. Native Filipino chickens are a tough sell commercially, he says, because they typically weigh in at only one kilogram apiece. But Inocencio's chickens are part native and part SASSO (a French breed), and grow to two kilos in just 63 days in a free-range system. They are also better adapted to the climate of the Philippines, unlike white chickens that are more vulnerable to heat. As a result, Inocencio's chickens not only are nutritious, but taste good. Raising white chickens, he says, forced small farmers to become "consumers of a chicken that doesn't taste like anything." Further, his chickens don't contain any antibiotics and are just 5 percent fat, compared to 35 percent in the white chicken. Because they are not raised in the very high densities of factory farms, these chickens actually enrich the environment with their manure. They also provide a reliable source of income for local farmers and give Filipinos a taste of how things used to be.

lations controlling air and water pollution from such farms are strengthened in one country, companies simply pack up and move to countries with more lenient rules. Western European nations now have among the strongest environmental regulations in the world; farmers can only apply manure during certain times of the year and they must follow strict controls on how much ammonia is released from their farms. As a result, a number of companies in the Netherlands and Germany are moving their factory farms—but to the United States, not to developing countries. According to a recent report in the *Dayton Daily News*, cheap land and less restrictive environmental regulations in Ohio are luring European livestock producers to the Midwest. There, dairies with fewer than 700 cows are not required to obtain permits, which would regulate how they control manure. But 700 cows can produce a lot of manure. In 2001, five Dutch-owned dairies were cited by the Ohio Environmental Protection Agency for manure spills. "Until there are international regulations controlling waste from factory farms," says William Weida, director of the Global Reaction Center for the Environment/Spira Factory Farm project, "it is impossible to prevent farms from moving to places with less regulation."

Mauricio Rosales of FAO's Livestock, Environment, and Development Project also stresses the need for siting farms where they will benefit both people and the environment. "Zoning," he says, "is necessary to produce livestock in the most economically viable places, but with the least impact." For instance, when livestock live in urban or peri-urban areas, the potential for nutrient imbalances is high. In rural areas manure can be a valuable resource because it contains nitrogen and phosphorous, which fertilize the soil. In cities, however, manure is a toxic, polluting nuisance.

The triumph of factory farming is not inevitable. In 2001, the World Bank released a new livestock strategy which, in a surprising reversal of its previous commitment to funding of large-scale livestock projects in developing nations, said that as the livestock sector grows "there is a significant danger that the poor are being crowded out, the environment eroded, and global food safety and security threatened." It promised to use a "people-centered approach" to livestock development projects that will reduce poverty, protect environmental sustainability, ensure food security and welfare, and promote animal welfare. This turnaround happened not because of pressure from environmental or animal welfare activists, but because the large-scale, intensive animal production methods the Bank once advocated are simply too costly. Past policies drove out smallholders because economies of scale for large units do not internalize the environmental costs of producing meat. The Bank's new strategy includes integrating livestock–environment interactions in to environmental impact assessments, correcting regulatory distortions that favor large producers, and promoting and developing markets for organic products. These measures are steps in the right direction, but more needs to be done by lending agencies, governments, non-governmental organizations, and individual consumers. Changing the meat economy will require a rethinking of our relationship with livestock and the price we're willing to pay for safe, sustainable, humanely-raised food.

Meat is more than a dietary element, it's a symbol of wealth and prosperity. Reversing the factory farm tide will require thinking about farming systems as more than a source of economic wealth. Preserving prosperous family farms and their landscapes and raising healthy, humanely treated animals, should also be viewed as a form of affluence.

Further Reading:
World Society for the Protection of Animals,
www.wspa.org.uk

Danielle Nierenberg *is a Staff Researcher at the Worldwatch Institute.*

From *World Watch*, May/June 2003. © 2003 by Worldwatch Institute, www.worldwatch.org.

Why race matters

in the fight for a healthy planet

By Jennifer Hattam

Sierra Club founder John Muir didn't have to worry about lead paint, asthma attacks, or unemployment rates. But John McCown does. The codirector of the Sierra Club's environmental-justice program, McCown is one of a growing number of Club staff and volunteers working in the inner cities and rural areas that most Americans—including environmentalists—either fear or forget.

These activists are cleaning up dirty air in Fresno, California; protecting neighborhoods and wetlands in Gulfport, Mississippi, fighting urban blight in Washington, D.C.; and preserving jobs and air quality in Middletown, Ohio. In doing so, they're also pioneering a different way of organizing. In communities of color and low-income neighborhoods, which are more frequently beset by polluting facilities and ecologically destructive development, environmental concerns vie for attention with worries about jobs, health, crime, and education. To succeed, an environmental campaign must respect these other priorities and engage in the larger struggle for social and economic improvement. To ignore those problems is to risk turning potential allies into opponents, as a group of anti-pesticide activists learned the hard way.

"When you're a new organizer, it's easy to think that you can come into a neighborhood and make things happen," says Julie Eisenhardt, a Sierra Club environmental-justice activist. "One thing that becomes clear very quickly is that people already know the solutions to their problems. They just need a little help to make it real."

Playing a support role means working with the people who bear an environmental burden disproportionate to their numbers—African Americans, Latinos, Native Americans, Asian Americans, and poor whites. These interactions can be fraught with distrust and discomfort. Environmentalists have a reputation as wealthy, white tree-huggers who don't care about people of color and the poor, while members of those communities are perceived as unconcerned about the environment. Overcoming those stereotypes requires expanding the definition of what is considered an environmental problem, and addressing how race and class issues color environmental ones. After all, as McCown points out, "Pollution does not respect the barriers that we ourselves put between us. It does not discriminate."

The Bridge Builder

Creating a more diverse movement, one community at a time.

Activism runs in John McCown's Georgia blood. His father fought for civil rights and in the late 1960s helped turn poverty-stricken Hancock County, a rural area dominated by its plantation past, into the first U.S. county since Reconstruction to be politically led by African Americans. The younger McCown participated as early as third grade, when he was among the first black students to integrate a local all-white school.

Some 35 years later, McCown is still working to overcome prejudice. In 1993, he was hired by the Sierra Club to help with out-

reach to at-risk communities. Once the sole Club employee devoted full-time to environmental-justice issues, he now oversees grassroots organizers around the country. McCown talked with Sierra about how low-income neighborhoods and communities of color are fighting corporate polluters and political neglect.

How did you get started with environmental organizing?

Back in 1986, the governor of Georgia was trying to put a $50 million hazardous-waste incinerator and an 887-acre solid-

JOINING TOGETHER FOR JUSTICE

In 1993, the Sierra Club officially adopted its first environmental-justice policy, recognizing that "to achieve our mission of environmental protection and a sustainable future for the planet, we must attain social justice and human rights at home and around the globe." Since then, Club staff and volunteers have:

- provided organizing assistance to over 250 low-income neighborhoods and communities of color nationwide
- hired full-time environmental-justice organizers in Detroit; Memphis; Washington, D.C.; the Southwest; and central Appalachia
- awarded some two dozen grants, including ones to help local groups lead "toxic tours" in Cincinnati, create community gardens in Tennessee, and educate the public about environmental damage caused by factory farms in rural Alabama and Mississippi
- helped shut down a polluting sewage-treatment facility and a medical-waste incinerator in Detroit
- helped African-American families in Pensacola, Florida, get funds to move away from the two Superfund sites in their community
- helped Native American groups in New Mexico and Arizona block construction of an 18,000-acre mine that threatened a sacred lake
- supported grassroots groups along the U.S.-Mexico border with grants and organizing assistance
- collaborated with Amnesty International to defend activists around the world who risk jail, exile, or even death for speaking out on environmental issues

ON THE WEB *For more on the Club's environmental-justice work, visit* **www.sierraclub.org/ environmental_justice.** *To learn about its international efforts, visit* **www.sierraclub.org/beyondtheborders.**

waste landfill in my town. We got organized and after a five-year battle, we eventually blocked both projects. At that time, I had heard about the Sierra Club, but it always seemed conspicuously absent from local struggles like the one in my hometown. I often wondered, since we're all fighting the same environmental foes, why aren't we talking to each other?

In the early 1990s, I finally met some Sierra Club activists from the Gulf region who were trying to make the Club more relevant in adversely impacted communities. Historically, the organization had not sought or valued the views of these communities. But these people realized that this was wrong, and that it was time to come together. They wanted to hire someone who could help make that happen. And they hired me.

The first time I brought Sierra Club people into an environmental-justice community was in Columbia, Mississippi, in 1994. We walked door to door, we stood in people's yards and around their porches, and we listened to them tell us what it's like to live in a neighborhood with a Superfund site. A chemical plant exploded there in 1977, and rather than rebuild, the company took most of the 4,000 drums of toxic chemicals and buried them there. When it rains, these chemicals wash through the community. People eat food from their gardens and get really sick.

The Sierra Club activists are hearing these stories, and they're realizing that, damn, these people are just like me! They've got the same hopes and aspirations. They want their children to grow up and be healthy, productive members of society. They're just catching hell because they happen to be African American and perceived as politically weak.

I've taken a lot of risks acting as a go-between for the Sierra Club and suffering communities; I've been called an Uncle Tom and all sorts of horrible names. But although I can remember lying on the floor of our house as a child, dodging bullets from the KKK, my father never taught us how to hate. I'm a spiritual person, and I know that God is not a racist. He created us all, so I want to do anything I can to help bring God's children together.

When you've seen multiracial coalitions succeed, what has been the key factor?

Some environmentalists will come into a community and say, "Don't worry! We'll do it for you! We'll speak for you!" Our approach is different. We believe that the real experts on environmental-justice issues are the people who are actually living beneath the smokestacks and on top of the landfills. These are your true leaders.

We go into a community only at its invitation, and we respect the right of the community to define its own agenda. We spend our resources on the challenges that *they* identify. We're giving workshops on fundraising and proposal-writing. We're hiring local organizers, we're renting offices with fax machines and computers that residents can use as a resource. This approach has been revolutionary.

What changes have you seen over the years in environmentalists' attitudes toward race and social justice?

This work has changed the entire direction of the Sierra Club. People are looking at the environmental-justice implications of everything they're working on—from sprawl to endangered species to clearcutting—and looking at ways to engage nontraditional constituencies. This Club will never be the same. It will be better.

At last year's environmental-justice conference, we sat down and had an honest discussion about race. Everybody shared their experiences—racial incidents that they may have suffered or participated in, instances where they stood by and did nothing—and we pledged as a result to confront institutional racism within the Sierra Club and get funding set aside to do racial-sensitivity training all over the country.

It's such a touchy topic—Americans don't like to talk about race.

But we've got to work through it. I've had time over the last 12 years to understand that Sierra Club volunteers are for real. I've stayed in their homes, eaten at their tables. I know their children, and I know these people are *serious* about working for justice. Without the benefit of that kind of interaction, other

Who Cares about the Environment

Flip through most magazines (including this one) and you might notice something about the people shown hiking, biking, or camping in ads celebrating the outdoors: They're almost always white. Indeed, while **minority groups make up 30 percent of the U.S. population,** they account for only 16 percent of all "outdoor enthusiasts," according to the Outdoor Industry Association.

But while they may engage less often in outdoor recreation, people of color do not care less about environmental problems. National opinion polls conducted between 1993 and 2000 consistently found that **a higher percentage of blacks than whites opt for pesticide-free produce,** and drive less for environmental reasons. The polls also showed that blacks join environmental groups—although not necessarily the nationally known ones—at similar rates as whites and are just as likely to call air and water pollution and global warming "very" or "extremely" dangerous to the environment. The actions of African-American leaders reflect these trends; **members of the Congressional Black Caucus are more likely to vote pro-environment** than their colleagues in the House.

In a 1998 survey by the Latino Issues Forum in California, **96 percent of Hispanics agreed that preserving the environment is important,** just behind reducing crime and improving public education—and ahead of protecting immigrant rights and bilingual education. The state's Hispanic voters voice these concerns at the ballot box, with 74 percent supporting a successful $2.6 billion parks-and-open-space bond measure in 2002, compared with 56 percent of whites. Another survey, conducted in 2002 in California, Arizona, and New Mexico, found that **Hispanics support wilderness protection more strongly than the general population.**

Despite these sentiments, the 1998 Latino Issues Forum survey found that "a large portion of the Latino community feels isolated from the environmental decision-making process," with 61 percent of respondents agreeing that white leaders make environmental policy. And **55 percent agreed that most environmentalists are "white, middle class, and live in the suburbs."** Recognizing our common ground is only the first step; diversifying our ranks must come next.

---

activists can't develop the same levels of trust. So we have to constantly create opportunities to come together.

There are misperceptions on both sides. You always hear that environmental groups don't care about communities of color and those communities don't care about the environment.

Absolutely. And who benefits from that type of garbage? The very people that we're fighting. We don't have the money that they have, so our only option is to start talking to each other—and that's their greatest fear.

When we first went to Columbia, Mississippi, the president of the company that had polluted this town lived in the same neighborhood as [former Sierra Club president] Robbie Cox. So when this CEO heard about our meeting, he called Robbie up and invited him to dinner. He thought he was going to have one of those good-old-boy conversations—you know, "Come on, Robbie, what's it going to take to get you guys out of there? You aren't serious about working with those po' black folks, are you?" But Robbie spoke up for people in the community, presented a letter with their concerns, and for the first time in ten years, that CEO made a visit to Columbia to talk with them about possible remedies. And he came down *twice*. Because the Sierra Club got involved. *[Editor's note: The community was*

eventually able to obtain funds for some residents to relocate, and healthcare for those who remained.]

To have white brothers and sisters stand in solidarity with us as we take on common foes, and say that this type of treatment of minority communities is not acceptable, adds credibility to the already credible struggle for environmental justice. We have to realize that our similarities greatly exceed our differences—we're all breathing the same air, drinking the same water, eating the same food, and fighting the same enemies. If I were a polluter or a corrupt politician, I would be deeply concerned about what I see happening in the Sierra Club.

Where do you hope we'll be on these issues in another decade?

I hope that when people think about conservation groups that are tearing down barriers and working to build relationships based on mutual respect and trust, the Sierra Club will readily come to mind. When they think of a national organization that is building broad coalitions to bring pain on polluting corporations around the country, the Sierra Club will readily come to mind. If the Club can get really serious about this, it will be the most powerful organization in the country, environmental or otherwise, because its membership will look like America. When the Sierra Club speaks, you will actually be hearing the voice of America.

Diversity at Work
From Urban Blight to Urban Might

White visitors to Washington, D.C., are often cautioned to stay away from the low-income, predominantly African-American communities in the southeastern part of the city. But in the late 1990s, Sierra Club members ventured into the Anacostia neighborhoods to help local res-

idents keep a private prison out of their public parkland. The collaboration led to a victory in 1999, when the zoning commission rejected the prison proposal. It also made it a little easier for Julie Eisenhardt, a white Wisconsinite with a degree in medieval history, to begin her work as the Sierra Club's first environmental-justice organizer in the nation's capital. Eisenhardt was introduced around the neighborhood by Eugene Dewitt Kinlow, an African-American politico from Anacostia who worked on the prison campaign. "I could tell people that when we fought the prison, an important organization that helped us succeed was the Sierra Club," says Kinlow, who recognized in Eisenhardt a fellow fighter for social justice.

"Eugene put his reputation on the line by getting involved with all this," says Eisenhardt. "If anyone asked, 'Who's this little blond girl?' he was the one who had to answer for me." The risk paid off. Eisenhardt forged relationships around the neighborhood, paving the way for an ongoing partnership with community groups. When Eisenhardt moved on to the Sierra Club's environmental-justice committee, the Club hired longtime Anacostia activist Linda Fennell to continue the fight against neighborhood blight, unequal levels of services, and air pollution from nearby highways and power plants.

Eugene Dewitt Kinlow
Washington, D.C., activist

"Our community has the largest percentage of people who are unemployed and on public assistance in D.C., and the largest population of kids in poverty. It also has little or no commerce. We had one grocery store, but that closed about five years ago. So, frankly, many environmental issues are subsumed under more immediate ones—how am I going to keep a roof over my head, how am I going to feed my kids and keep them safe from crime? It can be a difficult veil to lift.

"So it was incumbent upon Julie to look at the community and try to see it through their eyes—and then get them to look at things through an environmental lens. When you've got broken bottles in the neighborhood, illegal dumping, and trash in your river so it's not fishable or swimmable, those are environmental problems too.

"Of course it was difficult at first. Residents are very cautious about outsiders. People have come in and promised us many things and we don't ever receive them. They want access to our leaders and our political support, and then they just abandon us after they get what they want. But for the Sierra Club to put its office in the heart of Anacostia demonstrated a commitment right off the bat. Julie showed that she was real and that the commitment of the Club was true.

"Julie was dealing with local groups that had great ideas, but lacked resources. They could have a protest, but they didn't have a fax machine to get a press release out, or even a computer to make a flyer. The Sierra Club did, and they opened their doors to us. That meant a lot.

"About a year and a half after we defeated the prison, there was a proposal to put a trash-transfer station in our ward. They always

want to put the most negative things in our community. But since we had respectability from the prison fight, and the resources from the Sierra Club to assist us, we were able to say no. We've been accepting everyone else's sewage and junk for so long, and we will not accept any more."

Julie Eisenhardt
former Sierra Club EJ organizer
Washington, D.C.

"When I walked into a room in the neighborhood, it was easy to tell that I wasn't from there. It's important to be honest about who you are, and what your motivations are, and to not try to act like you understand what people are going through or what their experiences have been. Folks in these neighborhoods have had do-gooders come in for generations saying they're going to rescue them. So people perceived us with a certain amount of skepticism.

"Too many times, we're in such a hurry to get things done, we forget that we're human and we need to relate on a personal level. There are a lot of older women involved in community-building east of the river, and two or three times I made the mistake of assuming that I had to slow down or talk louder, and kind of explain things, but these women were just absolute fireballs. I learned very quickly that they don't need to be talked down to.

"We had a number of social events. At one, we had the garden-club presidents from both Anacostia and Georgetown. People associate Anacostia with poverty and drug-dealing, but here's an older woman putting planters out on the street and flowers in the corner park. She's trying to do the same thing as the wealthy people from Georgetown.

"It's not like we had to convert people. These folks already cared about environmental issues because their kids have asthma attacks, they're seeing lead poisoning affect the ability of students to learn. When you live in a city, that's really what the environment is. We just had to convince folks that the Sierra Club was there to help them on *their* issues.

"A lot of roadblocks come up in this kind of work, and we can't change the way we were raised or our different backgrounds, but we can discuss the preconceived notions that we bring to the table. We can also be geographically diverse in where we hold events and make our meetings transportation-accessible, with childcare and some snacks, so as many people as possible can get there. There are a lot of ways to get rid of the logistical things so we can get to the real stuff."

John McCown: "I knew that if I could link community groups with the resources of the Sierra Club, we could effect some real change."

WILL FRANKENFOOD SAVE THE PLANET?

BY JONATHAN RAUCH

*Over the next half century genetic engineering could
feed humanity and solve a raft of environmental ills—
if only environmentalists would let it*

That genetic engineering may be the most environmentally beneficial technology to have emerged in decades, or possibly centuries, is not immediately obvious. Certainly, at least, it is not obvious to the many U.S. and foreign environmental groups that regard biotechnology as a bête noire. Nor is it necessarily obvious to people who grew up in cities, and who have only an inkling of what happens on a modern farm. Being agriculturally illiterate myself, I set out to look at what may be, if the planet is fortunate, the farming of the future.

It was baking hot that April day. I traveled with two Virginia state soil-and-water-conservation officers and an agricultural-extension agent to an area not far from Richmond. The farmers there are national (and therefore world) leaders in the application of what is known as continuous no-till farming. In plain English, they don't plough. For thousands of years, since the dawn of the agricultural revolution, farmers have ploughed, often several times a year; and with ploughing has come runoff that pollutes rivers and blights aquatic habitat, erosion that wears away the land, and the release into the atmosphere of greenhouse gases stored in the soil. Today, at last, farmers are working out methods that have begun to make ploughing obsolete.

At about one-thirty we arrived at a 200-acre patch of farmland known as the Good Luck Tract. No one seemed to know the provenance of the name, but the best guess was that somebody had said something like "You intend to farm this? Good luck!" The land was rolling, rather than flat, and its slopes came together to form natural troughs for rainwater. Ordinarily this highly erodible land would be suitable for cows, not crops. Yet it was dense with wheat—wheat yielding almost twice what could normally be expected, and in soil that had grown richer in organic matter, and thus more nourishing to crops, even as the land was farmed. Perhaps most striking was the almost complete absence of any chemical or soil runoff. Even the beating administered in 1999 by Hurricane Floyd, which lashed the ground with nineteen inches of rain in less than twenty-four hours, produced no significant runoff or erosion. The land simply absorbed the sheets of water before they could course downhill.

At another site, a few miles away, I saw why. On land planted in corn whose shoots had only just broken the surface, Paul Davis, the extension agent, wedged a shovel into the ground and dislodged about eight inches of topsoil. Then he reached down and picked up a clump. Ploughed soil, having been stirred up and turned over again and again, becomes lifeless and homogeneous, but the clump that Davis held out was alive. I immediately noticed three squirming earthworms, one grub, and quantities of tiny white insects that looked very busy. As if in greeting, a worm defecated. "Plant-available food!" a delighted Davis exclaimed.

This soil, like that of the Good Luck Tract, had not been ploughed for years, allowing the underground ecosystem to return. Insects and roots and microorganisms had given the soil an elaborate architecture, which held the earth in place and made it a sponge for water. That was why erosion and runoff had been reduced to practically nil. Crops thrived because worms were doing the ploughing. Crop residue that was left on the ground, rather than ploughed under as usual, provided nourishment for the soil's biota and, as it decayed, enriched the soil. The farmer saved the fuel he would have used driving back and forth with a heavy plough. That saved money, and of course it also saved energy and reduced pollution. On top of all that, crop yields were better than with conventional methods.

The conservation people in Virginia were full of excitement over no-till farming. Their job was to clean up the James and York Rivers and the rest of the Chesapeake Bay watershed. Most of the sediment that clogs and clouds the rivers, and most of the fertilizer runoff that causes the algae blooms that kill fish, comes from farmland. By all but eliminating agricultural erosion and runoff—so Brian Noyes, the local conservation-district manager, told me—continuous no-till could "revolutionize" the area's water quality.

Even granting that Noyes is an enthusiast, from an environmental point of view no-till farming looks like a dramatic advance. The rub—if it is a rub—is that the widespread elimination of the plough depends on genetically modified crops.

The modern environmental movement was to a large extent founded on suspicion of markets and artificial substances. Markets exploit the earth; chemicals poison it. Biotech— seemingly the very epitome of the unnatural—touches both hot buttons.

It is only a modest exaggeration to say that as goes agriculture, so goes the planet. Of all the human activities that shape the environment, agriculture is the single most important, and it is well ahead of whatever comes second. Today about 38 percent of the earth's land area is cropland or pasture—a total that has crept upward over the past few decades as global population has grown. The increase has been gradual, only about 0.3 percent a year; but that still translates into an additional Greece or Nicaragua cultivated or grazed every year.

Farming does not go easy on the earth, and never has. To farm is to make war upon millions of plants (weeds, so-called) and animals (pests, so-called) that in the ordinary course of things would crowd out or eat or infest whatever it is a farmer is growing. Crop monocultures, as whole fields of only wheat or corn or any other single plant are called, make poor habitat and are vulnerable to disease and disaster. Although fertilizer runs off and pollutes water, farming without fertilizer will deplete and eventually exhaust the soil. Pesticides can harm the health of human beings and kill desirable or harmless bugs along with pests. Irrigation leaves behind trace elements that can accumulate and poison the soil. And on and on.

The trade-offs are fundamental. Organic farming, for example, uses no artificial fertilizer, but it does use a lot of manure, which can pollute water and contaminate food. Traditional farmers may use less herbicide, but they also do more ploughing, with all the ensuing environmental complications. Low-input agriculture uses fewer chemicals but more land. The point is not that farming is an environmental crime—it is not—but that there is no escaping the pressure it puts on the planet.

In the next half century the pressure will intensify. The United Nations, in its midrange projections, estimates that the earth's human population will grow by more than 40 percent, from 6.3 billion people today to 8.9 billion in 2050. Feeding all those people, and feeding their billion or so hungry pets (a dog or a cat is one of the first things people want once they move beyond a subsistence lifestyle), and providing the increasingly protein-rich diets that an increasingly wealthy world will expect— doing all of that will require food output to at least double, and possibly triple.

But then the story will change. According to the UN's midrange projections (which may, if anything, err somewhat on the high side), around 2050 the world's population will more or less level off. Even if the growth does not stop, it will slow. The crunch will be over. In fact, if in 2050 crop yields are still increasing, if most of the world is economically developed, and if population pressures are declining or even reversing—all of which seems reasonably likely—then the human species may at long last be able to feed itself, year in and year out, without putting any additional net stress on the environment. We might even be able to grow everything we need while reducing our agricultural footprint: returning cropland to wilderness, repairing damaged soils, restoring ecosystems, and so on. In other words, human agriculture might be placed on a sustainable footing forever: a breathtaking prospect.

The great problem, then, is to get through the next four or five decades with as little environmental damage as possible. That is where biotechnology comes in.

One day recently I drove down to southern Virginia to visit Dennis Avery and his son, Alex. The older Avery, a man in late middle age with a chinstrap beard, droopy eyes, and an intent, scholarly manner, lives on ninety-seven acres that he shares with horses, chickens, fish, cats, dogs, bluebirds, ducks, transient geese, and assorted other creatures. He is the director of global food issues at the Hudson Institute, a conservative think tank; Alex works with him, and is trained as a plant physiologist. We sat in a sunroom at the back of the house, our afternoon conversation punctuated every so often by dog snores and rooster crows. We talked for a little while about the Green Revolution, a dramatic advance in farm productivity that fed the world's burgeoning population over the past four decades, and then I asked if the challenge of the next four decades could be met.

"Well," Dennis replied, "we have tripled the world's farm output since 1960. And we're feeding twice as many people from the same land. That was a heroic achievement. But we have to do what some think is an even more difficult thing in this next forty years, because the Green Revolution had more land per person and more water per person—"

"— and more potential for increases," Alex added, "because the base that we were starting from was so much lower."

"By and large," Dennis went on, "the world's civilizations have been built around its best farmland. And we have used most of the world's good farmland. Most of the good land is already heavily fertilized. Most of the good land is already being planted with high-yield seeds. [Africa is the important exception.] Most of the good irrigation sites are used. We can't triple yields again with the technologies we're already using. And we might be lucky to get a fifty percent yield increase if we froze our technology short of biotech."

"Biotech" can refer to a number of things, but the relevant application here is genetic modification: the selective transfer of genes from one organism to another. Ordinary breeding can cross related varieties, but it cannot take a gene from a bacterium, for instance, and transfer it to a wheat plant. The organisms resulting from gene transfers are called "transgenic" by scientists—and "Frankenfood" by many greens.

Gene transfer poses risks, unquestionably. So, for that matter, does traditional crossbreeding. But many people worry that transgenic organisms might prove more unpredictable. One possibility is that transgenic crops would spread from fields into forests or other wild lands and there become environmental nuisances, or worse. A further risk is that transgenic plants might cross-pollinate with neighboring wild plants, producing "super-weeds" or other invasive or destructive varieties in the wild. Those risks are real enough that even most biotech enthusiasts—including Dennis Avery, for example—favor some government regulation of transgenic crops.

What is much less widely appreciated is biotech's potential to do the environment good. Take as an example continuous no-till farming, which really works best with the help of transgenic crops. Human beings have been ploughing for so long that we tend to forget why we started doing it in the first place. The short answer: weed control. Turning over the soil between plantings smothers weeds and their seeds. If you don't plough, your land becomes a weed garden—unless you use herbicides to kill the weeds. Herbicides, however, are expensive, and can be complicated to apply. And they tend to kill the good with the bad.

In the mid-1990s the agricultural-products company Monsanto introduced a transgenic soybean variety called Roundup Ready. As the name implies, these soybeans tolerate Roundup, an herbicide (also made by Monsanto) that kills many kinds of weeds and then quickly breaks down into harmless ingredients. Equipped with Roundup Ready crops, farmers found that they could retire their ploughs and control weeds with just a few applications of a single, relatively benign herbicide—instead of many applications of a complex and expensive menu of chemicals. More than a third of all U.S. soybeans are now grown without ploughing, mostly owing to the introduction of Roundup Ready varieties. Ploughless cotton farming has likewise received a big boost from the advent of bioengineered varieties. No-till farming without biotech is possible, but it's more difficult and expensive, which is why no-till and biotech are advancing in tandem.

In 2001 a group of scientists announced that they had engineered a transgenic tomato plant able to thrive on salty water—water, in fact, almost half as salty as seawater, and fifty times as salty as tomatoes can ordinarily abide. One of the researchers was quoted as saying, "I've already transformed tomato, tobacco, and canola. I believe I can transform any crop with this gene"—just the sort of Frankenstein hubris that makes environmentalists shudder. But consider the environmental implications. Irrigation has for millennia been a cornerstone of agriculture, but it comes at a price. As irrigation water evaporates, it leaves behind traces of salt, which accumulate in the soil and gradually render it infertile. (As any Roman legion knows, to destroy a nation's agricultural base you salt the soil.)

Every year the world loses about 25 million acres—an area equivalent to a fifth of California—to salinity; 40 percent of the world's irrigated land, and 25 percent of America's, has been hurt to some degree. For decades traditional plant breeders tried to create salt-tolerant crop plants, and for decades they failed.

Salt-tolerant crops might bring millions of acres of wounded or crippled land back into production. "And it gets better," Alex Avery told me. The transgenic tomato plants take up and sequester in their leaves as much as six or seven percent of their weight in sodium. "Theoretically," Alex said, "you could reclaim a salt-contaminated field by growing enough of these crops to remove the salts from the soil."

His father chimed in: "We've worried about being able to keep these salt-contaminated fields going even for decades. We can now think about *centuries.*"

One of the first biotech crops to reach the market, in the mid-1990s, was a cotton plant that makes its own pesticide. Scientists incorporated into the plant a toxin-producing gene from a soil bacterium known as *Bacillus thuringiensis*. With Bt cotton, as it is called, farmers can spray much less, and the poison contained in the plant is delivered only to bugs that actually eat the crop. As any environmentalist can tell you, insecticide is not very nice stuff—especially if you breathe it, which many Third World farmers do as they walk through their fields with backpack sprayers.

Transgenic cotton reduced pesticide use by more than two million pounds in the United States from 1996 to 2000, and it has reduced pesticide sprayings in parts of China by more than half. Earlier this year the Environmental Protection Agency approved a genetically modified corn that resists a beetle larva known as rootworm. Because rootworm is American corn's most voracious enemy, this new variety has the potential to reduce annual pesticide use in America by more than 14 million pounds. It could reduce or eliminate the spraying of pesticide on 23 million acres of U.S. land.

Over the next half century the world's population is projected to grow by more than 40 percent. The great problem is to get through that crunch with as little environmental damage as possible. That is where biotechnology comes in.

All of that is the beginning, not the end. Bioengineers are also working, for instance, on crops that tolerate aluminum, another major contaminant of soil, especially in the tropics. Return an acre of farmland to productivity, or double yields on an already productive acre, and, other things being equal, you reduce by an acre the amount of virgin forest or savannah that will be stripped and cultivated. That may be the most important benefit of all.

Of the many people I have interviewed in my twenty years as a journalist, Norman Borlaug must be the one who has saved the most lives. Today he is an unprepossessing eighty-nine-year-old man of middling height, with crystal-bright blue eyes and thinning white hair. He still loves to talk about plant breeding, the discipline that won him the 1970 Nobel Peace Prize: Borlaug led efforts to breed the staples of the Green Revolution. (See "Forgotten Benefactor of Humanity," by Gregg Easterbrook, an article on Borlaug in the January 1997 *Atlantic*.) Yet the renowned plant breeder is quick to mention that he began his career, in the 1930s, in forestry, and that forest conservation has never been far from his thoughts. In the 1960s, while he was working to improve crop yields in India and Pakistan, he made a mental connection. He would create tables detailing acres under cultivation and average yields—and then, in another column, he would estimate how much land had been saved by higher farm productivity. Later, in the 1980s and 1990s, he and others began paying increased attention to what some agricultural economists now call the Borlaug hypothesis: that the Green Revolution has saved not only many human lives but, by improving the productivity of existing farmland, also millions of acres of tropical forest and other habitat—and so has saved countless animal lives.

From the 1960s through the 1980s, for example, Green Revolution advances saved more than 100 million acres of wild lands in India. More recently, higher yields in rice, coffee, vegetables, and other crops have reduced or in some cases stopped forest-clearing in Honduras, the Philippines, and elsewhere. Dennis Avery estimates that if farming techniques and yields had not improved since 1950, the world would have lost an additional 20 million or so square miles of wildlife habitat, most of it forest. About 16 million square miles of forest exists today. "What I'm saying," Avery said, in response to my puzzled expression, "is that we have saved every square mile of forest on the planet."

Habitat destruction remains a serious environmental problem; in some respects it is the most serious. The savannahs and tropical forests of Central and South America, Asia, and Africa by and large make poor farmland, but they are the earth's storehouses of biodiversity, and the forests are the earth's lungs. Since 1972 about 200,000 square miles of Amazon rain forest have been cleared for crops and pasture; from 1966 to 1994 all but three of the Central American countries cleared more forest than they left standing. Mexico is losing more than 4,000 square miles of forest a year to peasant farms; sub-Saharan Africa is losing more than 19,000.

That is why the great challenge of the next four or five decades is not to feed an additional three billion people (and their pets) but to do so without converting much of the world's prime habitat into second- or third-rate farmland. Now, most agronomists agree that some substantial yield improvements are still to be had from advances in conventional breeding, fertilizers, herbicides, and other Green Revolution standbys. But it seems pretty clear that biotechnology holds more promise—probably much more. Recall that world food output will need to at least double and possibly triple over the next several decades. Even if production could be increased that much using conventional technology, which is doubtful, the required amounts of pesticide and fertilizer and other polluting chemicals would be immense. If properly developed, disseminated, and used, genetically modified crops might well be the best hope the planet has got.

If properly developed, disseminated, and used. That tripartite qualification turns out to be important, and it brings the environmental community squarely, and at the moment rather jarringly, into the picture.

Not long ago I went to see David Sandalow in his office at the World Wildlife Fund, in Washington, D.C. Sandalow, the organization's executive vice-president in charge of conservation programs, is a tall, affable, polished, and slightly reticent man in his forties who holds degrees from Yale and the University of Michigan Law School.

Some weeks earlier, over lunch, I had mentioned Dennis Avery's claim that genetic modification had great environmental potential. I was surprised when Sandalow told me he agreed. Later, in our interview in his office, I asked him to elaborate. "With biotechnology," he said, "there are no simple answers. Biotechnology has huge potential benefits and huge risks, and we need to address both as we move forward. The huge potential benefits include increased productivity of arable land, which could relieve pressure on forests. They include decreased pesticide usage. But the huge risks include severe ecological disruptions—from gene flow and from enhanced invasiveness, which is a very antiseptic word for some very scary stuff."

I asked if he thought that, absent biotechnology, the world could feed everybody over the next forty or fifty years without ploughing down the rain forests. Instead of answering directly he said, "Biotechnology could be part of our arsenal if we can overcome some of the barriers. It will never be a panacea or a magic bullet. But nor should we remove it from our tool kit."

Sandalow is unusual. Very few credentialed greens talk the way he does about biotechnology, at least publicly. They would readily agree with him about the huge risks, but they wouldn't be caught dead speaking of huge potential benefits—a point I will come back to. From an ecological point of view, a very great deal depends on other environmentalists' coming to think more the way Sandalow does.

Biotech companies are in business to make money. That is fitting and proper. But developing and testing new transgenic crops is expensive and commercially risky, to say nothing of politically controversial. When they decide how to invest their research-and-development money, biotech companies will naturally seek products for which farmers and consumers will pay top dollar. Roundup Ready products, for instance, are well suited to U.S. farming, with its high levels of capital spending on such things as herbicides and automated sprayers. Poor farmers in the developing world, of course, have much less buying power. Creating, say, salt-tolerant cassava suitable for growing on hardscrabble African farms might save habitat as well as lives—but commercial enterprises are not likely to fall over one another in a rush to do it.

If earth-friendly transgenics are developed, the next problem is disseminating them. As a number of the farmers and experts I talked to were quick to mention, switching to an unfamiliar new technology—something like no-till—is not easy. It requires capital investment in new seed and equipment, mastery of new skills and methods, a fragile transition period as farmer and ecology readjust, and an often considerable amount of trial and error to find out what works best on any given field. Such problems are only magnified in the Third World, where the learning curve is steeper and capital cushions are thin to nonexistent. Just handing a peasant farmer a bag of newfangled seed is not enough. In many cases peasant farmers will need one-on-one attention. Many will need help to pay for the seed, too.

Finally there is the matter of using biotech in a way that actually benefits the environment. Often the technological blade can cut either way, especially in the short run. A salt-tolerant or drought-resistant rice that allowed farmers to keep land in production might also induce them to plough up virgin land that previously was too salty or too dry to farm. If the effect of improved seed is to make farming more profitable, farmers may respond, at least temporarily, by bringing more land into production. If a farm becomes more productive, it may require fewer workers; and if local labor markets cannot provide jobs for them, displaced workers may move to a nearby patch of rain forest and burn it down to make way for subsistence farming. Such transition problems are solvable, but they need money and attention.

In short, realizing the great—probably unique—environmental potential of biotech will require stewardship. "It's a tool," Sara Scherr, an agricultural economist with the conservation group Forest Trends, told me, "but it's absolutely not going to happen automatically."

So now ask a question: Who is the natural constituency for earth-friendly biotechnology? Who cares enough to lobby governments to underwrite research—frequently unprofitable research—on transgenic crops that might restore soils or cut down on pesticides in poor countries? Who cares enough to teach Asian or African farmers, one by one, how to farm without ploughing? Who cares enough to help poor farmers afford high-tech, earth-friendly seed? Who cares enough to agitate for programs and reforms that might steer displaced peasants and profit-seeking farmers away from sensitive lands? Not politicians, for the most part. Not farmers. Not corporations. Not consumers.

Who is the natural constituency for earth-friendly biotechnology? The natural constituency is, of course, environmentalists. But very few credentialed greens would be caught dead speaking of biotechnology's huge potential benefits.

At the World Resources Institute, an environmental think tank in Washington, the molecular biologist Don Doering envisions transgenic crops designed specifically to solve environmental problems: crops that might fertilize the soil, crops that could clean water, crops tailored to remedy the ecological problems of specific places. "Suddenly you might find yourself with a virtually chemical-free agriculture, where your cropland itself is filtering the water, it's protecting the watershed, it's providing habitat," Doering told me. "There is still so little investment in what I call design-for-environment." The natural constituency for such investment is, of course, environmentalists.

B ut environmentalists are not acting as such a constituency today. They are doing the opposite. For example, Greenpeace declares on its Web site: "The introduction of genetically engineered (GE) organisms into the complex ecosystems of our environment is a dangerous global experiment with nature and evolution … GE organisms must not be released into the environment. They pose unacceptable risks to ecosystems, and have the potential to threaten biodiversity, wildlife and sustainable forms of agriculture."

Other groups argue for what they call the Precautionary Principle, under which no transgenic crop could be used until proven benign in virtually all respects. The Sierra Club says on its Web site,

> In accordance with this Precautionary Principle, we call for a moratorium on the planting of all genetically engineered crops and the release of all GEOs [genetically engineered organisms] into the environment, *including those now approved*. Releases should be delayed until extensive, rigorous research is done which determines the long-term environmental and health impacts of each GEO and there is public debate to ascertain the need for the use of each GEO intended for release into the environment.

Under this policy the cleaner water and healthier soil that continuous no-till farming has already brought to the Chesapeake Bay watershed would be undone, and countless tons of polluted runoff and eroded topsoil would accumulate in Virginia rivers and streams while debaters debated and researchers researched. Recall David Sandalow: "Biotechnology has huge potential benefits and huge risks, and we need to address both as we move forward." A lot of environmentalists would say instead, *before* we move forward." That is an important difference, particularly because the big population squeeze will happen not in the distant future but over the next several decades.

For reasons having more to do with politics than with logic, the modern environmental movement was to a large extent founded on suspicion of markets and artificial substances. Markets exploit the earth; chemicals poison it. Biotech touches both hot buttons. It is being pushed forward by greedy corporations, and it seems to be the very epitome of the unnatural.

Still, I hereby hazard a prediction. In ten years or less, most American environmentalists (European ones are more dogmatic) will regard genetic modification as one of their most powerful tools. In only the past ten years or so, after all, environmentalists have reversed field and embraced market mechanisms—tradable emissions permits and the like—as useful in the fight against pollution. The environmental logic of biotechnology is, if anything, even more compelling. The potential upside of genetic modification is simply too large to ignore—and therefore environmentalists will not ignore it. Biotechnology will transform agriculture, and in doing so will transform American environmentalism.

Jonathan Rauch is a correpondent for The Atlantic *and a senior writer for* National Journal. *He is also a writer in residence at the Brookings Institution and the author of several books, including* Government's End: Why Washington Stopped Working *(1999).*

UNIT 3
Energy: Present and Future Problems

Unit Selections

Key Points to Consider

- According to recent and still controversial projections on oil supplies, when will the world begin to run out of oil (that is, reach peak production after which the supply of oil will no longer be sufficient to meet the demand)? How can the world adjust to this potential shortfall within such a short period of time?

- What is the nature of the legal concept of "mineral rights" that can have negative impacts on the environmental quality of farm and ranch land? Should the owners of mineral rights have unlimited access to fossil fuels, despite the damage they may produce to the land's surface?

- What are some of the major benefits of such alternate energy sources as solar power and wind power? Do these energy alternatives really have a chance at competing with fossil fuels for a share of the global energy market?

- Why should the hydrogen fuel cell be considered as an important component of energy systems of the future? What limiting factors have stood in the way of development of hydrogen as an alternative fuel source?

 **Links: www.dushkin.com/online/**
These sites are annotated in the World Wide Web pages.

Alliance for Global Sustainability (AGS)
  http://globalsustainability.org/

Alternative Energy Institute, Inc.
  http://www.altenergy.org

Energy and the Environment: Resources for a Networked World
  http://zebu.uoregon.edu/energy.html

Institute for Global Communication/EcoNet
  http://www.igc.org/

Nuclear Power Introduction
  http://library.thinkquest.org/17658/pdfs/nucintro.pdf

U.S. Department of Energy
  http://www.energy.gov

There has been a tendency, particularly in the developed nations of the world, to view the present high standards of living as exclusively the benefit of a high-technology society. In the "techno-optimism" of the post World War II years, prominent scientists described the technical-industrial civilization of the future as being limited only by a lack of enough trained engineers and scientists to build and maintain it. This euphoria reached its climax in July 1969 when American astronauts walked upon the surface of the Moon, an accomplishment brought about solely by American technology—or so it was supposed. It cannot be denied that technology has been important in raising standards of living and permitting Moon landings, but how much of the growth in living standards and how many outstanding and dramatic feats of space exploration have been the result of technology alone? The answer is few—for in many of humankind's recent successes, the contributions of technology to growth have been no more important than the availability of incredibly cheap energy resources, particularly petroleum, natural gas, and coal.

As the world's supply of recoverable (inexpensive) fossil fuels dwindles and becomes, as evidenced by recent international events, more important as a factor in international conflict, it becomes increasingly clear that the energy dilemma is the most serious economic and environmental threat facing the Western world and its high standard of living. With the exception of the specter of global climate change, the scarcity and cost of con-

ventional (fossil fuel) energy is probably the most serious threat to economic growth and stability in the rest of the world as well. The economic dimensions of the energy problem are rooted in the instabilities of monetary systems produced by and dependent on inexpensive energy. The environmental dimensions of the problem are even more complex, ranging from the hazards posed by the development of such alternative sources as nuclear power to the inability of developing world farmers to purchase necessary fertilizer produced from petroleum, which has suddenly become very costly, and to the enhanced greenhouse effect created by fossil fuel consumption. The only answers to the problems of dwindling and geographically vulnerable, inexpensive energy supplies are conservation and sustainable energy technology. Both require a massive readjustment of thinking, away from the exuberant notion that technology can solve any problem. The difficulty with conservation, of course, is a philosophical one that grows out of the still-prevailing optimism about high technology. Conservation is not as exciting as putting a man on the Moon. Its tactical applications—caulking windows and insulating attics—are dog-paddle technologies to people accustomed to the crawl stroke. Does a solution to this problem entail the technological fixes of which many are so enamored? Probably not, as it appears that the accelerating energy demands of the world's developing nations will most likely be met first by increased reliance on the traditional (and still relatively cheap) fossil fuels. Although there is a need to reduce this reliance, there are few ready alternatives available to the poorer developing countries. It would appear that conservation is the only option.

But conservation is probably not enough to solve the environmental problems related to fossil fuel use. Even if we adopted all the available alternative energy sources immediately, there would still probably be significant shortfalls. Such a relatively pessimistic outlook could be tempered if promising new experimental strategies such as the hydrogen fuel cell prove useful in tests now being carried out in, among other places, Iceland.

Two approaches to the possibility of reducing dependence on foreign oil—and the problems that might be created thereby—are found in the unit's next two selections. In "Powder Keg" science writer Keith Kloor discusses some of the environmental and social consequences of increased natural gas development in Wyoming's Powder River Basin. Here, large energy companies own the mineral rights under the land owned and operated by livestock ranchers. The disjunctions between what the energy companies are allowed to do at the surface because they own the subsurface, and the best uses of the surface of the land by ranchers, are profound. A three-way struggle between big energy, the federal and state governments, and environmental advocates puts ranchers in the unusual position of siding with environmentalists. In "Personalized Energy," energy expert Stephen Millett suggests that, since fossil fuels have so many negatives for the environment and since alternative energy systems such as wind and solar are still a long ways from being capable of satisfying current demands, a new approach to energy shortages is needed. This new approach actually represents a return to an older system of energy acquisition and use in which local communities and individual households played a more important role in providing their own energy. He points to intriguing new developments in fuel cell technologies, for example, as mechanisms through which energy production may be redistributed from the huge producing plants to much smaller ones serving localities.

The final two selections in the section both deal with some of the most promising renewable or alternative energy strategies. In "Renewable Energy: A Viable Choice," authors Antonia Herzog, Timothy Lipman, Jennifer Edwards, and Daniel Kammen urge the development of energy sources such as solar, wind, and biomass energy to reduce U.S. reliance on fossil fuels and improve environmental quality in the world's largest energy consumer country. If this is going to happen, the authors suggest, then a strong national energy policy will be required, along with a reduction in American dependence on foreign oil. Another new energy source is touted by freelance writer Bill Keenan who, in "Hydrogen: Waiting for the Revolution," argues that hydrogen may well be the fuel of the future and that it has enormous benefits for both the economy and the environment. As ideal as hydrogen seems to be as a fuel source—and as technologically feasible as scientists suggest—there still is little concrete evidence that delivery of hydrogen as a cheap and readily available, non-polluting energy source is around the corner. This will not happen until large corporations, still linked to the fossil fuel energy system, become truly interested in the development and use of hydrogen and put R&D money into a hydrogen-based fuel system.

POWDER KEG

THE GAS INDUSTRY HAS BEEN BUSY IN WYOMING'S PRAIRIES AND GRASSLANDS, BUILDING THOUSANDS OF MILES OF ROADS AND SINKING MORE THAN 10,000 WELLS IN THE PAST THREE YEARS. BUT IN THE POWDER RIVER BASIN, RANCHERS ARE JOINING ENVIRONMENTALISTS TO TRY TO STILL THE DRILLS.

BY KEITH KLOOR

ED SWARTZ DOES NOT SEEM LIKE THE KIND OF guy you would threaten with bodily harm. He has spent almost all of his 62 years in a windswept corner of northeastern Wyoming, herding cattle, baling hay, and building waterlines to keep his ranch from going dry. He still works as long as the daylight lasts, even after suffering a near-fatal heart attack in 1995. But shortly after his ranch started dying in 1999, he stood up on a bus full of state officials from Wyoming and Montana and started fuming about his dried-up meadows and polluted creekbeds. Most audiences would have sympathized with Swartz. But this particular bus carried coal-bed methane executives and various state officials, including the governors of Wyoming and Montana. They were touring drilling sites in the Powder River basin; the Wyoming officials wanted to impress their counterparts in neighboring Montana with how smoothly the booming development was going.

Ed Swartz's property, however, which sits in the coal-rich basin, has been inundated with coal-bed methane discharge water from drilling sites near his ranch, contaminating his creek and preventing him from irrigating his alfalfa, the cattle ranch's lifeline. On this day he just wanted to make sure everyone on the bus was aware of it. And that he was not the only rancher with this problem. John Kennedy, a local coal-bed methane operator, was furious. "You're a liar!" he yelled at Swartz, cutting him off. Swartz, a strapping, leathery, third-generation Wyomingite whose grandfather homesteaded the family ranch in 1904, kept talking, unperturbed. "I'm going to hurt you!" Kennedy warned. Swartz said to go ahead and try. Inflamed, Kennedy again thundered, "I'm going to hurt you!" But the blustery driller never left his seat. Finally, Montana's governor, Judy Martz, settled things down, so the tour could continue.

Sitting around his kitchen table, chain-smoking low-tar cigarettes, Swartz tells me the story over a pot of turbo-charged coffee, a glinty grin on his face. Kennedy, I learn, has a reputation for harassing similarly aggrieved ranchers at public meetings and over the phone. Swartz, who has long been an active, bedrock Republican in what is a virtual one-party—Republican—state, is now considered a pariah for rocking the boat. "People think I'm just a rabble-rousing rancher around here," he says defiantly. That, undoubtedly, is because of the Clean Water Act lawsuit he has filed against Wyoming state officials and the drilling company (Denver-based Redstone Inc.) he asserts is responsible for the damages to his land.

Coal-bed methane development is a relatively new form of natural gas extraction that has exploded in the Powder River basin since 1997, with more than 10,000 gas wells already sunk on private and state lands. It is like mother's milk to state officials, because it produces both tax revenues and campaign contributions from the energy companies. "They are so oriented to the energy industry that they could give a red rat's ass about a rancher," Swartz bristles. "I'm as serious as I can be about that. There's not one of them that is concerned about a rancher around here."

ALONG THIS STORIED AND BLOODIED FRONTIER, INDIAN BATTLES RAGED, RANGE WARS WERE FOUGHT, AND BUFFALO BILL AND BUTCH CASSIDY SEALED THEIR MYTHS.

THE 8-MILLION-acre Powder River basin straddles the Wyoming–Montana border, offering a snapshot of the quintessential Old West, with its rolling hills and prairies nestled between the Bighorn Mountains and the Black Hills, 100 miles to the east. Along this storied and bloodied frontier, Indian battles raged, range wars first erupted between homesteaders and cattle barons, and Buffalo Bill and Butch Cassidy sealed their myths. It is the place, many claim, where the American cowboy was born.

The Powder River region is also where the Great Plains meet the Rocky Mountains, a bare, reddish, lunarlike landscape that skips between parched, stumpy buttes, green meadows, and unspoiled streams. It is a vast, mixed ecosystem of grassland and sagebrush, where, in that timeless Darwinian race, prairie dogs burrow deep and antelope hurtle fast and high, coyotes and mountain lions quick on their heels. The Powder River—described by settlers as "a mile wide and an inch deep, too thin to plow and too thick to drink"—is a 375-mile tributary of the Yellowstone River. Plying its shallow, muddy waters are the globally imperiled sturgeon chub, the channel catfish, and 23 other native fish.

During the 20th century the long boom-and-bust cycles of strip mines and oilfields left their own indelible imprint on the landscape, in the form of sawed-off hilltops and abandoned "orphan" wells. Even so, ranchers and environmentalists have always taken solace that the deep and large gashes scarring the land were mostly confined to a few areas.

No longer. The latest boom to hit the Powder River basin has spread out in a chaotic patchwork, pockmarking the historic landscape with thousands of miles of powerlines, pipelines, roads, compressor stations, and wellheads. Methane is a natural gas found in the region's plentiful coal seams. Water pressure holds the gas in the coal; pumping the water out in large volumes releases the gas. The process also produces wastewater laced with sodium, calcium, and magnesium—too saline to be used for irrigation, too tainted to be dumped in waterways. So in a semi-arid region where water is precious, energy companies are forced to store the methane water in "containment" pits, from which it often runs into water wells and into the tributaries of the Powder River.

"IT'S SO DAMN DISCOURAGING. EVERYTHING I WORKED FOR, THAT MY GRANDFATHER WORKED FOR, AND THAT MY SON IS WORKING FOR IS BEING WIPED OUT."

The resulting environmental damage in the Powder River basin has hit ranchers and the land equally hard. "It's so damn discouraging," Swartz tells me in a craggy voice tinged with resignation. "Everything I worked for, that my grandfather worked for, and that my son is working for is being wiped out." Over the years the family has endured many droughts (including the one the region is suffering today), its share of machinery breakdowns, and several diseases afflicting their cattle. But nothing compares with the poisonous runoff that is killing the ranch's vegetation. "They're [Redstone] using my place as a garbage dump," Swartz fumes.

About a year and a half ago, at the boom's peak, drillers were pumping 55 million gallons of water to the surface every day. Underground aquifers were being depleted and cottonwood trees flooded. The massive runoff of the methane-tainted water

has become so alarming—polluting creeks and streams and altering natural river flows—that earlier this year the conservation group American Rivers named Wyoming's Powder River as one of America's Most Endangered Rivers for 2002. "There could be 139,000 coal-bed methane wells in the Powder River basin by the end of the decade," says Rebecca Wodder, president of American Rivers, referring to energy-industry and government estimates. "Despite this, federal and state agencies have yet to formulate an adequate plan for minimizing the environmental consequences of drilling in the Powder River basin."

And they weren't about to until the U.S. Environmental Protection Agency (EPA) issued a report last May, slamming the Bureau of Land Management (BLM) for failing to assess the fallout from coal-bed methane development. At the time, drillers were already having their way on Wyoming's private and public lands, owing to lax state and federal environmental safeguards. They had just set their sights on a mother lode of rich coal-seam deposits on BLM lands in the Powder River region. Then, with the bureau poised to give the operators a quick go-ahead, the EPA's report stopped them in their tracks, throwing into doubt the development of gas wells on 8 million public acres in Wyoming. In particular, the EPA cited concerns about air quality from dust and compressor emissions, and the impact on wildlife and water quality. The EPA report forced the BLM to redo its environmental-impact statement on 51,000 new gas wells slated for development on federal lands.

The reassessment, scheduled to be released in January, stands to reverberate through out the Rocky Mountain West, where a gold rush mentality has taken hold. Energy officials have called the area the "Persian Gulf of natural gas." Gas companies have already struck hard and fast in Colorado's San Juan basin; now they're waiting for the green light on federal lands to expand there and across Montana and Wyoming.

Moreover, coal-bed methane development in the West is the cornerstone of President Bush's proposed domestic energy plan, which claims the area has enough natural gas to supply the energy needs of the United States for seven years. "The region is enormously rich in minerals," Ray Thomasson, a Colorado-based energy consultant, told a recent Denver gathering of energy experts and industry officials. "We just have to find out where the sweet spots are."

IN WYOMING, AS IN MOST OF THE WEST, SUBSURFACE RIGHTS SUPERSEDE SURFACE RIGHTS, SO UNLESS A LANDOWNER OWNS BOTH—FEW DO—WHOEVER OWNS THE MINERAL RIGHTS UNDER THE LAND HOLDS THE TRUMP CARD.

DESPITE THEIR PROUD BEARING, NANCY AND Robert Sorenson can't mask the sorrow and anguish in their voices as they describe what it's like to live in the first "sweet spot" of the coal-bed methane boom. Both natives of Wyoming, the Sorensons have spent the past 30 years—almost half their lives—on a 3,500-acre cattle ranch in the Powder River basin, 20 miles north of Ed Swartz's spread. They, too, have had methane-contaminated water run off onto their property, flooding their soil and boxwood elder trees.

The Sorensons have graciously invited me into their home for a lunch of stir-fried chicken and homemade nut bread and a discussion of their unwanted quandary. Robert is taciturn yet direct, and casts a sharp gaze behind his full mustache; Nancy has a warm, open smile but speaks softly and deliberately, always searching for the right words. Recently, rich coal seams have been discovered under their land. But since the Sorensons own only partial mineral rights, when the drillers came knocking, they had no power to turn them away. In Wyoming, as in most of the West, subsurface rights supersede surface rights, so unless a landowner owns both—few do—whoever owns the mineral rights under the land holds the trump card.

Though the Sorensons stand to collect royalties, they are agonizing over the repercussions. "I worry about our neighbors downstream, who will be hit hard by this," says Nancy, a retired schoolteacher and an active board member of the Powder River Basin Resource Council, a grassroots group that has united ranchers and local environmentalists. She also worries about the fate of the wildlife she sees every day on her morning walks around the ranch, such as antelope, mule deer, and foxes. (" I've seen a mountain lion twice, which was quite a thrill!") Living on the land all their lives, the Sorensons have become attuned to its natural rhythms, mindful of even the subtlest changes in predator-prey relationships. Over the years the couple has watched, with admiration, as ecological forces exerted their own balance. "When the rabbits become too much of problem, the bobcats take care of it," says Nancy. But, she adds in a doleful whisper, "I hate to see what happens to all these animals once this drilling starts, because you are fragmenting their environment."

Overall, the Powder River basin is home to more than 157,000 mule deer, 108,000 pronghorn antelope, and almost 12,000 elk. Even the BLM admitted in its first—albeit inadequate—environmental assessment that the proposed coal-bed methane development "may result in loss of viability of federal lands... and may result in trends toward federal listing" under the Endangered Species Act for 16 species, including the white-tailed prairie dog, the burrowing owl, and the Brewer's sparrow.

As clouds darken the early afternoon and a light, intermittent drizzle begins to fall, the Sorensons mourn the transformation—almost overnight—of their rural community into an industrial zone. "It's a total change of a way of life," says Robert, whose family homesteaded in the Powder River basin in 1881.

He's right. For two days I have zigzagged hundreds of miles west from the city of Gillette—the satellite home base of energy companies—to Sheridan, another energy outpost. In between antelope sightings, it seemed that for every 20 miles I rode along Highway 14-16, new dirt roads were being bulldozed in the rolling foothills for pipelines, and open pits were being dug for wastewater containment ponds. What's more, this was a quiet period, because the unseasonably wet, cold weather hampered drilling. In warmer temperatures, for instance, coal-bed methane operators use an atomizer to spray methane water at high pressure, so that most of it will evaporate—although the salt and minerals still coat the ground.

"IMAGINE IF YOU LIVED ACROSS THE STREET FROM A POWER PLANT. THAT'S WHAT IT'S LIKE FOR PEOPLE THAT LIVE NEAR A COAL-BED METHANE FIELD." SOME RANCHERS LIKEN THE SOUND TO THAT OF 747S TAKING OFF—CONSTANTLY.

During construction of a particular site, it's not uncommon, the Sorensons say, to have a hundred trucks trundling in 24 hours a day. Once the wellheads are sunk and the compressor stations built alongside—sheltered in houselike structures—the noisy, whirring process of methane gas extraction runs all day and all night. "Imagine if you lived across the street from a power plant," says Robert. "That's what it's like for people that live near a coal-bed methane field." Some ranchers liken the sound to that of 747s taking off—constantly.

Yet as Nancy admits, somewhat awkwardly, not everyone in the community is put off by the development. Though Wyomingites are deeply proud of their ties to the land and the ranching tradition, it's not an easy life. "I'm aware of neighbors who welcomed this [the gas development] with open arms because they were so far at the end of their rope," she says. "And because this is going to keep them on the land a little while longer, I'm absolutely understanding of their position." Her voice trails off, a faraway look in her eyes. "You know," she continues, a few seconds later, "we've been able to make a living off this place, which is quite unusual. Robert has made some smart moves, and we've both worked hard, but we've also been lucky in that we haven't had a major illness. So I can't blame other people for how they feel."

Like Ed Swartz, the Sorensons direct their anger at state officials, who they believe have permitted the coal-bed methane operators to run roughshod over ranchers and the land. For their ranch, the Sorensons signed 13 separate lease deals, like pipeline and powerline rights-of-ways, with energy companies. "Not once did we have a choice," says Nancy. What's more, Wyoming, unlike neighboring Montana and most states, doesn't have laws mandating that proposed industrial development—such as gas extraction—on nonfederal lands be assessed beforehand for environmental impact.

Shortly after the widespread environmental damages became evident several years ago, the Sorensons, Ed Swartz, and many other ranchers met with their state representatives to plead for tighter regulations on the drilling and for a set of rancher rights—to no avail.

"They've all just been bought and sold," says Robert. "Really, they all have." (Nearly 70 percent of all campaign contributions to Wyoming's state legislators come from the oil and gas industry.)

"They all know where the paychecks come from," Nancy chimes in. "There isn't anybody in local government that is going to fight this. And neither am I. We just need some tougher regulations."

The ranchers' biggest concern is the depletion of their underground aquifers. In Colorado, where coal-bed methane development has taken off in a number of areas, drillers are required by state law to clean the discharged methane water of all pollutants and "reinject" it into the ground. Not so in Wyoming. Drillers say that requirement would make their operations less cost-effective. Many ranchers, joining forces with the formidable Powder River Basin Resource Council, have also been petitioning state and federal lawmakers for a mandatory Surface Owner Agreement—which would allow landowners to have a say about where pipelines and roads are built. (The government broke its promise that powerlines would be buried and kept to minimum.) Above all, what pains the ranching community is the lack of adequate bonding—money the energy companies pay to cover land damages and reclamation. Just look around, Nancy says, at all the abandoned wells from previous booms. Undoubtedly, history will repeat itself, she asserts, if energy companies aren't required to pony up much more than the paltry $25,000 bond they pay for unlimited wellheads on Wyoming's federal lands (it's $75,000 for state and private lands).

Given this history, I'm surprised to hear the Sorensons say they are not against coal-bed methane development in principle (and that goes for Ed Swartz and the other Powder River basin ranchers I spoke with).

"No," Robert answers resolutely. "We just want to see it done right."

A FEW MONTHS AFTER MY TRIP TO THE Powder River basin last May, the bottom fell out of the natural gas market, dropping prices to less than a dollar for every thousand cubic feet, from a high of $12 two years earlier, a price that fueled the drilling frenzy. Some industry experts and plenty of environmentalists attribute the Powder River boom to the manipulation of energy prices in California by Enron and other energy companies. Whatever the reason, the plunging price has slowed development in the Powder River basin and given ranchers—and wildlife—some breathing room.

"We have a window of opportunity to get the problems addressed," says Jill Morrison, an organizer with the Powder River Basin Resource Council. "It's good we have this slowdown. Maybe now we can get better planning and development." Morrison cautions it could merely be a lull in a volatile energy market, and that in the meantime, drillers might use the depressed prices as an excuse to pay less money to repair damages to the land.

No matter the outcome, Ed Swartz, the Sorensons, and other Powder River ranchers under siege aren't counting on their politicians for help. In August, three months after the EPA released its critical report of the BLM's environmental assessment, Wyoming's governor, Jim Geringer, lambasted the EPA for its "aggressive and ill-informed approach" to coal-bed methane development in the Powder River region. And in September, perhaps to shore up their flagging spirits, Montana Governor Judy Martz told a roomful of oil and gas lobbyists that they were "the true environmentalists"—as opposed to, as one meeting attendee said, the "radical" critics of energy development.

After I heard both comments, I figured these officials had never seen Ed Swartz's polluted stream or had lunch with the Sorensons. "We are very environmentally conscious," Nancy told me during my visit, "and most ranchers we know are." Her tone was gentle but firm. "I'm not opposed to having development, but there should be some equilibrium."

It never occurred to me then, or months later, when Governor Martz and the gas-industry proponents made their barbs, that the Sorensons or Ed Swartz were "radicals." But I sure do know who the "true environmentalists" are in Wyoming.

WHAT YOU CAN DO

For more information, call the Powder River Basin Resource Council at 307-672-5809 or log on to www.powderriverbasin.org. You can also call or e-mail the Wyoming state office of the BLM at 307-775-6256; state_wyomail@blm.gov.

Personalized Energy
THE NEXT PARADIGM

In the future, energy will be more in the control of neighborhoods and home-owners. But for that to happen, new technologies need to be developed that bring efficiency and reliability up and costs down, says a technology futurist.

By Stephen M. Millett

Consumers want more control over their energy, and they want it cheaper, cleaner, more convenient, and more reliable. They want energy, like other commodities, to be more personalized. Technological improvements will focus on meeting those demands, but they won't happen quickly. Current forecasts for energy supply and demand are limited by a mind-set stuck in the past. But new technologies and new consumer imperatives will spawn new ideas about energy that could get us off the grid and bring power generation into neighborhoods and even into homes.

The primary question we need to ask about the future of energy is whether the old supply-and-demand paradigm of fossil fuels still applies. In that case, the key to solving our energy woes lies in finding ways to increase production of traditional hydrocarbon fuels (such as oil, natural gas, and coal) and promote consumer conservation. But if the old paradigm is out, there may be a whole new paradigm emerging, where new technologies, for in-

stance, could change the whole energy picture.

Right now, we hear too many discussions about drilling more oil, conserving energy, and other actions based on old-paradigm thinking. Indeed, statistics show a big gap between projected energy demand and supplies in the United States: Oil and natural gas consumption are going up and available quantities are going down, so we're going to have a big projected shortfall.

The biggest jump in American energy consumption in the twentieth century was the use of petroleum, and that's almost exclusively transportation. Transportation relies on petroleum to meet 95% of its energy needs, according to the U.S. Bureau of Transportation Statistics. The story on coal is a little bit different. Developed economies such as the United States used to use coal in homes for heating, but that's done almost nowhere anymore. Americans are using more and more coal, but it's to generate electricity in large power plants. Coal is now rarely used at the individual level.

The 2001 report of the president's National Energy Policy Development Group stated, "Renewable and alternative fuels offer hope for America's energy future but they supply only a small fraction of present energy needs. The day they fulfill the bulk of our [energy] needs is still years away. Until that day comes we must continue meeting the nation's energy requirements by the means available to us."

This assertion assumes no changes to the existing energy paradigm: no new technological breakthroughs, no shifts in people's values or consumers' demands, no surprising events—natural or manmade—to alter the energy picture. But this paradigm-blinder limits our thinking—and our forecasts. Paradigms and social systems are rarely permanent, and new technology often drives a transition to other paradigms.

My thesis is that we have just begun the shift away from what I call the "carbon-combustion paradigm" to a new "electro-hydrogen paradigm." The shift is going to be very dramatic in the next 20 years, but the

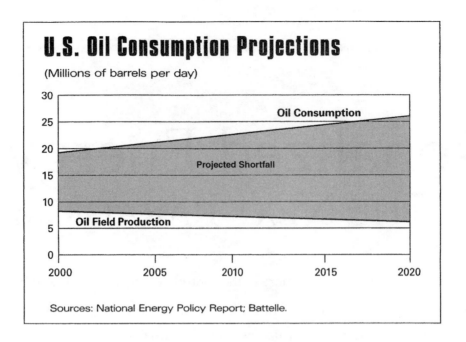

U.S. Oil Consumption Projections

(Millions of barrels per day)

Oil Consumption

Projected Shortfall

Oil Field Production

Sources: National Energy Policy Report; Battelle.

full integration is going to take easily a hundred years. We're going to see a lot of exciting technology innovations in the laboratory and in prototype systems in the next 10 to 20 years, but to go from our current paradigm and all its infrastructure to a new paradigm and all of its infrastructures is going to take a very long time.

Energy and the Consumer

Changes in consumer behavior are driving many trends. In the U.S. market, baby boomers seek convenience, while the elderly put heavy emphasis on the reliability and affordability of power. The question for policy makers and the energy industry is how reliable the electric grid will be in the future. Both baby boomers and Generation X'ers value customization—the personalization of products, especially computers and cell phones. Consumers also want more mobility and longevity in their products. And of course we all want inexpensive energy.

Along with consumer behavior, there are marketplace trends working toward this paradigm shift, including the effects of more-stringent environmental-quality regulations. The bad news is that no energy system will ever be 100% environmen-

tally friendly; the good news is that the next paradigm will be a lot friendlier than the last one was.

U.S. energy policy calls for greater energy self-sufficiency. An additional need, a new element since September 11, 2001, is security of the energy infrastructure. Before then, people didn't even worry about infrastructure security, but now it's a major issue. The electric grid system is absolutely vulnerable to weather and potential security compromises.

Another marketplace issue is the need for energy-cost stability and continued necessity for economic growth. If you take a strictly conservationist approach to this issue and say this cost stability and continued economic growth can be realized by voluntary simplicity, then you are limited by the old energy paradigm. This low-growth scenario has a low probability of occurring. We need more energy for continued economic growth, and that is still what most people in the world want.

Also impacting marketplace trends are emerging technologies such as those behind a gradual shift now beginning from central-station generation to decentralized, or distributed, generation of power from local sources. The paradigm shift to distributed generation, or distributed re-

sources, parallels the paradigm shift from fossil fuels to hydrogen. For example, one of the biggest needs we have is gasification of coal, though surprisingly little work is yet being done in this area. The current gasification technology has been in existence for at least 20 years, it isn't really very good, and it's very expensive. It's just not competitive.

The Real Hydrogen Future

It's easy to speculate about the hydrogen economy's potential and to go off into science-fiction scenarios. But because Battelle deals with the real-world challenges of governments and corporations, we spend a lot of time separating science from science fiction. So here's a little reality check: All forms of energy are going to have some negative environmental consequences. We need to recognize that fact and then try to make the energy system a lot better in the future than it is today. Another reality check is that no fuels will be free: There will always be costs for both the fuels and their infrastructure of production and distribution. The challenge is to find the new ways that improve the value (benefits/costs) relationship.

The challenge in this transitional period to the hydrogen future is to

About Battelle

The Battelle Memorial Institute was established in 1929 in Columbus, Ohio, and now manages or co-manages four of the 16 U.S. national labs. We are in the business of technology development, management, and commercialization. We do mostly government work, but we also have industrial clients. We're independent, meaning we have stakeholders but no stockholders. We are technically not for profit but, as we're reminded daily by our CEO, we are not for loss. We do more than $1.7 billion in business a year, and that's a lot of R&D. Battelle scientists have contributed to a wide range of breakthroughs, such as copy machines, optical digital recording, and bar codes; Battelle's R&D yields between 50 and 100 patented inventions a year.

All corporations and organizations face the challenge of keeping up with and anticipating change. At Battelle, we do futuring. I like to use the word *futuring* because the participle adds the action of making or doing something. We do trend analysis, expert focus groups, and expert judgment, and we have our own process of scenario analysis based on cross-impact analysis. We also have our own scenarios software, which we've been using for 20 years for our work with corporations.

Battelle studies "consumer value zones," where marketplace trends, new customer demands, and emerging technologies all converge. The study of energy's consumer value zone leads us to conclude, for example, that the future of energy is personal; that is, energy production will increasingly move from large, centralized power plants to distributed power.

Among its outreach projects, Battelle does an annual technology forecast and maintains a separate Web site for our scenarios and trends: www.dr-futuring.com.

For more information, see Battelle's Web site, www.battelle.org.

— *Stephen M. Millett*

extract hydrogen from hydrocarbon sources in an affordable way. There are many potential avenues being explored today. The approach at Battelle is to develop a universal reformer for the fuel cell, where we would take methane, methanol, and even gasoline and convert it into hydrogen at the point of burning it. The ability to extract sufficiently pure hydrogen from methane, methanol, or gasoline means that we could continue to use the existing infrastructure (such as all of those gas stations) to distribute safe liquid fuels without the expense and hazards of storing hydrogen today. Avoiding new infrastructure costs in the short run would greatly help the transition to the electrohydrogen paradigm of fuel cells for both transportation and stationary power generation applications.

Economics really favor the current, fading energy paradigm; economics do not yet favor the next one. We're going to have to see a lot of economic and regulatory changes, as well as technological changes. The challenge is cost. We can make fuel cells, we can produce hydrogen, but we can't do it at competitive prices relative to the existing hydrocarbon system, the carbon-combustion system. Electric utilities use a benchmark of $1,500 per kilowatt capacity; anything costing more than that is simply not competitive enough. Researchers at United Technologies, for example, are getting the cost of the fuel cell down below $3,000 per kilowatt—it has been as high as $15,000—but the price is still too high for general commercialization.

And what about solar cells? There's no question that there's a market today for solar cells, but it's largely a vanity technology for people who put it on their houses. If you take a pocket calculator that has solar panels, generating power measured in milliwatts, and normalize that to a kilowatt, the cost might be as high as $17,000. So, clearly, there's a long way to go before solar technology can replace the carbon-combustion system.

Alternative fuels like wind power, which is now growing, have been attractive because of a number of government incentives and subsidies at both the federal and the state levels.

Technologies to Watch

The technologies to watch include:

- Innovations in materials for batteries and fuel cells, especially PEM (polymer electrolyte membrane) and solid-oxide fuel cells.
- Breakthroughs in reducing diesel emissions.
- Innovations for reconfiguring backup and emergency power generation into distributed generation systems.
- Biofuel development.
- New approaches to the gasification of coal.
- Global warming and carbon-dioxide management.

For fuel cells and batteries, the biggest challenge is in materials development. Exciting new developments in battery technology include the sodium sulfur battery, which the American Electric Power Company is working on in Columbus, Ohio.

For PEM and solid-oxide fuel cells, the name of the game is the materials and getting their costs down. For instance, the current membrane material used in PEM fuel cells now costs as much as $800 per square meter. We need to get the cost down to $8 to make this transition to the next paradigm. So we need technology breakthroughs that bring costs down. In addition, the current use of platinum as the catalyst is obviously very expensive and needs to be changed.

Reducing diesel emissions is another significant area of research at Battelle and other institutions. Biofuel blending with diesel and other fuels is a very exciting growth area.

Bringing diesel emissions down will promote the transition of current backup generation to distributed resources coordinated with the power grid. Most utilities now dismiss customer-driven backup generation as simply being irrelevant to the grid, but if you can make all of those diesel generators environmentally compliant, and if you can coordinate them with the grid, then you've got a prototype distributed-generation system already in place.

Biofuel development, not just bioblending, is another breakthrough area. The DNA revolution in agriculture is very exciting because we could design plants—not just corn, but also chickweed or garbage grass, for instance—that could be engineered for high-starch content to be more easily converted into methanol.

Gasification of coal is a huge area of research and development. Affordable and efficient coal gasification would enable us to break down the constituent parts of coal and get the hydrogen atoms out of it. In an ideal process, we would be able to separate sulfur and other undesirable constituents out of the coal and extract pure hydrogen. We could also separate out the carbon content that produces carbon-dioxide emission from stacks. An innovative coal gasification technology would be a tremendous boon for the American economy—and the economies of Germany, Russia, China, and India, to mention just a few others. Hydrogen from coal would be a major step in the transition to fuel cells.

The new energy paradigm is also about the environment, and we at Battelle are concerned about global climate change and carbon-dioxide management. To that end, Battelle is actively pursuing approaches like carbon sequestration. We are currently working with the U.S. Department of Energy to evaluate how to capture carbon dioxide and store it underground so that it cannot escape into the atmosphere.

Toward a Distributed Power System

We're not suddenly going to do away with our coal-burning plants, but there are emerging opportunities to use large fuel cells and batteries in conjunction with central generation. This could produce emergency and peak power at the generation site as well as provide supplementary power at distributed sites. By "distributed," I don't mean we're going from the big power plant to the home all in one jump. Energy will be distributed first at the level of neighborhoods and districts, and then we'll work it on down to homes generating their own power. It'll go step by step, but the trend favors personalized energy.

We're going to see some exciting technologies developed in the next 10 years, but it's going to be a slow process toward full-blown commercialization. If we're on a low technology-development trajectory, it will take more time. If we get a couple of breakthroughs in technology or some regulatory changes, then we can be on a faster track, but no sooner than 2008 or even 2010 at the earliest unless a desperate need for power drives the trends faster. Slow progress favors the "tracker" and "adapter" companies and organizations, while fast progress favors the early innovators. Many companies are now agonizing over whether to be the progress leaders or the followers (or fast followers).

Who's going to lead this energy paradigm shift? Who really is going to provide the thought leadership and the breakthroughs? The Japanese are clearly ahead of the Americans in fuel cells. Honda and Toyota are ahead of the Big Three auto manufacturers in Detroit on energy breakthroughs for transportation. Honda in particular is the world leader in thinking through distributed power generation.

As for regulatory leadership, the question is who is going to provide the standards. There's a dearth of leadership for the new energy paradigm in the United States. Neither the federal government nor the states are showing signs of leadership, and there are very few progressive electric and gas utilities out there in the United States. Wherever the leadership comes from for the new energy paradigm, that's who will likely succeed and capture the largest market share.

But for now, those who should lead seem to be saying, "Change is good. You go first."

Stephen M. Millett is a thought leader at Battelle and co-author of *A Manager's Guide to Technology Forecasting and Strategy Analysis Methods*. His address is Battelle, 505 King Avenue, Columbus, Ohio 43201. E-mail milletts@battelle.org; Web site www.battelle.org.

Originally published in the July/August 2004 issue of *The Futurist,* pp. 44-48. Copyright © 2004 by World Future Society, 7910 Woodmont Avenue, Suite 450, Bethesda, MD 20814. Telephone: 301/656-8274; Fax: 301/951-0394; http://www.wfs.org. Used with permission from the World Future Society.

RENEWABLE ENERGY: A VIABLE CHOICE

By Antonia V. Herzog, Timothy E. Lipman, Jennifer L. Edwards, and Daniel M. Kammen

Renewable energy systems—notably solar, wind, and biomass—are poised to play a major role in the energy economy and in improving the environmental quality of the United States. California's energy crisis focused attention on and raised fundamental questions about regional and national energy strategies. Prior to the crisis in California, there had been too little attention given to appropriate power plant siting issues and to bottlenecks in transmission and distribution. A strong national energy policy is now needed. Renewable technologies have become both economically viable and environmentally preferable alternatives to fossil fuels. Last year the United States spent more than $600 billion on energy, with U.S. oil imports climbing to $120 billion, or nearly $440 of imported oil for every American. In the long term, even a natural gas–based strategy will not be adequate to prevent a buildup of unacceptably high levels of carbon dioxide (CO_2) in the atmosphere. Both the Intergovernmental Panel on Climate Change's (IPCC) recent Third Assessment Report and the National Academy of Sciences' recent analysis of climate change science concluded that climate change is real and must be addressed immediately—and that U.S. policy needs to be directed toward implementing clean energy solutions.[1]

Renewable energy technologies have made important and dramatic technical, economic, and operational advances during the past decade. A national energy policy and climate change strategy should be formulated around these advances. Despite dramatic technical and economic advances in clean energy systems, the United States has seen far too little research and development (R&D) and too few incentives and sustained programs to build markets for renewable energy technologies and energy efficiency programs.[2] Not since the late 1970s has there been a more compelling and conducive environment for an integrated, large-scale approach to renewable energy innovation and market expansion.[3] Clean, low-carbon energy choices now make both economic and environmental sense, and they provide the domestic basis for our energy supply that will provide security, not dependence on unpredictable overseas fossil fuels.

Energy issues in the United States have created "quick fix" solutions that, while politically expedient, will ultimately do the country more harm than good. It is critical to examine all energy options, and never before have so many technological solutions been available to address energy needs. In the near term, some expansion of the nation's fossil fuel (particularly natural gas) supply is warranted to keep pace with rising demand, but that expansion should be balanced with measures to develop cleaner energy solutions for the future. The best short-term options for the United States are energy efficiency, conservation, and expanded markets for renewable energy.

Traditional power plants based on fossil fuels emit pollutants that contribute substantially to climate change. Renewable energy sources are becoming economically viable and environmentally preferable alternatives.

For many years, renewables were seen as energy options that—while environmentally and socially attractive—occupied niche markets at best, due to barriers of cost and available infrastructure. In the last decade, however, the case for renewable energy has become economically compelling as well. There has been a true revolution in technological innovation, cost improvements, and our understanding and analysis of appropriate applications of renewable energy resources and technologies—notably solar, wind, small-scale hydro, and biomass-based

Figure 1. Capital cost forecasts for renewable energy technologies

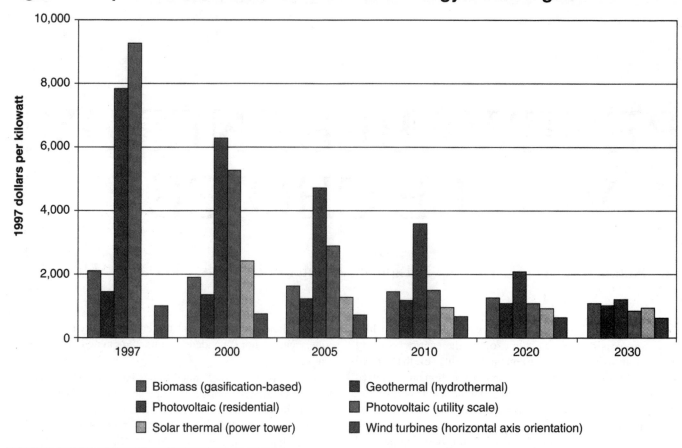

SOURCE: U.S. Department of Energy (DOE), *Renewable Energy Technology Characterizations*, Topical Report prepared by DOE Office of Utility Technologies and EPRI, TR-109496, December 1997.

energy, as well as advanced energy conversion devices such as fuel cells.[4] There are now a number of energy sources, conversion technologies, and applications that make renewable energy options either equal or better in price and services provided than the prevailing fossil-fuel technologies. For example, in a growing number of settings in industrialized nations, wind energy is now the least expensive option among all energy technologies—with the added benefit of being modular and quick to install and bring on-line. In fact, some farmers, notably in the U.S. Midwest, have found that they can generate more income per hectare from the electricity generated by a wind turbine than from their crop or ranching proceeds.[5] Also, photovoltaic (solar) panels and solar hot water heaters placed on buildings across the United States can help reduce energy costs, dramatically shave peak-power demands, produce a healthier living environment, and increase the overall energy supply.

The United States has lagged in its commitment to maintain leadership in key technological and industrial areas, many of which are related to the energy sector.[6] The United States has fallen behind Japan and Germany in the production of photovoltaic systems, behind Denmark in wind and cogeneration system deployment, and behind Japan, Germany, and Canada in the development of fuel-cell systems. Developing these industries within the United States is vital to the country's international competitiveness, commercial strength, and ability to provide for its own energy needs.

Renewable Energy Technologies

Conventional energy sources based on oil, coal, and natural gas have proven to be highly effective drivers of economic progress, but at the same time, they are highly damaging to the environment and human health. These traditional energy sources are facing increasing pressure on a host of environmental fronts, with perhaps the most serious being the looming threat of climate change and a needed reduction in greenhouse gas (GHG) emissions. It is now clear that efforts to maintain atmospheric CO_2 concentrations below even double the pre-industrial level cannot be accomplished in an oil- and coal-dominated global economy.

Theoretically, renewable energy sources can meet many times the world's energy demand. More important, renewable energy technologies can now be considered major components of local and regional energy systems. Solar, biomass, and wind energy resources, combined with new efficiency measures

Figure 2. Actual electricity costs in 2000

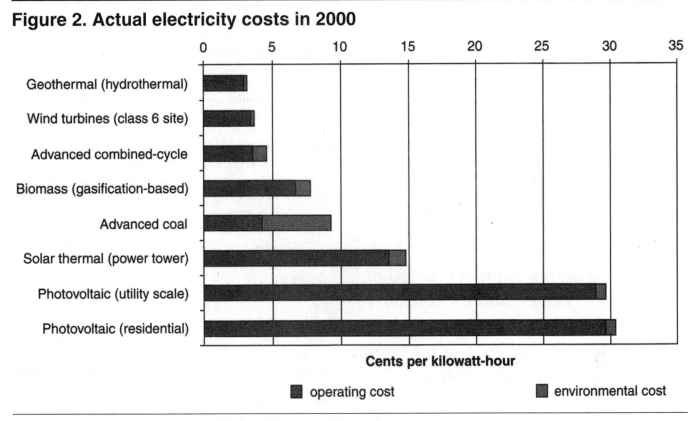

SOURCE: R.L. Ottinger et al., *Environmental Costs of Electricity* (New York: Oceana Publications, Inc., 1991); and U.S. Department of Energy, *Annual Energy Outlook 2000*, DOE/EIA-0383, Energy Information Administration, Washington, D.C., December 2000.

available for deployment in California today, could supply half of the state's total energy needs. As an alternative to centralized power plants, renewable energy systems are ideally suited to provide a decentralized power supply that could help to lower capital infrastructure costs. Renewable systems based on photovoltaic arrays, windmills, biomass, or small hydropower can serve as mass-produced "energy appliances" that can be manufactured at low cost and tailored to meet specific energy loads and service conditions. These systems have less of an impact on the environment, and the impact they do have is more widely dispersed than that of centralized power plants, which in some cases contribute significantly to ambient air pollution and acid rain.

There has been significant progress in cost reductions made by renewable technologies (see Figure 1).[7] In general, renewable energy systems are characterized by low or no fuel costs, although operation and maintenance costs can be considerable. Systems such as photovoltaics contain far fewer mechanically active parts than comparable fossil fuel combustion systems, and are therefore likely to be less costly to maintain in the long term.

Costs of solar and wind power systems have dropped substantially in the past 30 years and continue to decline. For decades, the prices of oil and natural gas have been, as one research group noted, "predictably unpredictable."[8] Recent analyses have shown that generating capacity from wind and solar energy can be added at low incremental costs relative to additions of fossil fuel–based generation. Geothermal and wind can be competitive with modern combined-cycle power plants—and geothermal, wind, and biomass all have lower total costs than advanced coal-fired plants, once approximate environmental costs are included (see Figure 2).[9] Environmental costs are based, conservatively, on the direct damage to the terrestrial and river systems from mining and pollutant emissions, as well as the impacts on crop yields and urban areas. The costs would be considerably higher if the damage caused by global warming were to be estimated and included.

The push to develop renewable and other clean energy technologies is no longer being driven solely by environmental concerns; these technologies are becoming economically competitive. According to Merrill Lynch's Robin Batchelor, the traditional energy sector has lacked appeal to investors in recent years because of heavy regulation, low growth, and a tendency to be cyclical.[10] The United States' lack of support for innovative new companies sends a signal that U.S. energy markets are biased against new entrants. The clean energy industry could, however, become a world-leading industry akin to that of U.S. semi-conductors and computer systems.

Renewable energy sources have historically had a difficult time breaking into markets that have been dominated by traditional, large-scale, fossil fuel–based systems. This is partly because renewable and other new energy technologies are only now being mass produced and have previously had high capital costs relative to more conventional systems, but also because coal-, oil-, and gas-powered systems have benefited from a range of subsidies over the years. These include military expenditures to protect oil exploration and production interests overseas, the costs of railway construction to enable economical delivery of coal to power plants, and a wide range of tax breaks.

One disadvantage of renewable energy systems has been the intermittent nature of some sources, such as wind and solar. A solution to this problem is to develop diversified systems that maximize the contribution of renewable energy sources but that also use clean natural gas and/or biomass-based power generation to provide base-load power (energy to meet the daily needs of society, leaving aside the peak in energy use associated, for example, with afternoon and evening air-conditioner or heating demands).

Solar energy can be harnessed by both industrial-scale and smaller residential solar power systems. Solar and other renewable energy sources could take over for fossil fuel–based power plants while saving energy and money.

Renewable energy systems face a situation confronting any new technology that attempts to dislodge an entrenched technology. For many years, the United States has been locked in to nuclear- and fossil fuel–based technologies, and many of its secondary systems and networks have been designed to accommodate only these sources. The U.S. administration's recent National Energy Policy plan focused on expanding the natural gas supply, without any attention to the benefits of building a diverse energy system.[11] The plan would add one to two new power plants each week for the next several years. The majority of these plants would be fired by natural gas, making the country far more dependent on natural gas than it ever was on oil—even at the height of the OPEC crisis in the 1970s.

Renewable energy technologies are characterized by low environmental costs, but many of these environmental costs are termed "externalities" and are not reflected in market prices. Only in certain areas and for certain pollutants do these environmental costs enter the picture. The international effort to limit the growth of GHG emissions through the Kyoto Protocol may lead to some form of carbon-based tax, which continues to face stiff political opposition in the United States. It is perhaps more likely that concern about emissions of particulate matter and ozone formation from fossil-fuel power plants will lead to expensive mitigation efforts by the plant operators, and this would help to tip the balance toward cleaner renewable systems.

There are two principal rationales for government support of R&D to develop clean energy technologies. First, conventional energy prices generally do not reflect the social and environmental cost of pollution. Second, private firms are generally unable to appropriate all the benefits of their R&D investments. The social rate of return for R&D exceeds the returns captured by individual firms, so they do not invest enough in R&D to maximize social benefits.[12] Public investment, however, would help spread innovation among clean energy companies, which would benefit the public.

Publicly funded market transformation programs (MTPs) for desirable clean energy technologies would provide an initial subsidy and incentive for market growth, thus stimulating long-term demand. A principal reason for considering MTPs is inherent in the production process. When a new technology is first introduced, it is more expensive than established substitutes. The unit cost of manufactured goods then tends to fall as a function of cumulative production experience. Cost reductions are typically very rapid at first and then taper off as the industry matures—resulting in an "experience curve". Gas turbines, photovoltaic cells and wind turbines have all exhibited this expected price-production relationship, with costs falling roughly 20 percent for each doubling of the number of units produced.[13]

If producers of clean energy consider the experience-curve effect when deciding how much to produce, they will "forward-price," producing at a loss initially to bring down their costs and thereby maximize profit over the entire production period. In practice, however, the benefits of production experience often spill over to competitor producers, and this potential problem discourages private firms from investing in bringing new products down the experience curve. Publicly funded MTPs can help correct the output shortfall associated with these experience effects.[14]

This suggests an important role for MTPs in national and international technology policies. MTPs are most effective with emerging technologies that have steep industry experience curves and a high probability of major long-term market penetration once subsidies are removed. The condition that these technologies be clean mitigates the risk of poor MTP performance, because the investments will alleviate environmental problems whose costs were not taken into account for the older, dirtier energy technologies. Renewable energy products are ideal candidates for support through MTPs, via federal policies that reward the early production of clean energy technologies.

Energy Efficiency

Energy efficiency improvements have contributed a great deal to economic growth and increased standard of living in the United States over the past 30 years, and there is much potential for further improvements in the decades to come. According to the U.S. Department of Energy (DOE), increasing energy efficiency could cut national energy use by 10 percent or more by 2010 and about 20 percent by 2020. The recent Interlaboratory Working Group study Scenarios for a Clean Energy Future estimates that cost-effective end-use technologies could reduce electricity consumption by about 1,000 billion kilowatt-hours (kWh) by 2020, almost entirely offsetting the projected growth in electricity use.[15] This level of energy savings would reduce U.S. carbon emissions by approximately 300 million metric tons, and many of these changes can actually be accomplished with an increase in profits. Still more benefits can be had for investments of only a few cents per kilowatt-hour, far less than the energy cost of new power plants.

Energy efficiency is the single greatest way to improve the U.S. energy economy. Based on data published by the Energy Information Administration (EIA), the American Council for an Energy Efficient Economy (ACEEE) estimates that total energy use per capita in the United States in 2000 was almost identical to that in 1973, while over the same period, economic output (measured by Gross Domestic Product or GDP) per capita increased 74 percent. Furthermore, national energy intensity (energy use per unit of GDP) fell 42 percent between 1973 and 2000, with about 60 percent of this decline attributable to real improvements in energy efficiency and about one-quarter due to structural changes and fuel switching. Between 1996 and 2000, GDP increased 19 percent while primary energy use increased just 5 percent. These statistics clearly indicate that energy use and GDP do not have to grow or decline in lock step with each other, but GDP can, in fact, increase while energy use does not.[16]

The federal government's energy efficiency programs have been a resounding success. Last year, DOE documented the results of 20 of its most successful energy efficiency and renewable energy technology initiatives over the past two decades.[17] These programs have already saved the nation 5.5 quadrillion Btu (British Thermal Units) of energy, equivalent to the amount of energy needed to heat every household in the United States for about a year, and worth about $30 billion in avoided energy costs. Over the last decade, the cost to taxpayers for these 20 programs has been $712 million, less than 3 percent of the energy bill savings that the programs created.[18]

In 1997, the President's Committee of Advisors on Science and Technology (PCAST), a panel that consisted mainly of distinguished academics and private-sector executives, conducted a detailed review of DOE's energy efficiency R&D programs. PCAST concluded, "R&D investments in energy efficiency are the most cost-effective way to simultaneously reduce the risks of climate change, oil import interruption, and local air pollution, and to improve the productivity of the economy." PCAST recommended that the DOE energy efficiency budget be doubled between the fiscal years of 1998 and 2003. They estimated that this could produce a 40-to-1 return on investment for the nation, including reductions in fuel costs of $15 billion to $30 billion by 2005 and $30 billion to $45 billion by 2010.[19] Funding for these DOE programs in the last several years has fallen far short of the PCAST recommendations.

Increasing the efficiency of homes, appliances, vehicles, businesses, and industries must be an important part of a sound national energy and climate-change policy. Increasing energy efficiency reduces energy waste (without forcing consumers to cut back on energy services or amenities), lowers GHG emissions, saves consumers and businesses money (because the energy savings more than pay for any increase in initial cost), protects against energy shortages, reduces energy imports, and reduces air pollution. Furthermore, increasing energy efficiency does not create a conflict between enhancing national security and energy reliability, on the one hand, and protecting the environment on the other.

Climate Change

The threat of global climate change is finally producing a growing understanding and acknowledgement by some in U.S. industry and government that a responsible national energy policy must include a sound global climate-change mitigation strategy. President George W. Bush has rejected the Kyoto Protocol, but the U.S. Congress, in particular the Senate, appears poised to take action to reduce domestic GHG emissions. For example, Senators Jim Jeffords (I-Vt.) and Joe Lieberman (D-Conn.) and Representatives Sherwood Boehlert (R-N.Y.) and Henry Waxman (D-Calif.) recently introduced legislation in Congress to reduce the emission of four pollutants from electricity generation. The legislation puts a national cap on power plants' emissions of nitrogen oxides, sulfur dioxide, mercury, and carbon dioxide, and requires every power plant to meet the most recent emission control standards. It allows market-oriented mechanisms such as the trading of emissions credits, which is widely seen as a way to control pollution and stimulate innovation at the lowest cost. In emissions trading, total emissions are capped and then a market is created involving those firms that have excess credits to sell (resulting from decreased emissions due to efficiency and other improvements) and those firms needing to purchase credits due to emissions exceeding an allocated baseline. In the United States, nitrogen oxides and sulfur dioxide markets have been highly successful. The CO_2 reductions required by the legislation would bring emissions back to 1990 levels by 2007, and the costs of implementing such measures would likely be dwarfed by the resulting benefits of industrial innovation.[20]

A tax credit that rewards fuel-efficient vehicles, including hybrid vehicles now on the market, could be used along with a tax penalty for inefficient vehicles to encourage purchases of fuel-efficient vehicles and the development of new technologies.

Legislation that controls the four major pollutants from power plants in an integrated package will help reduce regulatory uncertainties for electric generators and will be less costly than separate programs for each pollutant. Although voluntary action by companies is an attractive idea, in the last 10 years, voluntary actions have failed to reduce carbon dioxide emissions in the United States. Instead, emissions have increased by 15.5 percent since 1990, with an annual average increase of 1.5 percent since 1990, and they continue to increase.[21] EIA recently released data showing an increase of 2.7 percent in U.S. carbon dioxide emissions from 1999 to 2000. Solutions will become more costly and difficult if mandatory emissions reductions are not enacted now.

Policy Options

The ultimate solutions to meeting the nation's energy needs cost-effectively and reducing GHG emissions must be based on private-sector investment bolstered by well-targeted government R&D and incentives for emerging clean energy technologies. The United States now has the opportunity to build a sustainable energy future by engaging and stimulating the tremendous innovative and entrepreneurial capacity of the private sector. Advancing clean energy technologies requires a stable and predictable economic environment.

Research and Development Funding

Federal funding and leadership for renewable energy and energy efficiency projects has resulted in several notable successes, such as the U.S. Environmental Protection Agency's (EPA) Energy Star and Green Lights Programs, which have been emulated in a number of countries. Fifteen percent of the public-sector building space in the country has now signed up for the Energy Star buildings program, saving more than 21 billion kWh of energy in 1999 and reducing carbon emissions by about 4.4 million metric tons, which according to EPA, has resulted in $1.6 billion in energy bill savings. Despite these achievements, funding in this area has been scant and so uneven as to discourage private sector involvement. By increasing funding for these EPA programs, their scope could be considerably expanded.

The Bush administration's proposed cuts in its 2002 fiscal year budget for DOE's renewable energy and energy efficiency programs would harm existing public-private partnerships as well as R&D. This budgetary roller coaster harms all investments and sends mixed signals to industry.[22] Steadily increasing funding would transform the clean energy sector from a good idea to a pillar of the new economy.

Tax Incentives

The R&D tax credit, which goes to companies based on their R&D expenditures, has proven remarkably effective and popular with private industry, so much so that there is a strong consensus in both Congress and the administration to make this credit permanent. To complement this support of private-sector R&D, tax incentives directed toward those who use the technologies would provide the "demand pull" needed to accelerate the technology transfer process and the rate of market development.

Currently, non-R&D federal tax expenditures aimed at the production and use of energy have an unequal distribution across primary energy sources, distorting the market in favor of conventional energy technologies. Renewable fuels make up 4 percent of the United States' energy supply, yet they receive only 1 percent of federal tax expenditures and direct fiscal spending combined (see Table 1).[23] The largest single tax credit in 1999 was the Alternative Fuel Production Credit, which totaled more than $1 billion.[24] This income tax credit, which has gone primarily to the natural gas industry, was designed to reduce dependence on foreign energy imports by encouraging the production of gas, coal, and oil from unconventional sources (such as tight gas formations and coalbed methane) within the United States. Support for the production and further development of renewable fuels, all found domestically, would have a greater long-term effect on the energy system than any expansion of fossil-fuel capacity.

A production tax credit (PTC) of 1.7 cents/kWh now exists for electricity generated from wind power and "closed loop" biomass (biomass from dedicated energy crops and chicken litter). The wind power credit, in particular, has proven successful in encouraging strong growth of U.S. wind energy over the last several years—with a 30-percent increase in 1998 and a 40-percent increase in 1999. Approximately 2,000 megawatts (MW) of wind energy will be under development or proposed for completion before the end of 2001 (a 40-percent increase from 2000), when the federal wind energy PTC is scheduled to expire. Currently, Germany has twice the U.S. installed wind energy capacity, and the major wind-turbine manufacturers are now in Europe.[25]

This production credit should be expanded to include electricity produced by "open loop" biomass (including agricultural and forestry residues but excluding municipal solid waste), solar energy, geothermal energy, and landfill gas. The extension and expansion of PTC has recently been garnering strong and consistent support in the U.S. Congress. Investment tax incentives are also needed for smaller-scale renewable energy systems, such as residential photovoltaic panels and solar hot-water heaters, as well as small wind systems used in commercial and farm applications. In these cases, an investment credit in capital or installation expenditures is preferable to a production credit based on electricity generated, due to the relatively high capital cost of these smaller-scale renewable technologies and the fact that the electricity and heat produced is used directly.

Many new energy-efficient technologies have been commercialized in recent years or are nearing commercialization. Tax incentives can help manufacturers justify mass marketing and help buyers and manufacturers offset the relatively high initial capital and installation costs for new technologies. A key element in designing the credits is for only high-efficiency products to be eligible. If eligibility is set too low, there may not be enough energy savings to justify the credits. These tax credits should have limited duration and be reduced in value over time, because once these new technologies become widely available, costs should decline. In this manner, the credits will help innovative technologies get established in the marketplace but will not become permanent subsidies.

Recent federal tax credit legislation to encourage the use of high-efficiency technologies includes incentives for highly efficient clothes washers, refrigerators, and new homes; innovative building technologies such as furnaces, stationary fuel cells, gas-fired pumps, and electric heat-pump water heaters; and investments in commercial buildings that have reduced heating and cooling costs. The incentives currently being proposed in Congress and by the administration will have a relatively modest direct impact on energy use and CO_2 emissions. Savings may only amount to 0.3 quadrillion Btu of energy and 5 million metric tons of carbon emissions per year by 2012. How-

Table 1. U.S. energy consumption and federal expenditures

Fuel source	Primary energy supply 1998 consumption		Direct expenditures and tax expenditures (1999)	
	Value (quads, quadrillion Btu)	Percent	Value (millions of dollars)	Percent
Oil	36.57	40	263	16
Natural gas	21.84	24	1,048	65
Coal	21.62	24	85	5
Oil, gas, and coal combined			205	13
Nuclear	7.16	8	0	–
Renewables	3.48	4	19	1
Total	90.67	100	1620	100

NOTE: The Alternative Fuels Credit accounted for $1,030 of the $1,048 in expenditures for natural gas. Oil, gas, and coal combined includes expenditures that were not allocated to any one of the three individual fuels. Research and development are not included in direct expenditures and tax expenditures. Btu = British thermal units.

SOURCE: Energy Information Administration, *Federal Financial Interventions and Subsidies in Energy Markets 1999: Primary Energy*, U.S. Department of Energy (Washington, D.C., 1999).

ever, if these proposed tax credits help to establish innovative products in the marketplace and reduce the first-cost premium so that the products are viable after the credits are phased out, then the indirect impacts could be many times greater than the direct impacts. It has been estimated that total energy savings could reach 1 quadrillion Btu by 2010 and 2 quadrillion Btu by 2015 if these credits are successfully implemented.[26]

Vehicle Fuel Economy Standards

New vehicles with hybrid gasoline-electric power systems are now produced commercially, and fuel cell–electric vehicles are being produced in prototype quantities. These vehicles combine high-efficiency electric motors with revolutionary power systems to produce a new generation of motor vehicles that are vastly more efficient than today's simple cycle combustion systems. The potential for future hybrid and fuel-cell vehicles to achieve up to 100 miles per gallon (mpg) is believed to be both technically and economically viable in the near future. In the long term, fuel-cell vehicles running directly on hydrogen promise to allow motor vehicle use with very low fuel-cycle emissions.

The improvements in fuel economy that these new vehicles offer will help to slow growth in petroleum demand, reducing our oil import dependency and trade deficit. While the Partnership for a New Generation of Vehicles helped generate some vehicle technology advances, an increase in the Corporate Average Fuel Economy (CAFE) standard, which has been stagnant for 16 years now, is required to provide an incentive for companies to bring these new vehicles to market rapidly.

Recent analyses of the costs and benefits of motor vehicles with higher fuel economy have been conducted by the Union of Concerned Scientists, Massachusetts Institute of Technology, the Office of Technology Assessment, and Oak Ridge National Lab/ACEEE.[27] These studies have generally concluded that

with longer-term technologies, motor vehicle fuel economy can be raised to 45 mpg for cars with a retail price increase of $500 to $1,700 per vehicle, and to 30 mpg for light trucks with a retail price increase of $800 to $1,400 per vehicle.[28] These improvements could be the basis for a new combined fuel economy standard of 40 mpg for both cars and light trucks. The combined standard could be accomplished between 2008 and 2012. The net cost would be negligible once fuel savings are factored in, if the auto industry is given adequate time to retool for this new generation of vehicles. A lower combined standard could be implemented sooner and then raised incrementally each year to achieve the 40-mpg goal by 2012.

Tax credits for hybrid-electric vehicles, battery-electric vehicles, and fuel-cell vehicles are an important part of the puzzle. These funds could, in principle, be raised through a revision of the archaic "gas guzzler" tax, which does not apply to a significant percentage of the light-duty car and truck fleet. The tax penalty and tax credit in combination could be a revenue-neutral "fee-bate" scheme—similar to one recently proposed in California—that would simultaneously reward economical vehicles and penalize uneconomical ones.

Efficiency Standards

A critical strategy for effectively promoting energy efficiency is implementing new standards for buildings, appliances, and equipment. Tax credits do not necessarily remove all the market barriers that prevent clean energy technologies from spreading throughout the marketplace. These barriers include lack of awareness, rush purchases when an existing appliance breaks down, and purchases by builders and landlords who do not pay utility bills.

Significant advances in the efficiency of heating and cooling systems, motors, and appliances have been made in recent years, but more improvements are technologically and economically feasible. A clear federal statement of desired improvements in system efficiency would remove uncertainty about and reduce costs of implementing these changes. Under such a federal mandate, efficiency standards for equipment and appliances could be gradually increased, helping to expand the market share of existing high-efficiency systems.[29]

Extensive transmission and distribution networks waste a significant percentage of electricity generated by traditional power plants. Smaller-scale systems can be located closer to where the energy is used.

Standards remove inefficient products from the market and still leave consumers with a full range of products and features from which to choose. Building, appliance, and equipment efficiency standards have proven to be one of the federal government's most effective energy-saving programs. Analyses by DOE and others indicate that in 2000, appliance and equipment efficiency standards saved 1.2 quadrillion Btu of energy (1.3 percent of U.S. electric use) and reduced consumer energy bills by approximately $9 billion, with energy bill savings far exceeding any increase in product cost. By 2020, standards already enacted will save 4.3 quadrillion Btu per year (3.5 percent of projected U.S. energy use) and reduce peak electric demand by 120,000 MW (more than a 10-percent reduction). ACEEE estimates that by 2020, energy use could be reduced by 1.0 quadrillion Btu by quickly adopting higher standards for equipment that is currently covered under federal law, such as central air conditioners and heat pumps, and by adopting new standards for equipment not covered, such as torchiere (halogen) light fixtures, commercial refrigerators, and appliances that consume power while on standby.[30] Energy bills would decline by approximately $7 billion per year by 2020.[31]

A Renewable Portfolio Standard

The Renewable Portfolio Standard (RPS) is akin to the efficiency standards for vehicles and appliances that have proven successful in the past. A gradually increasing RPS is designed to integrate renewables into the marketplace in the most cost-effective fashion, and it ensures that a growing proportion of electricity sales is provided by renewable energy. An RPS provides the one true means to use market forces most effectively—the market picks the winning and losing technologies.

A number of studies indicate that a national renewable-energy component of 2 percent in 2002, growing to 10 percent in 2010 and 20 percent by 2020, that would include wind, biomass, geothermal, solar, and landfill gas, is broadly good for business and can readily be achieved.[32] States that decide to pursue more aggressive goals could be rewarded through an additional federal incentive program. In the past, federal RPS legislation has been introduced in Congress and it was proposed by the Clinton administration, but it has yet to be re-introduced by either this Congress or the Bush administration.

Including renewables in the United States' power-supply portfolio would protect consumers from fossil fuel price shocks and supply shortages by diversifying the energy options. A properly designed RPS will also create jobs at home and export opportunities abroad. To achieve compliance, a federal RPS should use market dynamics to stimulate innovation through a trading system. National renewable energy credit trading will encourage development of renewables in the regions of the country where they are the most cost-effective, while avoiding expensive long-distance transmission.

The coal, oil, natural gas, and nuclear power industries continue to receive considerable government subsidies, even though they are already well established. Without RPS or a similar mechanism, many renewables will not be able to survive in an increasingly competitive electricity market focused on producing power at the lowest direct cost. And while RPS is designed to deliver renewable that are most ready for the market, additional policies will still be needed to support emerging renewable technologies, like photovoltaics, that have enormous potential to become commercially competitive.

RPS is the surest market-based approach for securing the public benefits of renewables while supplying the greatest amount of clean power at the lowest price. It creates an ongoing incentive to drive down costs by providing a dependable and predictable market. RPS will promote vigorous competition among renewable energy developers and technologies to meet the standard at the lowest cost.

Analysis of the RPS target for 2020 shows renewable energy development in every region of the country, with most coming from wind, biomass, and geothermal sources. In particular, the Plains, Western, and mid-Atlantic states would generate more than 20 percent of their electricity from renewables.[33] Texas has become a leader in developing and implementing a successful RPS that then-Governor Bush signed into law in 1999. The Texas law requires electricity companies to supply 2,000 MW of new renewable resources by 2009, and the state is actually expected to meet this goal by the end of 2002, seven years ahead of schedule. Nine other states have signed an RPS into law: Arizona, Connecticut, Maine, Massachusetts, Nevada, New Jersey, New Mexico, Pennsylvania, and Wisconsin. Minnesota and Iowa have a minimum renewables requirement similar to RPS, and legislation that includes RPS is pending in several other states.

While the participation of 12 states signals a good start, this patchwork of state policies would not be able to drive down the costs of renewable energy technologies and move these technologies fully into the marketplace. Also, state RPS policies have differed substantially from each other thus far. These differences could cause significant market inefficiencies, negating the cost savings that a more comprehensive, streamlined, market-based federal RPS package would provide.

Small-Scale Distributed Energy Generation and Cogeneration

Small-scale distributed electricity generation has several advantages over traditional central-station utility service. Distributed generation reduces energy losses incurred by sending electricity long distances through an extensive transmission and distribution network (often an 8- to 10-percent loss of energy). In addition, generating equipment located close to the end use allows waste heat to be utilized (a process called cogeneration) to meet heating and hot water demands, significantly boosting overall system efficiency.

Distributed generation has faced several barriers in the marketplace, most notably from complicated and expensive utility interconnection requirements. These barriers have led to a push for national safety and power quality standards, now being finalized by the Institute of Electrical and Electronics Engineers (IEEE). Although the adoption of these standards would significantly decrease the economic burden on manufacturers, installers, and customers, the utilities are allowed discretion in adopting or rejecting them.

In designing credits, highest priority should go to renewable or fossil fuel systems that utilize waste heat through combined heat and power (CHP) designs. While a distributed generation system may achieve 35- to 45-percent electrical efficiency, the addition of heat utilization can raise overall efficiency to 80 percent. Industrial CHP potential is estimated to be 88,000 MW, the largest sectors being in the chemicals and paper industries. Commercial CHP potential is estimated to be 75,000 MW, with education, health care, and office building applications making up the most significant percentages.[36]

A National Public Benefits Fund

Electric utilities have historically funded programs to encourage the development of a host of clean energy technologies. Unfortunately, increasing competition and deregulation have led utilities to cut these discretionary expenditures in the last several years. Total utility spending on demand-side management programs fell more than 50 percent from 1993 to 1999. Utilities should be encouraged to invest in the future through rewards (such as tax incentives) for companies that reinvest profits and invigorate the power sector.[37] A national public benefits fund could be financed through a national, competitively neutral wires charge of $0.002 per kWh.

Cost and Benefit Analyses

A range of recent studies are all coming to the same conclusions: that simple but sustained standards and investments in a clean energy economy are not only possible but would also be highly beneficial to future prosperity in the United States.[37] If energy policies proceed as usual, the nation is expected to increase its reliance on coal and natural gas to meet strong growth in electricity use (42 percent by 2020). To meet this demand, it is estimated that 1,300 300-MW power plants would need to be built, with electricity generated by non-hydro renewables only

increasing from 2 percent today to 2.4 percent of total generation in 2020.[38] A set of clean energy polices could meet a much larger share of our future energy needs, with energy efficiency measures projected to almost completely offset the projected growth in electricity use.[39] A clean energy strategy would significantly reduce emissions from utilities. In fact, through a steady shift to clean energy production, power plant carbon dioxide reductions (as proposed in the current legislation before Congress), would not be difficult or expensive to meet.[40]

The United States is becoming increasingly dependent on oil—$120 billion was spent on oil imports last year. Making renewable energy sources a larger part of the energy economy would enable the nation to provide for its own energy needs.

Recent analysis by the Union of Concerned Scientists focused on the costs and environmental impacts of a package of clean energy polices and how the package would affect fossil fuel prices and consumer energy bills. They found that using energy more efficiently and switching from fossil fuels to renewable energy sources will save consumers money by decreasing energy use.[41] A whole-economy analysis carried out by the International Project for Sustainable Energy Paths has also shown that Kyoto-type targets can easily be met, with a net increase of 1 percent in the nation's 2020 GDP, by implementing the right policies.[42]

One of the greatest advantages that energy efficiency and renewable energy sources offer over new power plants, transmission lines, and pipelines is the ability to deploy these technologies very quickly. They can be installed—and benefits can be reaped—immediately.[43] In addition, reductions in CO_2 emissions will have a "clean cascade" effect on the economy because many other pollutants are emitted during fossil fuel combustion.

The renewable and energy-efficient technologies and policies described here have already proven successful and cost-effective at the national and state levels. Supporting them would allow the United States to cost-effectively meet GHG emission targets while providing a sustainable, clean energy future.[44]

We stand at a critical point in the energy, economic, and environmental evolution of the United States. Renewable energy and energy efficiency are now not only affordable, but their expanded use will also open new areas of innovation. Creating opportunities and a fair marketplace for a clean energy economy requires leadership and vision. The tools to implement this evolution are now well known. We must recognize and overcome the current road blocks and create the opportunities needed to put these renewable and energy-efficient measures into effect.

This article is based on testimony provided by D. M. Kammen to the U. S. Senate Commerce, Science and Transportation (July 10, 2001) and U. S. Senate Finance (July 11, 2001) Committees. Antonia V. Herzog and Timothy E. Lipman are postdoctoral researchers and Jennifer L. Edwards is a research assistant at the Renewable and Appropriate Energy Laboratory (RAEL), Energy and Resources Group (ERG), at the University of California at Berkeley. Daniel M. Kammen is a professor of Energy and Society with ERG and a professor of Public Policy with the Goldman School of Public Policy. Address correspondence to D. M. Kammen, 310 Barrows Hall, University of California, Berkeley, CA 94720-3050, or dkammen@socrates.berkeley.edu. Additional material can be found at http://socrates.berkeley.edu/~rael.

NOTES

1. Intergovernmental Panel on Climate Change (IPCC), *Climate Change 2001: The Scientific Basis* (Working Group I of the IPCC, World Meteorological Organization - U.N. Environment Program, Geneva), January 2001; and National Research Council (NRC), *Climate Change Science: An Analysis of Some Key Questions*, Committee on the Science of Climate Change, (National Academy Press, Washington, D.C., 2001). For more information on energy and climate change, see J. P. Holdren, "The Energy-Climate Challenge: Issues for the New U.S. Administration," *Environment*, June 2001, 8–21.

2. D. M. Kammen and R. M. Margolis, "Evidence of Under-Investment in Energy R&D Policy in the United States and the Impact of Federal Policy," *Energy Policy* 27 (1999), 575–84; and R. M. Margolis and D. M. Kammen, "Underinvestment: The Energy Technology and R&D Policy Challenge," *Science* 285 (1999), 690–92.

3. This work appeared in two influential forms that reached dramatically different audiences: A. B. Lovins, "Energy Strategy: The Road Not Taken," *Foreign Affairs* (1976), 65–96; and A. B. Lovins, *Soft Energy Paths: Toward a Durable Peace* (New York: Harper Colophon Books, 1977).

4. A. V. Herzog, T. E. Lipman, and D. M. Kammen, "Renewable Energy Sources," in *Our Fragile World: Challenges and Opportunities for Sustainable Development*, forerunner to the Encyclopedia of Life Support Systems (EOLSS), Volume 1, Section 1 (UNESCO-EOLSS Secretariat, EOLSS Publishers Co. Ltd., 2001).

5. P. Mazza, *Harvesting Clean Energy for Rural Development: Wind*, Climate Solutions Special Report, January 2001.

6. Kammen and Margolis, note 2 above.

7. U.S. Department of Energy (DOE), *Renewable Energy Technology Characterizations*, Topical Report Prepared by DOE Office of Utility Technologies and EPRI, TR-109496, December 1997.

8. B. Haevner and M. Zugel, *Predictably Unpredictable: Volatility in Future Energy Supply and Price From California's Over Dependence on Natural Gas*, CALIPIRG Charitable Trust Research Report, September, 2001).

9. R. L. Ottinger et al., *Environmental Costs of Electricity* (New York: Oceana Publications, Inc., 1991); U.S. Department of Energy (DOE), *Annual Energy Outlook 2000*, DOE/EIA-0383 (00), Energy Information Administration, Washington, D. C., December 2000; and U.S. Department of Energy, 1997.

10. Reuters News Service, "Fuel Cells and New Energies Come of Age Amid Fuel Crisis," 11 September 2000.

11. National Energy Policy, "Reliable, Affordable, and Environmentally Sound Energy for America's Future," Report of the National Energy Policy Development Group, Office of the President, May 2001.

12. Kammen and Margolis, note 2 above.

13. International Institute for Applied Systems Analysis/World Energy Council, *Global Energy Perspectives to 2050 and Beyond* (Laxenburg, Austria, and London, 1995).

14. R. D. Duke and D. M. Kammen, "The Economics of Energy Market Transformation Initiatives," *The Energy Journal* 20 (1999): 15–64.

15. Interlaboratory Working Group, *Scenarios for a Clean Energy Future* (Oak Ridge, Tenn.: Oak Ridge National Laboratory; and Berkeley, Calif.: Lawrence Berkeley National Laboratory), ORNL/CON-476 and LBNL-44029, November 2000.

16. S. Nadel and H. Geller, "Energy Efficiency Polices for a Strong America," American Council for an Energy-Efficient Economy (ACEEE), May 2001 (draft).

17. Clean Energy Partnerships, *A Decade of Success*, Office of Energy Efficiency and Renewable Energy, DOE/EE-0213 (Washington, D.C., 2000).

18. Nadel and Geller, note 16 above.

19. President's Committee of Advisors on Science and Technology (PCAST), *Federal Energy Research and Development for the Challenges of the Twenty-First Century*, Washington, D.C., Energy Research and Development Panel, November 1997.

20. F. Krause, S. DeCanio, and P. Baer, "Cutting Carbon Emissions at a Profit: Opportunities for the U.S." (El Cerrito, Calif.: International Project for Sustainable Energy Paths), May 2001; and A. P. Kinzig and D. M. Kammen, "National Trajectories of Carbon Emissions: Analysis of Proposals to Foster the Transition to Low-Carbon Economies," *Global Environmental Change* 8 (3) (1998): 183–208.

21. Energy Information Administration (EIA), *U.S. Carbon Dioxide Emissions from Energy Sources 2000 Flash Estimate*, based on data from the Monthly Energy Review (May 2001) and the Petroleum Supply Annual 2000, DOE (Washington, D.C.), June 2001.

22. Kammen and Margolis, note 2 above.

23. This does not include revenue outlays for the Alcohol Fuels Excise Tax, which reduces the tax paid on ethanol-blended gasoline. Most ethanol used in the United States is produced from corn, and the GHG emissions impact is uncertain and has been shown to be negligible (M. Delucchi, *A Revised Model of Emissions of Greenhouse Gases from the Use of Transportation Fuels and Electricity*, Institute of Transportation Studies, UCD-ITS-RR-97-22, (Davis, Calif., 1997)); and EIA, *Federal Financial Interventions and Subsidies in*

Energy Markets 1999: Primary Energy, DOE (Washington, D.C., 1999).

24. Established by the Windfall Profit Tax Act of 1980, this tax credit is $3 per barrel of oil equivalent produced, and it phases out when the price of oil rises to $29.50 per barrel (1979 dollars).

25. American Wind Energy Association web site, available at http://www.awea.org, accessed on September 8, 2001.

26. Nadal and Geller, note 16 above.

27. J. Mark, "Greener SUVs: A Blueprint for Cleaner, More Efficient Light Trucks," Union of Concerned Scientists, 1999; M. A. Weiss, J. B. Heywood, E. M. Drake, A. Schafer, and F. F. AuYeung, "On the Road in 2020: A Lifecycle Analysis of New Automobile Technologies," Energy Laboratory, Massachusetts Institute of Technology, MIT EL 00-003 (Cambridge, Mass., October 2000); Office of Technology Assessment, *Advanced Vehicle Technology: Visions of a Super-Efficient Family Car*, U.S. Congress, OTA-ETI-638 (Washington, D.C., September 1995); and D. L. Greene and J. Decicco, "Engineering-Economic Analyses of Automotive Fuel Economy Potential In The United States," *Annual Review of Energy and the Environment*, 25: (2000) 477–536.

28. Greene and Decicco, note 27 above; and Interlaboratory Working Group, note 15 above.

29. S. L. Clemmer, D. Donovan, and A. Nogee, "Clean Energy Blueprint: A Smarter National Energy Policy for Today and the Future, Phase I," Union of Concerned Scientists and Tellus Institute, June 2001.

30. K. B. Rosen and A. K. Meier, *Energy Use of Televisions and Videocassette Recorders in the U.S.*, DOE, LBNL-42393, (Berkeley, Calif.: Lawrence Berkeley National Laboratory), March 1999.

31. Nadal and Geller, note 16 above.

32. Clemmer, Donovan, and Nogee, note 29 above; S. L. Clemmer, A. Nogee, and M. Brower, "A Powerful Opportunity: Making Renewable Electricity the Standard," Union of Concerned Scientists, January 1999; and A. Nogee, S. Clemmer, B. Paulos, and B. Haddad, "Powerful Solutions: 7 Ways to Switch America to Renewable Energy," Union of Concerned Scientists, January 1999.

33. Clemmer, Nogee, and Brower, note 32 above.

34. Distributed generation reflects a new way to manage energy supply and demand. Instead of the old system of large capital-intensive central-station power plants, the improvements in energy efficiency, renewable energy, and small, modular5 "micro-turbines" that burn gas, as well as hydro and other resources, energy supplies could be located closer to users, reducing transmission losses, improving system reliability, and energy security.

35. R. K. Dixon, Office of Power Technologies, U.S. Department of Energy, Second International CHP Symposium, Amsterdam, Netherlands, May 2001.

36. Kammen and Margolis, note 2 above.

37. Interlaboratory Working Group, note 15 above; Krause, DeCanio, and Baer, note 20 above; and Clemmer, Donovan, and Nogee, note 29 above.

38. National Energy Policy, note 11 above.

39. Clemmer, Donovan, and Nogee, note 29 above.

40. Ibid.

41. Ibid.

42. Krause, DeCanio, and Baer, note 20 above.

43. Kinzig and Kammen, note 20 above.

44. P. Baer, et al. "Equity and Greenhouse Gas Responsibility," *Science* 289 (2000), 2287.

From *Environment*, December 2001, pp. 8-20. Reprinted with permission of the Helen Dwight Reid Educational Foundation. Published by Heldref Publications, 1319 Eighteenth St., NW, Washington, DC 20036-1802. © 2001.

Hydrogen: Waiting For the Revolution

Everybody agrees it's the future fuel of choice. Why hasn't the future arrived?

By Bill Keenan

Here's how you'll live in the Hydrogen Age: Your car, powered by hydrogen fuel cells and electric motors, quietly drives along smog-free highways. At night, when you return your vehicle to the garage, you hook up its fuel cell to a worldwide distributed-energy network; the central power grid automatically purchases your battery's leftover energy, offsetting your overall energy costs.

In the garage, you also have a suitcase-sized electrolyzer, or other conversion device, plugged into the electrical system to pump a fresh batch of hydrogen into your car. (The fuel cell uses hydrogen to produce electricity, which powers the motor.) If you need a refill as you're driving along one of the nation's highways, you pull up to a clean, quiet hydrogen fueling station to top off in less time than it takes today to fill a car with gasoline.

The electricity in your home will also come from hydrogen, either via small local fuel-cell power plants or residential fuel cells in your basement. "Moreover," says Jeremy Rifkin, president of the Foundation on Economic Trends and author of *The Hydrogen Economy*, "sensors attached to every appliance or machine powered by electricity—refrigerators, air-conditioners, washing machines, security alarms—will provide up-to-the-minute information on energy prices, as well as on temperature, light, and other environmental conditions, so that factories, offices, homes, neighborhoods, and whole communities can continuously and automatically adjust their energy consumption to one another's needs and to the energy load flowing through the system."

The U.S. Department of Energy is only slightly less enthusiastic, maintaining in a report that in the hydrogen economy, "America will enjoy a secure, clean, and prosperous energy sector that will continue for generations to come. It will be produced cleanly, with near-zero net carbon emissions, and it will be transported and used safely. [Hydrogen] will be the fuel of choice for American businesses and consumers."

The new energy regime will have economic and political ramifications as well. Oil companies and utility companies will merge and morph into "energy companies" with a focus on generating renewable energy and local power distribution, including purchasing power from residential customers. Distributed energy production will also result in a worldwide "democratization of energy," bringing low-cost power to underdeveloped areas.

Oil and Hydrogen Don't Mix

Driving the interest in a hydrogen-based energy system: threats to the economy, the environment, and national security. Oil production, by current estimates, will likely peak sometime between 2020 and 2040. At this point, the world's economies will have consumed half of the known oil reserves, with two-thirds of the remaining oil in the volatile Middle East. As a result, prices will rise dramatically, and global consumers will experience increasingly frequent shortages.

Global warming is another significant threat that a shift to hydrogen might ameliorate. The release of carbon dioxide into the atmosphere from the burning of fossil fuels such as coal, oil, and natural gas makes up about 85 percent of greenhouse-gas emissions in the United States. This increase has resulted in an unprecedented rate of global warming, according to most scientific experts. The thinning of the polar ice caps, the retreat of glaciers around the world, the spread of tropical diseases to more temperate climates, and the rising of global sea levels are all evidence of global warming. Says

The Mecahnics of Hydrogen

While still untested on a large scale, the promise of a hydrogen economy is based on a number of undeniable realities. Hydrogen can be burned or converted into electricity in a way that creates virtually no pollution. It is also Earth's most abundant element, available everywhere in the world. While hydrogen is scarce naturally in pure form, it can be generated easily by reforming gasoline, methanol, natural gas, and other readily available resources. It can also be created by electrolysis, a process by which electricity is run through water to separate the oxygen and hydrogen molecules.

The fuel cell, which combines oxygen in the air with hydrogen to create electricity and water, is the vital link in the hydrogen vision. It closes the energy loop and allows electricity to be stored and transported via hydrogen and then reconverted back into electricity.

In an ideal future, renewable-energy sources such as wind, solar, or water power will be used to create hydrogen through electrolysis. The hydrogen can be converted again to electricity locally by means of a fuel cell to power a car, provide energy for a home, power a laptop, or operate any number of other products.

—B.K

Rifkin: "Weaning the world away from a fossil-fuel energy regime will limit carbon-dioxide emissions to only twice their pre-industrial levels and mitigate the effects of global warming on the Earth's already beleaguered biosphere."

Add to these threats the burden of growing world populations, an increasingly unstable political situation in the Middle East, and the likelihood of longer and more frequent blackouts and brownouts resulting from an aging and vulnerable power grid in the United States, and the promise of a safe, pollution-free, and distributed power system based on hydrogen becomes increasingly attractive.

Pathways and Roadblocks

Does all this sound too good to be true? It is: The hydrogen economy faces serious obstacles. More than 90 percent of the hydrogen produced today comes from reformulated natural gas generated through a process that creates a significant amount of carbon dioxide. Energy for this process, or for electrolysis, a more expensive way of generating hydrogen would also come from power plants fueled by oil or natural gas. So in the near term, a shift to hydrogen will not greatly reduce the world's dependence on fossil fuels and, in fact, may well hasten the greenhouse effect and global warming by increasing carbon-dioxide emissions.

Consequently, a lot of discussion about the hydrogen economy revolves around the various "pathways," or means of producing hydrogen. Atakan Ozbek, director of energy research at ABI Research, a technology-research think tank, points out that while hydrogen can come from virtually any fuel, energy from oil and gas is currently cheaper and more efficient than energy from renewable resources such as wind, sun, or water. Then, too, in the event of an oil crisis and resultant electricity shortage, coal will likely be pressed into service, regardless of the environmental cost. Nuclear power plants can also provide electricity to create hydrogen, but nuclear energy's high cost—plus the still-hot controversy over waste disposal—make such a pathway less than certain.

"What we're trying to find out right now," Ozbek says, "is how to get hydrogen to the fuel cell in a way that is economically feasible and makes sense engineering-wise."

Environmental considerations are paramount: If coal is reintroduced in a large way into our "energy portfolio"— whether to produce hydrogen or as part of our existing energy plan to replace oil—carbon-dioxide emissions will rise significantly.

The Department of Energy roadmap anticipates this, and the DOE is funding research into the "sequestration" of carbon-dioxide gases created by coal processing and natural-gas reformation. This would involve capturing these gases at some point in the energy process and permanently storing them underground or in the ocean.

To many, this is unrealistic. Jon Ebacher, vice president of power-systems technology for GE Energy, won't say that sequestration is impossible, but his comments fall short of an endorsement. Even a fairly efficient coal plant, Ebacher says, produces millions of tons of carbon dioxide each year. "So if you're going to sequester carbon dioxide from all of the plants that use hydrocarbon fuels," he says, "that's a pretty massive undertaking."

Only a hydrogen economy based 100 percent on renewable power would result in zero emissions—the vision that has captured so many imaginations. And that vision remains decades away. In the meantime, Ebacher says, "natural gas can see us through a transition period until we get solar and other renewable-energy efficiencies up to a much higher level." That transition period, he suggests, might last twenty-five to fifty years.

Another potential roadblock: transport and storage of hydrogen. Less dense than other fuels, the gas must be compressed or liquefied to be stored or moved efficiently, adding to costs and inconvenience. While the existing natural-gas infrastructure would seem to offer a convenient pathway to hydrogen delivery, this can't be done without a major retrofit. Indeed, Ebacher says, almost all of the country's existing natural-gas pipeline would have to be modified to handle hydrogen.

Finally, fuel-cell researchers must make significant advances. The power produced by a fuel cell is significantly more expensive per unit than that produced by an internal-combustion engine. Fuel-cell vehicle development is also beset by problems and costs related to type of fuel, storage, and performance. A number of prototype and "concept car" fuel-cell vehicles have been produced and displayed at auto shows and fuel-cell conferences around the world—but at a development cost of about $250,000 or more per vehicle. GM estimates that it

spent between $1 million and $2 million to develop its Hy-wire fuel-cell concept car. A consumer version would cost far less, obviously, but likely would still take sticker shock to a whole new dimension.

Putting a Brake on Hydrogen Cars

Linking the hydrogen age to cars could be a critical policy mistake, according to Joseph Romm, former acting assistant secretary for the DOE's Office of Energy Efficiency and Renewable Energy and author of *The Hype About Hydrogen*. Despite car-company promises to have fuel-cell vehicles in dealer showrooms by 2010, if not sooner, Romm argues that the cost of fuel cells, problems with onboard storage of hydrogen in vehicles, and the issues related to creating a hydrogen delivery infrastructure are likely to push the market for hydrogen fuel-cell vehicles well into the future.

The focus on hydrogen as an immediate goal in the transportation sector amounts to confusing a means (hydrogen) with an end (greenhousegas reduction), Romm explains. This could have harmful consequences, since, he estimates, it will take thirty to fifty years for hydrogen vehicles to have a significant impact on greenhouse gases. A recent National Academy of Sciences study seconds this point, stating, "In the best-case scenario, the transition to a hydrogen economy would take many decades, and any reductions in oil imports and carbon-dioxide emissions are likely to be minor during the next twenty-five years."

"If the goal is to reduce greenhouse gases," Romm argues, "then there are technologies available right now that can have a more immediate effect"—hybrid vehicles, for instance. And diverting existing (and limited) natural-gas supplies to create hydrogen for vehicles "would make that fuel less available where its use could result in a more immediate reduction in greenhouse-gas emissions— in replacing existing oil and coal-burning electric-power plants in the nation's energy grid with cleaner natural-gas power plants."

As a fuel, hydrogen is "simply a better mousetrap."

In fact, some hydrogen-technology companies have back-burnered research

As GE Goes, So Goes The Nation?

Fuel cells will probably not be a viable market until a company like General Electric gets into the business in a big way, say critics of the hydrogen economy.

GE is indeed researching fuel cells, albeit cautiously, in keeping with its approach to most other energy markets. "We do have an investment in fuel cells," says Jon Ebacher, the company's vice president of power-systems technology." I don't know if it will ever get to the dimensions where it will work at the huge volumes that were once forecast, but I think it's quite viable in niche markets." Right now, he sees a possible market in "industrial facilities that have isolated power needs, where you have a maintenance crew that deals with heating, ventilating, and air-conditioning." But, he says, "there's a distance between where they are today and the huge potential in consumer markets that was forecast at one time."

GE has also invested substantially in researching a type of fuel-cell system that would employ a gas turbine and hydrogen system working as a combined cycle. Right now, Ebacher says, GE is considering creating a power plant based on this system by 2013. It could be sooner, depending on external factors, including political developments around both fuel and the environment, the price of fuel, and the types of fuel that are available. But there are still unresolved technical challenges that could push that back.

Natural-gas prices in particular are an important barometer. "During the California energy crisis, the price of gas spiked up to $7 per million BTUs," Ebacher says. Higher gas prices, he says, "could spark some other research efforts that may come in front of fuel cells—coal, for instance. It's possible to run a combined-cycle system on coal. You put a chemical plant beside a combined-cycle power plant to process the coal into a gaseous fuel to run the electrical plant." But right now, Ebacher says, "the capital cost of doing that doesn't cross the goal line. However, if the price of natural gas or its availability gets in a bad place, all of a sudden the capital cost of doing that might not look so bad."

"It all revolves around availability, economics, and the environment—where the pressures are, what are the levers. But if you talk about running out of hydrocarbon fuels, then you would have to say that hydrogen had better be in the cards."

—B. K

and development on transportation applications. "The horizons for fuel-cell vehicles keep getting pushed out further and further, and it's unlikely that somebody's going to license and commit to a uniform, standardized hydrogen technology for at least ten to fifteen years," says Stephen Tang, an industry consultant and former president and CEO of Eatontown, NJ.-based Millennium Cell, which makes a system called Hydrogen on Demand that supplies hydrogen to fuel cells.

To pay the bills in the meantime, Tang says, "Millennium Cell has targeted markets that it believes can tolerate the price of hydrogen and fuel cells, such as consumer electronic devices, standby power, and military portables. In all of those markets, you're competing with an incumbent technology that is rather expensive in its own right and also has some limitations in performance. In these markets, then, we can focus on hydrogen as a performance fuel and not focus so much on the environmental benefits or the energy-independence benefits—attributes that buyers have difficulty valuing. It's simply a better mousetrap: Hydrogen allows you to run your cell phone much longer, or your laptop much longer, without being a slave to the energy grid or inferior batteries."

Who Will Lead?

Despite the limitations, there is growing momentum for hydrogen vehicles.

Hybrid vehicles may be a "bridging technology" toward the hydrogen age, but it's one that "doesn't at all curb the nation's appetite for oil," says Chris Borroni-Bird, GM's director of design-technology fusion. Therefore, the automaker directs about a third of its R&D—over $1 billion thus far and involving more than six hundred people—toward fuel cells. The company insists that it will have a commercially viable fuel-cell vehicle available by the end of the decade.

In other business sectors, investment in hydrogen technology is slowly returning after the boom in hydrogen technology stocks in 1999-2000 and the subsequent bust that lasted until last year. "Behind a lot of the hype, there was tremendous capital inflow in the mid-1990s going into 2000," Tang says. Unfortunately, the number of commercial products—and the resulting revenue—in the industry have been "underwhelming" relative to investment dollars. That has made the investment community more cautious so far, but things are changing. "Right now there is a much more realistic view of the possibilities," Tang says. "The investor today is looking more toward interesting niche strategies and early market penetration rather than the hope of the mass market, the home run where fifty million cars are going to be sold with your product in it."

With the investment community poised and the technology issues coming together, says ABI Research's Ozbek, "Everything is feeding into this giant equation—you can consider it a giant chemical reaction—and once everything has been fed in and the equation solved, it's going to change the whole energy infrastructure." Federal support and direction will be especially important. While Ozbek considers President Bush's $1.7 billion State of the Union pledge for energy research a good start, he would like the government to provide such research incentives as Japan and the European Union have in recent years.

And though the president disappointed many hydrogen proponents by making no specific mention of hydrogen-energy R&D in his 2004 address, his proposed 2005 budget did increase funding for hydrogen research. The federal government, Ozbek argues, should provide enhanced tax credits for buyers of fuel-cell vehicles and fuel credits for energy companies and other investing in building a hydrogen infrastructure. Jeremy Rifkin agrees, urging the federal government to take the lead by establishing benchmarks—mandating tougher fuel-efficiency standards and requiring a greater use of renewable energy sources by power companies—as the European Union currently does.

Is GM's Hy-Wire The Car of the Future?

General Motors has had a reputation for being rather conservative when it comes to both new technological developments and vehicle design, but it seems to have leapt ahead of other carmakers with its concept car, the Hy-wire.

The idea, says Chris Borroni-Bird, director of design-technology fusion for GM, is that "if you design a vehicle around the fuel cell and hydrogen tanks, you might be able to create a better vehicle than if you just put those same systems in a car designed for an internal-combustion engine."

The Hy-wire design puts the fuel cell and hydrogen storage tanks into a skateboard-like chassis that allows for greater flexibility and interchangeability of body types. Customized car bodies are then effectively "docked into" the uniform chassis.

And because the fuel cell can provide much greater electrical output than today's batteries, GM's designers have replaced mechanical and hydraulic systems for steering and braking with an electronically controlled one. "This system provides more design freedom, because those electrical wires can be routed in numerous ways, replacing a fixed steering column," Borroni-Bird says.

The Hy-wire prototype has no gas engine, no brake pedals, and no instrument panel. The fuel cell enables you to operate everything by wire. The electronic controls are included in a compact handgrip console that extends from the floor from between the front seats of the vehicle. Drivers can steer, brake, or accelerate with the controls built into the handgrips.

Because GM puts the hydrogen directly on board the vehicle, there is no need for the car to convert fossil fuels or other renewable sources into hydrogen. As a result, it can claim to offer a zero-emission vehicle and market the car to be compatible with a network of hydrogen fueling stations.

To that end, GM "applauds any hydrogen infrastructure projects, anywhere in the world," says Tim Vail, GM's director of business development for fuel-cell activities. Yet it will take a lot of applause to get the government to invest the estimated $11 billion to get a sufficient mass of hydrogen refueling stations to support 1 million vehicles, in proximity to 70 percent of the nation's population. "But," says the optimistic Vail, "$11 billion is nothing compared to past infrastructure projects such as the highways or the railroads. So it's not that big an issue to overcome. You just have to have the will to do it."

—B. K.

California's Hydrogen Highway

One state isn't waiting for action from companies or the federal government. In California, the new Schwarzenegger administration has committed to an energy plan that aims to create a "hydrogen highway" in the state by 2010. The ambitious plan proposes the construction of hydrogen fueling stations every twenty miles along the state's twenty-one major interstate highways. By taking this step to break the chicken-and-egg dilemma (which comes first, the vehicle or the fueling infrastructure?) and by continuing to impose strict mandates on automakers for fuel efficiencies, California could jumpstart the hydrogen economy.

"The pieces are all on the table," says Terry Tamminen, secretary of California's Environmental Protection Agency, "and there have been demonstration projects, but they have not been pulled together into any kind of unified vision, something that average people can use and where we can more fully commercialize the technology. So we're taking a lot of this work that's already been done, bringing it together, adding some timetables and leadership, and then of course asking for some federal money to help

with the pieces that aren't paid for by private industry or other investments."

California could jumpstart the hydrogen economy.

California already has several hydrogen fueling stations, serving research projects and some municipal fleets, and about a dozen more are in the works. For instance, SunLine Transit Agency, a local public-transit company, now operates a hydrogen fueling station that it uses to test its hydrogen-powered buses. And AC Transit, which provides public transportation in the San Francisco Bay area, expects to have three fuel-cell-powered buses later this year.

The state's goal is to provide an infrastructure of fueling stations to support a consumer market for fuel-cell vehicles. "If we can deliver such a network by a certain date," Tamminen explains, "we can then ask car companies to deliver on their promises to start delivering cars to showrooms."

One of the things driving California's plan is a California Energy Comission report that, Tamminen says, "includes credible evidence that in three to five years we are going to have serious shortages of refined fuels in

the state. Not because there's not enough petroleum under the sands of Iraq but, rather, because we don't have enough refinery capacity in the state—or in the country—to keep up with the demand created by longer commutes, poorer fuel economy, and a growing population. The report predicts a likelihood of $3 to $5 per gallon gasoline prices and periodic shortages.

"During the oil embargo of the mid-1970s, we had twenty-four thousand retail gasoline outlets in the state, compared to ten thousand today. If there are shortages, not only will there be gas lines—they will be twice as long."

Consequently, it's not a question of if but when we move toward a hydrogen economy. Even Romm, who is dubious about short-term prospects for hydrogen, concludes: "The longer we wait to act, and the more inefficient, carbon-emitting infrastructure that we lock into place, the more expensive and the more onerous will be the burden on all segments of society when we finally do act."

BILL KEENAN *is a freelance business writer and former editor of* Selling *magazine.*

UNIT 4

Biosphere: Endangered Species

Unit Selections

Key Points to Consider

- Are there ways to assess the value or worth of living organisms other than those from whom we derive direct benefits (our domesticated plant and animals species)? What are the relationships between economic assessments of the biosphere and moral or value judgments on the preservation of species?

- Why is the spread of invasive species through the global trading network so difficult to control and what kinds of damage do invasive species produce? What suggestions would you make to remedy the problem?

- How can market forces be used to increase the effectiveness of biodiversity? Can the market function more rapidly and efficiently than government agencies in controlling or influencing the rate of environmental disruption?

- What are some of the primary reasons behind the rate of species loss that some scientists believe represents a mass extinction? Why do decision-makers dispute scientific claims about rates of species extinction?

 Links: www.dushkin.com/online/
These sites are annotated in the World Wide Web pages.

Endangered Species
http://www.endangeredspecie.com/
Friends of the Earth
http://www.foe.co.uk/index.html
Smithsonian Institution Web Site
http://www.si.edu
World Wildlife Federation (WWF)
http://www.wwf.org

Tragically, the modern conservation movement began too late to save many species of plants and animals from extinction. In fact, even after concern for the biosphere developed among resource managers, their effectiveness in halting the decline of herds and flocks, packs and schools, or groves and grasslands has been limited by the ruthlessness and efficiency of the competition. Wild plants and animals compete directly with human beings and their domesticated livestock and crop plants for living space and for other resources such as sunlight, air, water, and soil. As the historical record of this competition in North America and other areas attests, since the seventeenth century human settlement has been responsible—either directly or indirectly—for the demise of many plant and wildlife species. It should be noted that extinction is a natural process—part of the evolutionary cycle—and not always created by human activity; but human actions have the capacity to accelerate a natural process that might otherwise take a millennia.

In the opening article of this unit, one of the world's best-known writers on biological issues asks the central question that will control the future of the biosphere. In "What Is Nature Worth?" Edward O. Wilson notes that there are powerful economic reasons for implementing plans to preserve the world's natural biological diversity. Wilson notes that present losses of Earth's plants and animals from human impact are progressing at rates from 1,000 to 10,000 times those of the average over the last half-billion years. If for no other reason than the fact that many of those lost species have incalculable economic value, humankind needs to fully understand just what it is doing to other inhabitants of Earth. But Wilson notes that there are moral arguments as well as economic ones for slowing the rate of extinction.

The second selection in the unit addresses one of the most pervasive biological issues throughout human history—that of invasive species—and one that has contributed significantly to the extinction rates that Wilson discusses. In "Strangers in Our Midst: The Problem of Invasive Alien Species," scientist Jeffrey McNeely of the World Conservation Union notes that "invasive" or non-native species that become established in environments and spread rapidly because they lack natural enemies are not only very damaging to human economic and other interests but loom large as one of the world's most serious biological threats. While the migration of people, plants, and animals has been a feature of human history for a millennia, as globalization and an increasing emphasis on international trade has dominated the late 20th and early 21st centuries, natural barriers to the ready movement of plants and animals without human aid have disappeared. But no nation on earth has, as yet, developed an effective coordinated strategy to deal with this more pervasive problem. The next article deals with some similar problems but from an entirely different perspective. In "Markets for Biodiversity

Services: Potential Roles and Challenges," authors Michael Jenkins, Sara Scherr, and Mira Inbar (all with non-profit NGOs) argue that while historically it may have been the role of government to deal with such problems as invasive species and the loss of biodiversity, an alternative based in the market exists. Public sector financing for protection of the biosphere is facing increasingly severe challenges, particularly in countries where the prevailing political philosophies may deny that problems exist. In such situations, the authors contend, the lower cost approach to protecting and preserving biodiversity may be to pay landowners to manage their lands in ways that will preserve natural systems and conserve native species. At least some of this payment could come from conservation organizations such as the Nature Conservancy, but other logical sources include private corporations, research institutes, and private individuals.

What is essential is to find market-based mechanisms such as eco-tourism that will convince resource owners and managers that good stewardship creates economic value.

The unit's final selection deals with the essential issue regarding the biosphere: species extinction. In "On the Termination of Species," senior *Scientific American* writer, W. Wayt Gibbs, tells us that the scientific evidence of present mass extinction of species is being challenged by skeptics and ignored by politicians. He suggests that part of the problem is the difficulty of defining the dimension of the extinction problem. It is also important that people understand why species loss is important and how it can be dealt with. Part of the solution is figuring out ways to make biodiversity pay in economic as well as ecological terms.

What Is Nature Worth?

There's a powerful economic argument for preserving our living natural environment:
The biosphere promotes the long-term material prosperity and health of the human race
to a degree that is almost incalculable. But moral reasons, too, should compel us
to take responsibility for the natural world.

by Edward O. Wilson

In the early 19th century, the coastal plain of the southern United States was much the same as in countless millenniums past. From Florida and Virginia west to the Big Thicket of Texas, primeval stands of cypress and flatland hardwoods wound around the corridors of longleaf pine through which the early Spanish explorers had found their way into the continental interior. The signature bird of this wilderness, a dweller of the deep bottomland woods, was the ivory-billed woodpecker, *Campephilus principalis*. Its large size, exceeding a crow's, its flashing white primaries, visible at rest, and its loud nasal call—*kent!... kent!... kent!*—likened by John James Audubon to the false high note of a clarinet, made the ivory-bill both conspicuous and instantly recognizable. Mated pairs worked together up and down the boles and through the canopies of high trees, clinging to vertical surfaces with splayed claws while hammering their powerful, off-white beaks through dead wood into the burrows of beetle larvae and other insect prey. The hesitant beat of their strikes—*tick tick... tick tick tick... tick tick*—heralded their approach from a distance in the dark woods. They came to the observer like spirits out of an unfathomed wilderness core.

Alexander Wilson, early American naturalist and friend of Audubon, assigned the ivorybill noble rank. Its manners, he wrote in *American Ornithology*

(1808–14), "have a dignity in them superior to the common herd of woodpeckers. Trees, shrubbery, orchards, rails, fence posts, and old prostrate logs are all alike interesting to those, in their humble and indefatigable search for prey; but the royal hunter before us scorns the humility of such situations, and seeks the most towering trees of the forest, seeming particularly attached to those prodigious cypress swamps whose crowded giant sons stretch their bare and blasted or moss-hung arms midway to the sky."

A century later, almost all of the virgin bottomland forest had been replaced by farms, towns, and second-growth woodlots. Shorn of its habitat, the ivorybill declined precipitously in numbers. By the 1930s, it was down to scattered pairs in the few remaining primeval swamps of South Carolina, Florida, and Louisiana. In the 1940s, the only verifiable sightings were in the Singer Tract of northern Louisiana. Subsequently, only rumors of sightings persisted, and even these faded with each passing year.

The final descent of the ivorybill was closely watched by Roger Tory Peterson, whose classic *A Field Guide to the Birds* had fired my own interest in birds when I was a teenager. In 1995, the year before he died, I met Peterson, one of my heroes, for the first and only time. I asked him a question common in conversations among American naturalists: What of

the ivory-billed woodpecker? He gave the answer I expected: "Gone."

I thought, surely not gone *everywhere*, not *globally*! Naturalists are among the most hopeful of people. They require the equivalent of an autopsy report, cremation, and three witnesses before they write a species off, and even then they would hunt for it in séances if they thought there were any chance of at least a virtual image. Maybe, they speculate, there are a few ivorybills in some inaccessible cove, or deep inside a forgotten swamp, known only to a few close-mouthed cognoscenti. In fact, several individuals of a small Cuban race of ivorybills were discovered during the 1960s in an isolated pine forest of Oriente Province. Their current status is unknown. In 1996, the Red List of the World Conservation Union reported the species to be everywhere extinct, including Cuba. I have heard of no further sightings, but evidently no one at this writing knows for sure.

Why should we care about *Campephilus principalis*? It is, after all, only one of 10,000 bird species in the world. Let me give a simple and, I hope, decisive answer: because we knew this particular species, and knew it well. For reasons difficult to understand and express, it became part of our culture, part of the rich mental world of Alexander Wilson and all those afterward who cared about it. There is no way to make a full and final valuation of the ivorybill or any other species in the natural world. The measures we use increase in number and magnitude with no predictable limit. They rise from scattered, unconnected facts and elusive emotions that break through the surface of the subconscious mind, occasionally to be captured by words, though never adequately.

We, *Homo sapiens*, have arrived and marked our territory well. Winners of the Darwinian lottery, bulge-headed paragons of organic evolution, industrious bipedal apes with opposable thumbs, we are chipping away the ivorybills and other miracles around us. As habitats shrink, species decline wholesale in range and abundance. They slide down the Red List ratchet, and the vast majority depart without special notice. Over the past half-billion years, the planet lost perhaps one species per million species each year, including everything from mammals to plants. Today, the annual rate of extinction is 1,000 to 10,000 times faster. If nothing more is done, one-fifth of all the plant and animal species now on earth could be gone or on the road to extinction by 2030. Being distracted and self-absorbed, as is our nature, we have not yet fully understood what we are doing. But future generations, with endless time to reflect, will understand it all, and in painful detail. As awareness grows, so will their sense of loss. There will be thousands of ivorybilled woodpeckers to think about in the centuries and millenniums to come.

Is there any way now to measure even approximately what is being lost? Any attempt is almost certain to produce an underestimate, but let me start anyway with macroeconomics. In 1997, an international team of economists and environmental scientists put a dollar amount on all the ecosystems services provided to humanity free of charge by the living natural environment. Drawing from multiple databases, they estimated the contribution to be $33 trillion or more each year. This amount is nearly twice the 1997 combined gross national product (GNP) of all the countries in the world—$18 trillion. *Ecosystems services* are defined as the flow of materials, energy, and information from the biosphere that support human existence. They include the regulation of the atmosphere and climate; the purification and retention of fresh water; the formation and enrichment of the soil; nutrient cycling; the detoxification and recirculation of waste; the pollination of crops; and the production of lumber, fodder, and biomass fuel.

> IF HUMANITY WERE TO TRY TO REPLACE THE FREE SERVICES OF THE NATURAL ECONOMY WITH SUBSTITUTES OF ITS OWN MANUFACTURE, THE GLOBAL GNP WOULD HAVE TO BE INCREASED BY AT LEAST $33 TRILLION.

The 1997 megaestimate can be expressed in another, even more cogent, manner. If humanity were to try to replace the free services of the natural economy with substitutes of its own manufacture, the global GNP would have to be increased by at least $33 trillion. The exercise, however, cannot be performed except as a thought experiment. To supplant natural ecosystems entirely, even mostly, is an economic—and even physical—impossibility, and we would certainly die if we tried. The reason, ecological economists explain, is that the *marginal value*, defined as the rate of change in the value of ecosystems services relative to the rate of decline in the availability of these services, rises sharply with every increment in the decline. If taken too far, the rise will outpace human capacity to sustain the needed services by combined natural and artificial means. Hence, a much greater dependence on artificial means—in other words, environmental prostheses—puts at risk not just the biosphere but humanity itself.

Most environmental scientists believe that the shift has already been taken too far, lending credit to the folk injunction "Don't mess with Mother Nature." The lady is our mother all right, and a mighty dispensational force as well. After evolving on her own for more than three billion years, she gave birth to us a mere million years ago, the blink of an eye in evolutionary time. Ancient and vulnerable, she will not tolerate the undisciplined appetite of her gargantuan infant much longer.

Abundant signs of the biosphere's limited resilience exist all around. The oceanic fish catch now yields $2.5 billion to the U.S. economy and $82 billion worldwide. But it will not grow further, simply because the amount of ocean is fixed and the number of organisms it can generate is static. As a result, all of the world's 17 oceanic fisheries are at or below sustainable yield. During the 1990s, the annual global catch leveled off around 30 million tons. Pressed by ever-growing global demand, it can be expected eventually to drop. Already, fisheries of the western North Atlantic, the Black Sea, and portions of the Caribbean have largely collapsed. Aquaculture, or the farming of fish, crustaceans, and mollusks, takes up part of the slack, but at rising environmental cost. This "fin-and-shell revolution" necessitates the conversion of valuable wetland habitats, which are nurseries for marine life. To feed the captive populations, fodder must be diverted from crop production. Thus, aquaculture competes with other human activities for productive land while reducing natural habitat. What was once free for the taking must now be manufactured. The ultimate result will be an upward inflationary pressure across wide swaths of the world's coastal and inland economies.

Another case in point: Forested watersheds capture rainwater and purify it before returning it by gradual runoffs to the lakes and sea, all for free. They can be replaced only at great cost. For generations, New York City thrived on exceptionally clean water from the Catskill Mountains. The watershed inhabitants were proud that their bottled water was once sold throughout the Northeast. As their population grew, however, they converted more and more of the watershed forest into farms, homes, and resorts. Gradually, the sewage and agricultural runoff adulterated the water, until it fell below Environmental Protection Agency standards. Officials in New York City now faced a choice: They could build a filtration plant to replace the Catskill watershed, at a $6 billion to $8 billion capital cost, followed by $300 million annual running costs, or they could restore the watershed to somewhere near its original purification capacity for $1 billion, with subsequently very low maintenance costs. The decision was easy, even for those born and bred in an urban environment. In 1997, the city raised an environmental bond issue and set out to purchase forested land and to subsidize the upgrading of septic tanks in the Catskills. There is no reason the people of New York City and the Catskills cannot enjoy the double gift from nature in perpetuity of clean water at low cost and a beautiful recreational area at no cost.

There is even a bonus in the deal. In the course of providing natural water management, the Catskill forest region also secures flood control at very little expense. The same benefit is available to the city of Atlanta. When 20 percent of the trees in the metropolitan area were removed during its rapid development, the result was an annual increase in stormwater runoff of 4.4 billion cubic feet. If enough containment facilities were built to capture this volume, the cost would be at least $2 billion. In contrast, trees replanted along streets and in yards, and parking areas are a great deal cheaper than concrete drains and revetments. Their maintenance cost is near zero, and, not least, they are more pleasing to the eye.

In conserving nature, whether for practical or aesthetic reasons, diversity matters. The following rule is now widely accepted by ecologists: The more numerous the species that inhabit an ecosystem, such as a forest or lake, the more productive and stable is the ecosystem. By "production," the scientists mean the amount of plant and animal tissue created in a given unit of time. By "stability," they mean one or the other, or both, of two things: first, how narrowly the summed abundances of all species vary through time; and, second, how quickly the ecosystem recovers from fire, drought, and other stresses that perturb it. Human beings understandably wish to live in the midst of diverse, productive, and stable ecosystems. Who, if

given a choice, would build a home in a wheat field instead of a parkland?

Ecosystems are kept stable in part by the insurance principle of biodiversity: If a species disappears from a community, its niche will be more quickly and effectively filled by another species if there are many candidates for the role instead of few. Example: A ground fire sweeps through a pine forest, killing many of the understory plants and animals. If the forest is biodiverse, it recovers its original composition and production of plants and animals more quickly. The larger pines escape with some scorching of their lower bark and continue to grow and cast shade as before. A few kinds of shrubs and herbaceous plants also hang on and resume regeneration immediately. In some pine forests subject to frequent fires, the heat of the fire itself triggers the germination of dormant seeds genetically adapted to respond to heat, speeding the regrowth of forest vegetation still more.

WHO, IF GIVEN A CHOICE, WOULD BUILD A HOME IN A WHEAT FIELD INSTEAD OF A PARKLAND?

A second example of the insurance principle: When we scan a lake, our macroscopic eye sees only relatively big organisms, such as eelgrass, pondweeds, fishes, water birds, dragonflies, whirligig beetles, and other things big enough to splash and go bump in the night. But all around them, in vastly greater numbers and variety, are invisible bacteria, protistans, planktonic single-celled algae, aquatic fungi, and other microorganisms. These seething myriads are the true foundation of the lake's ecosystem and the hidden agents of its stability. They decompose the bodies of the larger organisms. They form large reservoirs of carbon and nitrogen, release carbon dioxide, and thereby damp fluctuations in the organic cycles and energy flows in the rest of the aquatic ecosystem. They hold the lake close to a chemical equilibrium, and, to a point, they pull it back

from extreme perturbations caused by silting and pollution.

In the dynamism of healthy ecosystems, there are minor players and major players. Among the major players are the ecosystems engineers, which add new parts to the habitat and open the door to guilds of organisms specialized to use them. Biodiversity engenders more biodiversity, and the overall abundance of plants, animals, and microorganisms increases to a corresponding degree.

By constructing dams, beavers create ponds, bogs, and flooded meadows. These environments shelter species of plants and animals that are rare or absent in free-running streams. The submerged masses of decaying wood forming the dams draw still more species, which occupy and feed on them.

Elephants trample and tear up shrubs and small trees, opening glades within forests. The result is a mosaic of habitats that, overall, contains larger numbers of resident species.

Florida gopher tortoises dig 30-foot-long tunnels that diversify the texture of the soil, altering the composition of its microorganisms. Their retreats are also shared by snakes, frogs, and ants specialized to live in the burrows.

Euchondrus snails of Israel's Negev Desert grind down soft rocks to feed on the lichens growing inside. By converting rock to soil and releasing the nutrients photosynthesized by the lichens, the snails multiply niches for other species.

TO EVALUATE INDIVIDUAL SPECIES SOLELY BY THEIR KNOWN PRACTICAL VALUE AT THE PRESENT TIME IS BUSINESS ACCOUNTING IN THE SERVICE OF BARBARISM.

Overall, a large number of independent observations from differing kinds of ecosystems point to the same conclusion: The greater the number of species that live together, the more stable and productive the ecosystems these species compose. On the other hand, mathematical models that attempt to describe the interactions of species in ecosystems show that the apparent opposite also occurs: High levels of diversity can reduce the stability of individual species. Under certain conditions, including random colonization of the ecosystem by large numbers of species that interact strongly with one another, the separate but interlocking fluctuations in species populations can become more volatile, thus making extinction more likely. Similarly, given appropriate species traits, it is mathematically possible for increased diversity to lead to decreased production.

When observation and theory collide, scientists turn to carefully designed experiments for resolution. Their motivation is especially strong in the case of biological systems, which are typically far too complex to be grasped by observation and theory alone. The best procedure, as in the rest of science, is first to simplify the system, then to hold it more or less constant while varying the important parameters one or two at a time to see what happens. In the 1990s a team of British ecologists, in an attempt to approach these ideal conditions, devised the *ecotron*, a growth chamber in which artificially simple ecosystems can be assembled as desired, species by species. Using multiple ecotrons, they found that productivity, measured by the increase of plant bulk, rose with an increase in species numbers. Simultaneously, ecologists monitoring patches of Minnesota grassland—outdoor equivalents of ecotrons—during a period of drought found that patches richer in species diversity underwent less decline in productivity and recovered more quickly than patches with less diversity.

These pioneering experiments appeared to uphold the conclusion drawn earlier from natural history, at least with reference to production. Put more precisely, ecosystems tested thus far do not possess the qualities and starting conditions allowed by theory that can reduce production and produce instability as a result of large species numbers.

But—how can we be sure, the critics asked (pressing on in the best tradition of science), that the increase in production in particular is truly the result of just an increase in the number of species?

Maybe the effect is due to some other factor that just happens to be correlated with species numbers. Perhaps the result is a statistical artifact. For example, the larger the number of plant species present in a habitat, the more likely it is that at least one kind among them will be extremely productive. If that occurs, the increase in the yield of plant tissue—and in the number of the animals feeding on it—is only a matter of luck of the draw, and not the result of some pure property of biodiversity itself. At its base, the distinction made by this alternative hypothesis is semantic. The increased likelihood of acquiring an outstandingly productive species can be viewed as just one means by which the enrichment of biodiversity boosts productivity. (If you draw on a pool of 1,000 candidates for a basketball team, you are more likely to get a star than if you draw on a pool of 100 candidates.)

Still, it is important to know whether other consequences of biodiversity enrichment play an important role. In particular, do species interact in a manner that increases the growth of either one or both? This is the process called *overyielding*. In the mid-1990s, a massive study was undertaken to test the effect of biodiversity on productivity that paid special attention to the presence or absence of overyielding. Multiple projects of BIODEPTH, as the project came to be called, were conducted during a two-year period by 34 researchers in eight European countries. This time, the results were more persuasive. They showed once again that productivity does increase with biodiversity. Many of the experimental runs also revealed the existence of overyielding.

Over millions of years, nature's ecosystems engineers have been especially effective in the promotion of overyielding. They have coevolved with other species that exploit the niches they build. The result is a harmony within ecosystems. The constituent species, by spreading out into multiple niches, seize and cycle more materials and energy than is possible in similar ecosystems. *Homo sapiens* is an ecosystems engineer too, but a bad one. Not having coevolved with the majority of life forms we now encounter around the world, we elimi-

nate far more niches than we create. We drive species and ecosystems into extinction at a far higher rate than existed before, and everywhere diminish productivity and stability.

I will grant at once that economic and production values at the ecosystem level do not alone justify saving every species in an ecosystem, especially those so rare as to be endangered. The loss of the ivory-billed woodpecker has had no discernible effect on American prosperity. A rare flower or moss could vanish from the Catskill forest without diminishing the region's filtration capacity. But so what? To evaluate individual species solely by their known practical value at the present time is business accounting in the service of barbarism. In 1973, the economist Colin W. Clark made this point persuasively in the case of the blue whale, *Balaenopterus musculus*. A hundred feet in length and 150 tons in weight at maturity, the species is the largest animal that ever lived on land or sea. It is also among the easiest to hunt and kill. More than 300,000 blue whales were harvested during the 20th century, with a peak haul of 29,649 in the 1930–31 season. By the early 1970s, the population had plummeted to several hundred individuals. The Japanese were especially eager to continue the hunt, even at the risk of total extinction. So Clark asked, What practice would yield the whalers and humanity the most money: Cease hunting and let the blue whales recover in numbers, then harvest them sustainably forever, or kill the rest off as quickly as possible and invest the profits in growth stocks? The disconcerting answer for annual discount rates over 21 percent: Kill them all and invest the money.

Now, let us ask, what is wrong with that argument?

Clark's implicit answer is simple. The dollars-and-cents value of a dead blue whale was based only on the measures relevant to the existing market—that is, on the going price per unit weight of whale oil and meat. There are many other values, destined to grow along with our knowledge of living *Balaenopterus musculus* and as science, medicine, and aesthetics grow and strengthen, in dimensions and magnitudes still unforeseen. What was the value of the blue whale in A.D. 1000? Close to zero. What will be its value in A.D. 3000? Essentially limitless—to say nothing of the measure of gratitude the generation then alive will feel to those who in their wisdom saved the whale from extinction.

No one can guess the full future value of any kind of animal, plant, or microorganism. Its potential is spread across a spectrum of known and as yet unimagined human needs. Even the species themselves are largely unknown. Fewer than two million are in the scientific register, with a formal Latinized name, while an estimated five to 100 million—or more—await discovery. Of the species known, fewer than one percent have been studied beyond the sketchy anatomical descriptions used to identify them.

> OF THE SPECIES KNOWN,
> FEWER THAN ONE
> PERCENT HAVE BEEN
> STUDIED BEYOND THE
> SKETCHY ANATOMICAL
> DESCRIPTIONS USED TO
> IDENTIFY THEM.

Agriculture is one of the vital industries most likely to be upgraded by attention to the remaining wild species. The world's food supply hangs by a slender thread of biodiversity. Ninety percent is provided by slightly more than 100 plant species out of a quarter-million known to exist. Twenty species carry most of the load, of which only three—wheat, maize, and rice—stand between humanity and starvation. For the most part, the premier 20 are those that happened to be present in the regions where agriculture was independently invented some 10,000 years ago, namely the Mediterranean perimeter and southwestern Asia; Central Asia; the Horn of Africa; the rice belt of tropical Asia; and the uplands of Mexico, Central America, and Andean South America. Yet some 30,000 species of wild plants, most occurring outside

these regions, have edible parts consumed at one time or other by hunter-gatherers. Of these species, at least 10,000 can be adapted as domestic crops. A few, including the three species of New World amaranths, the carrotlike arracacha of the Andes, and the winged bean of tropical Asia, are immediately available for commercial development.

In a more general sense, all the quarter-million plant species—in fact, all species of organisms—are potential donors of genes that can be transferred by genetic engineering into crop species in order to improve their performance. With the insertion of the right snippets of DNA, new strains can be created that are, variously, cold resistant, pest resistant, perennial, fast growing, highly nutritious, multipurpose, sparing in their consumption of water, and more easily sowed and harvested. And compared with traditional breeding techniques, genetic engineering is all but instantaneous.

The method, a spinoff of the revolution in molecular genetics, was developed in the 1970s. During the 1980s and 1990s, before the world quite realized what was happening, it came of age. A gene from the bacterium *Bacillus thuringiensis*, for example, was inserted into the chromosomes of corn, cotton, and potato plants, allowing them to manufacture a toxin that kills insect pests. No need to spray insecticides; the engineered plants now perform this task on their own. Other transgenes, as they are called, were inserted from bacteria into soybean and canola plants to make them resistant to chemical weed killers. Agricultural fields can now be cheaply cleared of weeds with no harm to the crops growing there. The most important advance of all, achieved in the 1990s, was the creation of golden rice. This new strain is laced with bacterial and daffodil genes that allow it to manufacture beta-carotene, a precursor of vitamin A. Because rice, the principal food of three billion people, is deficient in vitamin A, the addition of beta-carotene is no mean humanitarian feat. About the same time, the almost endless potential of genetic engineering was confirmed by two circus tricks of the trade: A bacterial gene was implanted into a monkey, and a

jellyfish bioluminescence gene into a plant.

But not everyone was dazzled by genetic engineering, and inevitably it stirred opposition. For many, human existence was being transformed in a fundamental and insidious way. With little warning, genetically modified organisms (GMOs) had entered our lives and were all around us, changing incomprehensibly the order of nature and society. A protest movement against the new industry began in the mid-1990s and exploded in 1999, just in time to rank as a millennial event with apocalyptic overtones. The European Union banned transgenic crops, the Prince of Wales compared the methodology to playing God, and radical activists called for a global embargo of all GMOs. "Frankenfoods," "super-weeds," and "Farmageddon" entered the vocabulary: GMOs were, according to one British newspaper, the "mad forces of genetic darkness." Some prominent environmental scientists found technical and ethical reasons for concern.

As I write, public opinion and official policy toward genetic engineering have come to vary greatly from one country to the next. France and Britain are vehemently opposed. China is strongly favorable, and Brazil, India, Japan, and the United States cautiously so. In the United States particularly, the public awoke to the issue only after the transgenie (so to speak) was out of the bottle. From 1996 to 1999, the amount of U.S. farmland devoted to genetically modified crops had rocketed from 3.8 million to 70.9 million acres. As the century ended, more than half the soybeans and cotton grown, and nearly a third (28 percent) of the corn, were engineered.

> WITH LITTLE WARNING,
> GENETICALLY MODIFIED
> ORGANISMS HAD ENTERED
> OUR LIVES AND WERE ALL
> AROUND US, CHANGING
> INCOMPREHENSIBLY THE
> ORDER OF NATURE
> AND SOCIETY.

There are, actually, several sound reasons for anxiety over genetic engineering, which I will now summarize and evaluate.

Many people, not just philosophers and theologians, are troubled by the ethics of transgenic evolution. They grant the benefits but are unsettled by the reconstruction of organisms in bits and pieces. Of course, human beings have been creating new strains of plants and animals since agriculture began, but never at the sweep and pace inaugurated by genetic engineering. And during the era of traditional plant breeding, hybridization was used to mix genes almost always among varieties of the same species or closely similar species. Now it is used across entire kingdoms, from bacteria and viruses to plants and animals. How far the process should be allowed to continue remains an open ethical issue.

The effects on human health of each new transgenic food are hard to predict, and certainly never free of risk. However, the products can be tested just like any other new food products on the market, then certified and labeled. There is no reason at this time to assume that their effects will differ in any fundamental way. Yet scientists generally agree that a high level of alertness is essential, and for the following reason: All genes, whether original to the organism or donated to it by an exotic species, have multiple effects. Primary effects, such as the manufacture of a pesticide, are the ones sought. But destructive secondary effects, including allergenic or carcinogenic activity, are also at least a remote possibility.

Transgenes can escape from the modified crops into wild relatives of the crop where the two grow close together. Hybridization has always occurred widely in agriculture, even before the advent of genetic engineering. It has been recorded at one or another time and place in 12 of the 13 most important crops used worldwide. However, the hybrids have not overwhelmed their wild parents. I know of no case in which a hybrid strain outcompetes wild strains of the same or closely related species in the natural environment. Nor has any hybrid turned into a superweed, in the same class as the worst wild nonhybrid weeds that afflict the planet. As a rule, domesticated species and strains are less competitive than their wild counterparts in both natural and human-modified environments, Of course, transgenes could change the picture. It is simply too early to tell.

Genetically modified crops can diminish biological diversity in other ways. In a now famous example, the bacterial toxin used to protect corn is carried in pollen by wind currents for distances of 60 meters or more from the cultivated fields. Then, landing on milkweed plants, the toxin is capable of killing the caterpillars of monarch butterflies feeding there. In another twist, when cultivated fields are cleared of weeds with chemical sprays against which the crops are protected by transgenes, the food supply of birds is reduced and their local populations decline. These environmental secondary effects have not been well studied in the field. How severe they will become as genetic engineering spreads remains to be seen.

Many people, having become aware of the potential threats of genetic engineering in their food supply, understandably believe that yet another bit of their freedom has been taken from them by faceless corporations (who can even name, say, three of the key players?) using technology beyond their control or even understanding. They also fear that an industrialized agriculture dependent on high technology can by one random error go terribly wrong. At the heart of the anxiety is a sense of helplessness. In the realm of public opinion, genetic engineering is to agriculture as nuclear engineering is to energy.

The problem before us is how to feed billions of new mouths over the next several decades and save the rest of life at the same time—without being trapped in a Faustian bargain that threatens freedom and security. No one knows the exact solution to this dilemma. Most scientists and economists who have studied both sides of it agree that the benefits outweigh the risks. The benefits must come from an evergreen revolution that has as its goal to lift food production well above the level attained by the green revolution of the 1960s, using technology and regulatory policy more advanced, and even safer, than that now in existence.

Genetic engineering will almost certainly play an important role in the evergreen revolution. Energized by recognition of both its promise and its risk, most countries have begun to fashion policies to regulate the marketing of transgenic crops. The ultimate driving force in this rapidly evolving process is international trade. More than 130 countries took an important first step in 2000 to address the issue by tentatively agreeing to the Cartagena Protocol on Biosafety, which provides the right to block imports of transgenic products. The protocol also sets up a joint "biosafety clearing house" to publish information on national policy. About the same time, the U.S. National Academy of Sciences, joined by the science academies of five other countries (Brazil, China, India, Mexico, and the United Kingdom) and the Third World Academy of Sciences, endorsed the development of transgenic crops. They made recommendations for risk assessment and licensing agreements and stressed the needs of the developing countries in future research programs and capital investment.

Medicine is another domain that stands to gain enormously from the world's store of biodiversity, with or without the impetus of genetic engineering. Pharmaceuticals in current use are already drawn heavily from wild species. In the United States, about a quarter of all prescriptions dispensed by pharmacies are substances extracted from plants. Another 13 percent originate from microorganisms, and three percent more from animals—making a total of about 40 percent derived from wild species. What's even more impressive is that nine of the 10 leading prescription drugs originally came from organisms. The commercial value of the relatively small number of natural products is substantial. The over-the-counter cost of drugs from plants alone was estimated in 1998 to be $20 billion in the United States and $84 billion worldwide.

But only a tiny fraction of biodiversity has been utilized in medicine, despite its obvious potential. The narrowness of the base is illustrated by the dominance of ascomycete fungi in the control of bacterial diseases. Although only about 30,000 species of ascomycetes—two percent of the total known species of organisms—have been studied, they have yielded 85 percent of the antibiotics in current use. The underutilization of biodiversity is still greater than these figures alone might suggest—because probably fewer than 10 percent of the world's ascomycete species have even been discovered and given scientific names. The flowering plants have been similarly scanted. Although it is likely that more than 80 percent of the species have received scientific names, only some three percent of this fraction have been assayed for alkaloids, the class of natural products that have proved to be among the most potent curative agents for cancer and many other diseases.

There is an evolutionary logic in the pharmacological bounty of wild species. Throughout the history of life, all kinds of organisms have evolved chemicals needed to control cancer in their own bodies, kill parasites, and fight off predators. Mutations and natural selection, which equip this armamentarium, are processes of endless trial and error. Hundreds of millions of species, evolving by the life and death of astronomical numbers of organisms across geological stretches of time, have yielded the present-day winners of the mutation-and-selection lottery. We have learned to consult them while assembling a large part of our own pharmacopoeia. Thus, antibiotics, fungicides, antimalarial drugs, anesthetics, analgesics, blood thinners, blood-clotting agents, agents that prevent clotting, cardiac stimulants and regulators, immunosuppressive agents, hormone mimics, hormone inhibitors, anticancer drugs, fever suppressants, inflammation controls, contraceptives, diuretics and antidiuretics, antidepressants, muscle relaxants, rubefacients, anticongestants, sedatives,

and abortifacients are now at our disposal, compliments of wild biodiversity.

Revolutionary new drugs have rarely resulted from the pure insights of molecular and cellular biology, even though these sciences have grown very sophisticated and address the causes of disease at the most fundamental level. Rather, the pathway of discovery has usually been the reverse: The presence of the drug is first detected in whole organisms, and the nature of its activity subsequently tracked down to the molecular and cellular levels. Then the basic research begins.

THE PROBLEM BEFORE US IS HOW TO FEED BILLIONS OF NEW MOUTHS OVER THE NEXT SEVERAL DECADES AND SAVE THE REST OF LIFE AT THE SAME TIME.

The first hint of a new pharmaceutical may lie among the hundreds of remedies of Chinese traditional medicine. It may be spotted in the drug-laced rituals of an Amazonian shaman. It may come from a chance observation by a laboratory scientist unaware of its potential importance for medicine. More commonly nowadays, the clue is deliberately sought by the random screening of plant and animal tissues. If a positive response is obtained—say, a suppression of bacteria or cancer cells—the molecules responsible can be isolated and tested on a larger scale, using controlled experiments with animals and then (cautiously!) human volunteers. If the tests are successful, and the atomic structure of the molecule is also in hand, the substance can be synthesized in the laboratory, then commercially, usually at lower cost than by extraction from harvested raw materials. In the final step, the natural chemical compounds provide the prototype from which new classes of organic chemicals can be synthesized, adding or taking away atoms and double bonds here and there. A few of the novel substances may prove more efficient than the natural prototype. And of equal importance to the

pharmaceutical companies, these analogues can be patented.

Serendipity is the hallmark of pharmacological research. A chance discovery can lead not only to a successful drug but to advances in fundamental science, which in time yield other successful drugs. Routine screening, for example, revealed that an obscure fungus growing in the mountainous interior of Norway produces a powerful suppressor of the human immune system. When the molecule was isolated from the fungal tissue and identified, it proved to be a complex molecule of a kind never before encountered by organic chemists. Nor could its effect be explained by the contemporary principles of molecular and cellular biology. But its relevance to medicine was immediately obvious, because when organs are transplanted from one person to another, the immune system of the host must be prevented from rejecting the alien tissue. The new agent, named cyclosporin, became an essential part of the organ transplant industry. It also served to open new lines of research on the molecular events of the immune response itself.

The surprising events that sometimes lead from natural history to medical breakthrough would make excellent science fiction—if only they were untrue. The protagonists of one such plot are the poison dart frogs of Central and South America, which belong to the genera *Dendrobates* and *Phyllobates* in the family Dendrobatidae. Tiny, able to perch on a human fingernail, they are favored as terrarium animals for their beautiful colors: The 40 known species are covered by various patterns of orange, red, yellow, green, or blue, usually on a black background. In their natural habitat, dendrobatids hop about slowly and are relatively unfazed by the approach of potential predators. For the trained naturalist their lethargy triggers an alarm, in observance of the following rule of animal behavior: If a small and otherwise unknown animal encountered in the wild is strikingly beautiful, it is probably poisonous, and if it is not only beautiful but also easy to catch, it is probably deadly. And so it is with dendrobatid frogs, which, it turns out, secrete a powerful toxin from glands on their backs. The potency varies according to species. A single individual of one (perfectly named) Colombian species, *Phyllobates horribilis*, for example, carries enough of the substance to kill 10 men. Indians of two tribes living in the Andean Pacific slope forests of western Colombia, the Emberá Chocó and the Noanamá Chocó, rub the tips of their blowgun darts over the backs of the frogs, very carefully, then release the little creatures unharmed so they can make more poison.

COLLECTING SAMPLES OF VALUABLE SPECIES FROM RICH ECOSYSTEMS AND CULTIVATING THEM IN BULK ELSEWHERE IS NOT ONLY PROFITABLE BUT THE MOST SUSTAINABLE OF ALL.

In the 1970s a chemist, John W. Daly, and a herpetologist, Charles W. Myers, gathered material from a similar Ecuadorian frog, *Epipedobates tricolor*, for a closer look at the dendrobatid toxin. In the laboratory, Daly found that very small amounts administered to mice worked as an opiumlike painkiller, yet otherwise lacked the properties of typical opiates. Would the substance also prove nonaddictive? If so, it might be turned into the ideal anesthetic. From a cocktail of compounds taken from the backs of the frogs, Daly and his fellow chemists isolated and characterized the toxin itself, a molecule resembling nicotine, which they named epibatidine. This natural product proved 200 times more effective in the suppression of pain than opium, but was also too toxic, unfortunately, for practical use. The next step was to redesign the molecule. Chemists at Abbott Laboratories synthesized not only epibatidine but hundreds of novel molecules resembling it. When tested clinically, one of the products, code-named ABT-594, was found to combine the desired properties: It depressed pain like epibatidine, including pain from nerve damage of a kind usually impervious to opiates, and it was nonaddictive.

ABT-594 had two additional advantages: It promoted alertness instead of sleepiness, and it had no side effects on respiration or digestion.

The full story of the poison dart frogs also carries a warning about the conservation of tropical forests. The destruction of much of the habitat in which populations of *Epipedobates* live almost prevented the discovery of epibatidine and its synthetic analogues. By the time Daly and Myers set out to collect enough toxin for chemical analysis, after their initial visit to Ecuador, one of the two prime rainforest sites occupied by the frogs had been cleared and replaced with banana plantations. At the second site, which fortunately was still intact, they found enough frogs to harvest just one milligram of the poison. From that tiny sample, chemists were able, with skill and luck, to identify epibatidine and launch a major new initiative in pharmaceutical research.

It is no exaggeration to say that the search for natural medicinals is a race between science and extinction, and will become critically so as more forests fall and coral reefs bleach out and disintegrate. Another adventure dramatizing this point began in 1987, when the botanist John Burley collected samples of plants from a swamp forest near Lundu in the Malaysian state of Sarawak, on the northwestern corner of the island of Borneo. His expedition was one of many launched by the National Cancer Institute (NCI) to search for new natural substances to add to the fight against cancer and AIDS. Following routine procedure, the team collected a kilogram of fruit, leaves, and twigs from each kind of plant they encountered. Part was sent to the NCI laboratory for assay, and part was deposited in the Harvard University Herbarium for future identification and botanical research.

One such sample came from a small tree at Lundu about 25 feet high. It was given the voucher code label Burley-and-Lee 351. Back at the NCI laboratories, an extract made from it was tested routinely against human cancer cells grown in culture. Like the majority of such preparations, it had no effect. Then it was run through screens designed to test its potency against the AIDS virus.

The NCI scientists were startled to observe that Burley-and-Lee 351 gave, in their words, "100 percent protection against the cytopathic effects of HIV-I infection," having "essentially halted HIV-I replication." In other words, while the substance the sample contained could not cure AIDS, it could stop cold the development of disease symptoms in HIV-positive patients.

The Burley-and-Lee 351 tree was determined to belong to a species of *Calophyllum*, a group of species belonging to the mangosteen family, or Guttiferae. Collectors were dispatched to Lundu a second time to obtain more material from the same tree, with the aim of isolating and chemically identifying the HIV inhibitor. The tree was gone, probably cut down by local people for fuel or building materials. The collectors returned home with samples from other *Calophyllum* trees taken in the same swamp forest, but their extracts were ineffective against the virus.

Peter Stevens, then at Harvard University, and the world authority on *Calophyllum*, stepped in to solve the problem. The original tree, he found, belonged to a rare strain named *Calopsyllum lanigerum*, variety *austrocoriaceum*. The trees sampled on the second trip were another species, which explained their inactivity. No more specimens of *austrocoriaceum* could be found at Lundu. The search for the magic strain widened, and finally a few more specimens were located in the Singapore Botanic Garden. Thus supplied with enough raw material, chemists and microbiologists were able to identify the anti-HIV substance as (+)-calanolide A. Soon afterward the molecule was synthesized, and the synthetic proved as effective as the raw extract. Additional research revealed calanolide to be a powerful inhibitor of reverse transcriptase, an enzyme needed by the HIV virus to replicate itself within the human host cell. Studies are now underway to determine the suitability of calanolide for market distribution.

The exploration of wild biodiversity in the search for useful resources is called *bioprospecting*. Propelled by venture capital, it has in the past 10 years grown into a respectable industry within a global market hungry for new pharmaceuticals. It is also a means for discovering new food sources, fibers, petroleum substitutes, and other products. Sometimes bioprospectors screen many species of organisms in search of chemicals with particular qualities, such as antisepsis or cancer suppression. On other occasions bioprospecting is opportunistic, focusing on one of a few species that show signs of yielding a valuable resource. Ultimately, entire ecosystems will be prospected as a whole, and all of the species assayed for most or all of the products they can yield.

The extraction of wealth from an ecosystem can be destructive or benign. Dynamiting coral reefs and clearcutting forests yield fast profits but are unsustainable. Fishing coral reefs lightly and gathering wild fruit and resins in otherwise undisturbed forest are sustainable. Collecting samples of valuable species from rich ecosystems and cultivating them in bulk elsewhere, in biologically less favored areas, is not only profitable but the most sustainable of all.

Bioprospecting with minimal disturbance is the way of the future. Its promise can be envisioned with the following matrix for a hypothetical forest: To the left, make a list of the thousands of plant, animal, and microbial species, as many as you can, recognizing that the vast majority have not yet been examined, and many still lack even a scientific name. Along the top, prepare a horizontal row of the hundreds of functions imaginable for all the products of these species combined. The matrix itself is the combination of the two dimensions. The spaces filled within the matrix are the potential applications, whose nature remains almost wholly unknown.

The richness of biodiversity's bounty is reflected in the products already extracted by native peoples of the tropical forests, using local knowledge and low technology of a kind transmitted solely by demonstration and oral teaching. Here, for example, is a small selection of the most common medicinal plants used by tribes of the upper Amazon, whose knowledge has evolved from their combined experience with the more than 50,000 species of flowering plants native to the region: motelo sanango, *Abuta grandifolia* (snakebite, fever); dye plant, *Arrabidaea chica* (anemia, conjunctivitis); monkey ladder, *Bauhinia guianensis* (amoebic dysentery); Spanish needles, *Bidens alba* (mouth sores, toothache); firewood tree, or capirona, species of *Calycophyllum* and *Capirona* (diabetes, fungal infection); wormseed, *Chenopodium ambrosioides* (worm infection); caimito, *Chrysophyllum cainito* (mouth sores, fungal infection); toad vine, *Cissus sicyoides* (tumors); renaquilla, *Clusia rosea* (rheumatism, bone fractures); calabash, *Crescentia cujete* (toothache); milk tree, *Couma macrocarpa* (amoebic dysentery, skin inflammation); dragon's blood, *Croton lechleri* (hemorrhaging); fer-de-lance plant, *Dracontium loretense* (snakebite); swamp immortelle, *Erythrina fusca* (infections, malaria); wild mango, *Grias neuberthii* (tumors, dysentery); wild senna, *Senna reticulata* (bacterial infection).

Only a few of the thousands of such traditional medicinals used in tropical forests around the world have been tested by Western clinical methods. Even so, the most widely used already have commercial value rivaling that of farming and ranching. In 1992 a pair of economic botanists, Michael Balick and Robert Mendelsohn, demonstrated that single harvests of wild-grown medicinals from two tropical forest plots in Belize were worth $726 and $3,327 per hectare (2.5 acres) respectively, with labor costs thrown in. By comparison, other researchers estimated per hectare yield from tropical forest converted to farmland at $228 in nearby Guatemala and $339 in Brazil. The most productive Brazilian plantations of tropical pine could yield $3,184 per hectare from a single harvest.

In short, medicinal products from otherwise undisturbed tropical forests can be locally profitable, on condition that markets are developed and the extraction rate is kept low enough to be sustainable. And when plant and animal food products, fibers, carbon credit trades, and ecotourism are added to the mix, the commercial value of sustainable use can be boosted far higher.

Examples of the new economy in practice are growing in number. In the Petén region of Guatemala, about 6,000 families live comfortably by sustainable extraction of rainforest products. Their combined annual income is $4 million to $6 million, more than could be made by converting the forest into farms and cattle ranches. Ecotourism remains a promising but largely untapped additional resource.

Nature's pharmacopoeia has not gone unnoticed by industry strategists. They are well aware that even a single new molecule has the potential to recoup a large capital investment in bioprospecting and product development. The single greatest success to date was achieved with extremophile bacteria living in the boiling-hot thermal springs of Yellowstone National Park. In 1983 Cetus Corporation used one of the organisms, *Thermus aquaticus*, to produce a heat-resistant enzyme needed for DNA synthesis. The manufacturing process, called *polymerase chain reaction* (PCR), is today the foundation of rapid genetic mapping, a stanchion of the new molecular biology and medical genetics. By enabling microscopic amounts of DNA to be multiplied and typed, PCR also plays a key role in crime detection and forensic medicine. Cetus's patents on PCR technology, which have been upheld by the courts, are immensely profitable, with annual earnings now in excess of $200 million—and growing.

Bioprospecting can serve both mainstream economics and conservation when put on a firm contractual basis. In 1991, Merck signed an agreement with Costa Rica's National Institute of Biodiversity (INBio) to assist the search for new pharmaceuticals in Costa Rica's rainforests and other natural habitats. The first deposit was $1 million dispensed over two years, with two similar consecutive grants to follow. During the first period, the field collectors concentrated on plants, in the second on insects, and in the third on microorganisms. Merck is now working through the immense library of materials it gathered during the field program and testing and refining chemical extracts made from them.

Also in 1991, Syntex signed a contract with Chinese science academies to receive up to 10,000 plant extracts a year for pharmaceutical assays. In 1998, Diversa Corporation signed on with Yellowstone National Park to continue bioprospecting the hot springs for biochemicals from thermophilic microbes. Diversa pays the park $20,000 yearly to collect the organisms for study, as well as a fraction of the profits generated by commercial development. Funds returning to Yellowstone will be used to promote conservation of the unique microbes and their habitat, as well as basic scientific research and public education.

Still other agreements have been signed between NPS Pharmaceuticals and the government of Madagascar, between Pfizer and the New York Botanical Garden, and between the international company GlaxoSmithKline and a Brazilian pharmaceutical company, with part of the profits pledged to the support of Brazilian science.

Perhaps it is enough to argue that the preservation of the living world is necessary to our long-term material prosperity and health. But there is another, and in some ways deeper, reason not to let the natural world slip away. It has to do with the defining qualities and self-image of the human species. Suppose, for the sake of argument, that new species can one day be engineered and stable ecosystems built from them. With that distant prospect in mind, should we go ahead and, for short-term gain, allow the original species and ecosystems to be lost? Yes? Erase Earth's living history? Then also burn the art galleries, make cordwood of the musical instruments, pulp the musical scores, erase Shakespeare, Beethoven, and Goethe, and the Beatles too, because all these—or at least fairly good substitutes—can be re-created.

The issue, like all great decisions, is moral. Science and technology are what we can do; morality is what we agree we should or should not do. The ethic from which moral decisions spring is a norm or standard of behavior in support of a value, and value in turn depends on purpose. Purpose, whether personal or global, whether urged by conscience or graven in sacred script, expresses the image we hold of ourselves and our society. A conservation ethic is that which aims to pass on to future generations the best part of the nonhuman world. To know this world is to gain a proprietary attachment to it. To know it well is to love and take responsibility for it.

EDWARD O. WILSON *is Pellegrino University Research Professor and Honorary Curator in Entomology at Harvard University's Museum of Comparative Zoology. His books include* Sociobiology: The New Synthesis *(1975),* Consilience: The Unity of Knowledge *(1998), and two Pulitzer Prize winners,* On Human Nature *(1978) and* The Ants *(1990, with Burt Holldobler). This essay is taken from his latest book,* The Future of Life *(2002). Published by arrangement with Alfred A. Knopf, a division of Random House, Inc. Copyright: © 2002 By Edward O. Wilson.*

From *The Wilson Quarterly*, Winter 2002, pp. 20-39. © 2002 by Edward O. Wilson. Reprinted by permission of the author.

Strangers in Our Midst:
The Problem of Invasive Alien Species

By Jeffrey A. McNeely

Invasive alien species—non-native species that become established in a new environment then proliferate and spread in ways that damage human interests—are now recognized as one of the greatest biological threats to our planet's environmental and economic well-being.[1]

Most nations are already grappling with complex and costly invasive-species problems: Zebra mussels (*Dreissena polymorpha*) from the Caspian and Black Sea region affect fisheries, mollusk diversity, and electric-power generation in Canada and the United States; water hyacinth (*Eichornia crassipes*) from the Amazon chokes African and Asian waterways; rats originally carried by the first Polynesians exterminate native birds on Pacific islands; and deadly new disease organisms (such as the viruses causing SARS, HIV/AIDS, and West Nile fever) attack human, animal, and plant populations in temperate and tropical countries. For all animal extinctions where the cause is known, invasive alien species are the leading culprits, contributing to the demise of 39 percent of species that have become extinct since 1600.[2] The 2000 IUCN Red List of Threatened Species reported that invasive alien species harmed 30 percent of threatened birds and 15 percent of threatened plants.[3] Addressing the problem of these invasive alien species is urgent because the threat is growing daily and the economic and environmental impacts are severe.

A key question is whether the global reach of modern human society can be matched by an appropriate sense of responsibility. One critical element of this question is the definition of "native," a concept with challenging spatial and temporal dimensions. While every species is native to a particular geographic area, this is just a snapshot in time, because species are constantly expanding and contracting their ranges, sometimes with human help. For example, Britain has nearly 40 more species of birds today than were recorded 200 years ago. About a third of these are deliberate introductions, such as the Little Owl (*Athene noctua*), while the others are natural colonizations that may be taking advantage of climate change.[4]

> **An invasive alien species is not a "bad" species but rather one "behaving badly" in a particular context.**

According to one view, local biological "enrichment" by non-native species always harms native species at some level, so any introduction should be regarded, at least in principle, as undesirable. An opposing view is that because species are constantly expanding or contracting their range, new species—especially those that are beneficial to people, such as crops, ornamental plants, and pets—should be welcomed as "increasing biodiversity" unless they are clearly harmful. According to this perspective, in the case of British birds noted above, only those introduced by people and that are causing ecological or economic damage, such as pigeons, are considered to be invasive.

All continental areas have suffered from invasions of alien species, losing biological diversity as a result, but the problem is especially acute on islands in general and for small islands in particular. The physical isolation of islands over millions of years has favored the evolution of unique species and ecosystems, so islands often have a high proportion of endemic species. The evolutionary processes associated with isolation have also meant that island species are especially vulnerable to predators, pathogens, and parasites from other areas. More than 90 percent of the 115 birds known to have become extinct over the past 400 years were endemic to islands.[5] Most of these evolved in the absence of mammalian predators, so the arrival of rats and cats carried by people has had a devastating impact.

Island plants are also affected. For example, the tree *Miconia calvescens* replaced the forest canopy on more than 70 percent of the island of Tahiti over a 50-year time span, starting with a few trees in two botanical gardens. Some 40–50 of the 107 plant species endemic to the island of Tahiti are believed to be on the verge of extinction primarily due to this invasion.[6] Introduced animals also can affect plants. For example, goats introduced on St. Clemente Island, California, have caused the extinction of eight endemic species of plants and have endangered eight others.[7]

An invasive alien species is not a "bad" species but rather one "behaving badly" in a particular context, usually due to inappropriate human agency or intervention. A species may be so threatened in its natural range that it is given legal protection, yet it may generate massive ecological and other damage elsewhere.

The degradation of natural habitats, ecosystems, and agricultural lands (through loss of vegetation and soil and pollution of land and waterways) that has occurred throughout the world has made it easier for non-native species to become invasive, opening up new possibilities for them. For all of these reasons, and others that will become apparent below, the issue of invasive alien species is receiving growing international attention.

The Vectors: How Species Move Around the World

The natural barriers of oceans, mountains, rivers, and deserts have provided the isolation that has enabled unique species and ecosystems to evolve. But in just a few hundred years, these barriers have been overcome by technological changes that helped people move species vast distances to new habitats, where some of them became invasive. The growth in the volume of international trade, from US$192 billion in 1960 to almost $6 trillion in 2003,[8] provides more opportunities than ever for species to be spread either accidentally or deliberately.

Some movement seems accidental, or at least incidental, in that transporting the species was not the purpose of the transporter. For example, ballast water is now regarded as the most important vector for transoceanic movements of shallow-water coastal organisms, dispersing fish, crabs, worms, mollusks, and microorganisms from one ocean to another. Enclosed water bodies like San Francisco Bay are especially vulnerable. The bay already has at least 234 invasive alien species, causing significant economic damage. California has one of the toughest ballast water laws in the nation, requiring ships from foreign ports to exchange their ballast water 200 miles from the California coastline, but enforcement remains spotty at best.

Ballast water may also be important in the epidemiology of waterborne diseases affecting plants and animals. One study measured the concentration of the bacteria *Vibrio cholerae*—which cause human epidemic cholera—in the ballast water of vessels arriving to the Chesapeake Bay from foreign ports, finding the bacteria in plankton samples from all ships.[9]

Other invasives are hitchhikers on global trade. For example, the Asian long-horned beetle (*Anoplophora glabripennis*) is one of the newest and most harmful invasive species in the United States. Originating in northeastern Asia, it finds its way to the United States through packing crates made of low-quality timber (that which is too infested for other uses). The number of insects found in materials imported from China increased from 1 percent of all interceptions in 1987 to 20 percent in 1996.[10] Outbreaks were reported in and around Chicago as early as 1992, in Brooklyn in August 1996, and in California in 1997. The beetle finds a congenial home among native maples, elders, elms, horse chestnuts, and others. The U.S. Department of Agriculture predicted that if the beetle becomes established, it could denude Main Street, USA, of shade trees, affect lumber and maple sugar production, threaten tourism in infested areas, and reduce biological diversity in forests.[11]

Another dangerous trade-related species for North America is the Asian gypsy moth (*Lymantria dispar*), which was first reported in the United States in 1991, entering as egg masses attached to ships or cargo from eastern Siberia. The caterpillars of this species are known to feed on more than 600 species of trees, and as moths, the females can disperse themselves over long distances. Scientists fear that this species could cause vastly more damage than the European gypsy moth, which already defoliates 1.5 million hectares of forest per year in North America.

With almost 700 million people crossing international borders as tourists each year, the opportunities for them to carry potential invasive species, either knowingly or unknowingly, are profound and increasing. Many tourists return with living plants that may become invasive, or carry exotic fruits that may be infested with invasive insects that can plague agriculture back home. Travelers may also carry diseases between countries, as apparently happened with the SARS virus. Tourism is considered an especially efficient pathway for invasive alien species

on subAntarctic islands such as South Georgia. Visitors to the island reached 15,000 in 1999. Part of the problem is that many tourists are visiting similar islands on the same trip, increasing the chances of a seed, fruit, or insect being carried, more than would be expected from a single landing of a few people who spend an extended time on one island.[12]

Many species are introduced on purpose but have unintended consequences. One example of purposeful introduction gone wrong is the extensive stocking program that introduced African tilapia *Oreochromis* into Lake Nicaragua in the 1980s, resulting in the decline of native populations of fish and the imminent collapse of one of the world's most distinctive freshwater ecosystems. The alteration of Lake Nicaragua's ecosystem is likely to have effects on the planktonic community and primary productivity of the entire lake—Central America's largest—destroying native fish populations and likely leading to unanticipated consequences.[13]

Sport fishers have also had an influence, importing their favorite game fish into new river systems, where they can have significant negative impacts on native species. For example, the northern pike (*Esox lucius*) has invaded rivers in Alaska and is replacing native species of salmon. While the northern pike occurs naturally in some parts of Alaska, it was introduced to the salmon-rich south-central area in the 1950s, probably by a fisherman who brought it to Bulchitna Lake. Flooding in the 1980s subsequently spread the pike into the streams of the Susitna and Matanuska river basins. Pike have now occupied at least a dozen lakes and four rivers in some of the richest salmon and trout habitat in the Pacific Northwest. Rainbow trout are an even greater threat. Originating in western North America, they have been introduced into 80 new countries, often with devastating impacts on native fish.

Pets are also a problem. Domestic cats can plunder ecosystems that they did not previously inhabit. On Marion Island in the sub-Antarctic Indian Ocean, cats were estimated to kill about 450,000 seabirds annually.[14] Exotic pets may escape—or be released when they have outlived their novelty—and become established in their new home. Stories of crocodiles in the

Manhattan's sewer system are probably fanciful, but many former pets are becoming established in the wild. For example, Monk parakeets (*Myiopsitta monachus*), descended from former pets that were released possibly in the 1960s, have invaded some 76 localities in 15 U.S. states.[15] Native to southern South America, they are the only parrots that build their own nests, some of which support several hundred individuals and have separate families living in different chambers. Some believe that they soon will become widespread throughout the lower 48 states, posing a significant threat to at least some agricultural lands by feeding on ripening crops. And Burmese pythons (*Python molurus*) have become established in Everglades National Park, where they reach a very large size and prey on many native species, even alligators.

Pet stores often advertise invasive species that are legally controlled. For example, the July 2000 issue of the magazine *Tropical Fish Hobbyist* recommended several species of the genus *Salvinia* as aquarium plants, even though they are considered noxious weeds in the United States and prohibited by Australian quarantine laws.

The globalization of trade and the power of the Internet offer new challenges, as sales of seeds and other organisms by mail order or over the Internet pose new and very serious risks to the ecological security of all nations. Controls on harvest and export of species are required as part of a more responsible attitude of governments toward the potential of spreading genetic pollution through invasive species. Further, all receiving countries want to ensure that they are able to control what is being imported. Virtually all countries in the world have serious problems in this regard, an issue that some countries are calling "biosecurity."

The Science of Understanding Invasions

Biodiversity is dynamic, and the movement of species around the world is a continuing process that is accelerating through expanding global trade. By trying to identify which species are especially likely to become invasive, and

hence harmful to people, ecologists are improving the quality of invasion biology as a predictive science so that people can continue to benefit from global biodiversity without paying the costs resulting from species that later become harmful.

Previous examples indicate the characteristics that can make a species invasive. For instance, coastal ecosystems are frequently invaded by microorganisms from ballast water for three main reasons. First, concentrations of bacteria and viruses exceed those reported for other taxonomic groups in ballast water by 6 to 8 orders of magnitude, and the probability of successful invasion increases with inoculation concentration. Second, the biology of many microorganisms combines a high capacity for increase, asexual reproduction, and the ability to form dormant resting stages. Such flexibility in life history can broaden the opportunity for successful colonization, allowing rapid population growth when suitable environmental conditions occur. And third, many microorganisms can tolerate a broad range of environmental conditions, such as salinity or temperature, so many sites may be suitable for colonization.[16] Insects are a major problem because they can lay dormant or travel as egg masses and are difficult to detect. The African tilapia introduced to Lake Nicaragua adapted well, because they are able to grow rapidly; feed on a wide range of plants, fish, and other organisms; and form large schools that can migrate long distances. Further, they are maternal mouth brooders, so a single female can colonize a new environment by carrying her young in her mouth.[17] Rapid growth, generalized diet, ability to move large distances, and prolific breeding are all characteristics of successful invaders.

It is not always simple, however, to distinguish a beneficial non-native species from one at significant risk of becoming invasive. A non-native species that is useful in one part of a landscape may invade other parts of the landscape where its presence is undesirable, and some species may behave well for decades before suddenly erupting into invasive status. The Nile Perch (*Lates niloticus*), for example, was introduced

to Lake Victoria in the 1950s but did not become a problem until the 1980s, when it was a key factor in the extinction of as many as half of the lake's 500 species of endemic fish, attractive prey for the perch.[18] That said, ecologists over the past several decades have agreed on some broad principles for guiding risk assessment. First, the probability of a successful invasion increases with the initial population size and with the number of attempts at introduction. While it is possible for a species to invade with a single gravid female or fertile spore, the odds of doing so are very low. Second, among plants, the longer a non-native plant has been recorded in a country and the greater the number of seeds or other propagules that it produces, the more likely it will become invasive. Third, species that are successful invaders in one situation are likely to be successful in other situations; rats, water hyacinth, microorganisms, and many others fall into this category. Fourth, intentionally introduced species may be more likely to become established than are unintentionally introduced species, at least partly because the vast majority of these have been selected for their ability to survive in the environment where they are introduced. Fifth, plant invaders of croplands and other highly disturbed areas are concentrated in herbaceous families with rapid growth and a wide range of environmental tolerances, while invaders of undisturbed natural areas are usually from woody families, especially nitrogen-fixing species that can live in nitrogen-poor soils.[19] And sixth, fire, like disturbance in general, increases invasion by introduced species. So ecosystems that are naturally prone to fire, such as the fynbos of South Africa, coastal chaparral in California, and maquis in the Mediterranean,[20] can be heavily invaded if fire-liberated seeds of invasive species are available. (These are all shrub communities adapted to cool, wet winters and hot, dry summers, where fire is a regular phenomenon. They are also rich in species: Fynbos have about 8,500 species that include many endemic *Proteaceae*; chaparral have about 5,000 species; and maquis have 25,000—of which about 60 percent are endemic to the Mediterranean region.[21])

INVASIVE ALIEN SPECIES AND PROTECTED AREAS

Protected areas are widely perceived as being devoted to conserving natural ecosystems. Ironically, protected areas are in fact heavily damaged by invasive alien species, and many protected-area managers consider this their biggest problem. Some examples:

- Galapagos National Park, a World Heritage site, is being affected by numerous invasive alien species, including pigs, goats, feral cats, fire ants, and mosquitoes.
- Kruger National Park, South Africa's largest, has recorded 363 alien plant species, including water weeds that pose a serious threat to the park's rivers.
- In the Wadden Sea, a biosphere reserve and Ramsar site protected by the Netherlands, Germany, and Denmark, the Pacific oyster has invaded, having escaped captive management. It is disrupting tourism because of its sharp shells. It has also carried with it numerous other invasive alien species.
- The Wet Tropic World Heritage Area of North Queensland, Australia, is infested by numerous invasive alien species, of which the worst is the pond apple from Florida, which has invaded creeks and riverbanks, wetlands, melaleuca swamps, and mangrove communities. Feral pigs, another invasive species, help to spread the species. The pond apple is now rare in its native range in the Florida Everglades.
- Everglades National Park in Florida, another World Heritage site, is threatened by the invasion of melaleuca from Queensland, demonstrating that species that may behave well in their natural habitat can be a serious problem when they invade somewhere else.
- Tongariro National Park, New Zealand, is also a World Heritage site, but a third of its territory has been infested by heather, a European plant deliberately introduced into New Zealand by an early park warden in 1912 in an attempt to reproduce the moors of Scotland.

These are just a few examples among many that could be cited that demonstrate that even the most strictly protected areas can be extremely vulnerable to invasion by non-native species.

Other ecological factors that may favor nonindigenous species include a lack of controlling natural enemies, the ability of an alien parasite to switch to a new host, an ability to be an effective predator in the new ecosystem, the availability of artificial or disturbed habitats that provide an ecosystem the aliens can easily invade, and high adaptability to novel conditions.[22]

It is sometimes argued that systems with great species diversity are more resistant to new species invading. However, a study in a California riparian system found that the most diverse natural assemblages are in fact the most invaded by non-native plants, and protected areas worldwide are heavily invaded by non-native plants and animals.[23] Dalmatian toadflax (*Linaria dalmatica*) is invading relatively undisturbed shrub-steppe habitat in the Pacific Northwest, wetland nightshade (*Solanum tampicense*) is invading cypress wetlands in central and south Florida, and garlic mustard (*Allilaria officinalis*) is often found in relatively undisturbed systems in the northern parts of North America.

This work helps resolve the controversy over the relationship between biodiversity and invasions, suggesting that the scale of investigation its a critical factor. Theory suggests that non-native species should have a more difficult time invading a diverse ecosystem, because the web of species interactions should be more efficient in using resources such as nutrients, light, and water than would fewer species, leaving fewer resources available for the nonnative species. But even in well-protected landscapes such as national parks, invaders often seem to be more successful in diverse ecosystems. Even though diversity does matter in fending off invasives, its effects are negated by other factors at larger scales. The most diverse ecosystems might be at the greatest risk of invasion, while losses of species, if they affect community-scale diversity, may erode invasion resistance.[24]

The Economic Impacts of Invasion

One reason invasive alien species are attracting more attention is that they are having substantial negative impacts on numerous economic sectors, even beyond the obvious impacts on agriculture (weeds), forestry (pests), and health (diseases or disease vectors). The probability that any one introduced species will become invasive may be low, but the damage costs and costs of control of the species that do become invasive can be extremely high (such as the recent invasion of eastern Canada by the European brown spruce longhorn beetle (*Tetropium fuscum*), which threatens the Canadian timber industry).

Estimates of the economic costs of invasive alien species include considerable uncertainty, but the costs are profound—and growing (see Table 1).

Most of these examples come from the industrialized world, but developing countries are experiencing similar, and perhaps proportionally greater, damage. Invasive alien insect pests—such as the white cassava mealybug (*Phenacoccus herreni*) and larger grain borer (*Prostephanus truncates*) in Africa—pose direct threats to food security. Alien weeds constrain efforts to restore degraded land, regenerate forests, and improve utilization of water for irrigation and fisheries. Water hyacinth and other alien water weeds that choke waterways currently cost developing countries in Africa and Asia more than US$100 million annually. Invasive alien species pose a threat to more than $13 billion of current and planned World Bank funding to projects in the irrigation, drainage, water supply, sanitation, and power sectors.[25] And a study of three developing nations (South Africa, India, and Brazil) found annual losses to introduced pests of $138 billion per year.[26]

Table 1. Indicative costs of some invasive alien species (in U.S. dollars)

Species	Economic variable	Economic impact
Introduced disease organisms	Annual cost to human, plant, and animal health in the United States	$41 billion per year[a]
A sample of alien species of plants and animals	Economic costs of damage in the United States	$137 billion per year[b]
Salt cedar	Value of ecosystem services lost in western United States	$7–16 billion over 55 years[c]
Knapweed and leafy spurge	Impact on economy in three U.S. states	Direct costs of $40.5 million per year; indirect costs of $89 million[d]
Zebra mussel	Damages to U.S. industry	Damage of more than $2.5 billion to the Great Lakes fishery between 1998–2000;[e] $5 billion to U.S. industry by 2000[f]
Most serious invasive alien plant species	Costs 1983–1992 of herbicide control in England	$344 million per year for 12 species[g]
Six weed species	Costs in Australia agroecosystems	$105 million per year[h]
Pinus, Hakeas, and Acacia	Costs on South African floral kingdom to restore to pristine state	$2 billion total for impacts felt over several decades[i]
Water hyacinth	Costs in seven African countries	$20–50 million per year[j]
Rabbits	Costs in Australia	$373 million per year (agricultural losses)[k]
Varroa mite	Economic cost to beekeeping in New Zealand	An estimated $267–602 million over the next 35 years[l]

[a] P. Daszak, A. Cunningham, and A. D. Hyatt, "Emerging Infectious Diseases of Wildlife: Threats to Biodiversity and Human Health," *Science*, 21 January 2000, 443–49.
[b] D. Pimentel, L. Lach, R. Zuniga, and D. Morrison, "Environmental and Economic Costs of Non-indigenous Species in the United States," *BioScience* 50 (2000): 53–65.
[c] E. Zavaleta, "Valuing Ecosystem Services Lost to Tamarix Invasion in the United States," in H. A. Mooney and R. J. Hobbs, eds., *Invasive Species in a Changing World* (Washington DC: Island Press, 2000).
[d] D. A. Bangsund, F. L. Leistritz, and J. A. Leitch, "Assessing Economic Impacts of Biological Control of Weeds: The Case of Leafy Spurge in the Northern Great Plains of the United States," *Journal of Environmental Management* 56 (1999): 35–43; and D. A. Bangsund, S. A. Hirsch, and J. A. Leitch, *The Impact of Knapweed on Montana's Economy* (Fargo, ND: Department of Agricultural Economics, North Dakota State University, 1996).
[e] F. C. Focazio, "Coordinated Issue Area: Aquatic Nuisance, Non-Indigenous, and Invasive Species," *Coastlines* 30, no.1 (2001): 4–5.
[f] "Coatings to Repel Zebra Mussels," U.S. Army Construction Engineering Research Laboratory fact sheet, http://www.cecer.army.mil/facts/sheets/FL10.html.
[g] M. Williamson, "Measuring the Impact of Plant Invaders in Britain," in S. Starfinger, K. Edwards, I. Kowarik, and M. Williamson, eds., *Plant Invasions: Ecological Mechanisms and Human Responses* (Leiden, Netherlands: Backhuys, 2000), 57–70.
[h] A. Watkinson, R. Freckleton, and P. Dowling, "Weed Invasions of Australian Farming Systems: From Ecology to Economics," in C. Perrings, M. Williamson, and S. Dalmazzone, eds., *The Economics of Biological Invasions* (Cheltenham, UK: Edward Elgar, 2000), 94–116.
[i] J. Turpie and B. Heydenrych, "Economic Consequences of Alien Infestation of the Cape Floral Kingdom's Fynbos Vegetation," in Perrings, Williamson, and Dalmazzone, ibid., pages 152–82.
[j] S. Joffe and S. Cook, *Management of the Water Hyacinth and Other Aquatic Weeds: Issues for the World Bank* (Cambridge, UK: Commonwealth Agriculture Bureau International (CABI) Bioscience, 1997).
[k] P. White and G. Newton-Cross, "An Introduced Disease in an Invasive Host: The Ecology and Economics of Rabbit Carcivirus Disease (RCD) in Rabbits in Australia," in Perrings, Williamson, and Dalmazzone, note h above, pages 117–37.
[l] R. Wittenberg and M. J. W. Cock, eds., *Invasive Alien Species: A Tool Kit of Best Prevention and Management Practices* (Wallingford, UK: Global Invasive Species Programme and CABI, 2001).

SOURCE: J. A. McNeely.

In addition to the direct costs of managing invasives, the economic costs also include their indirect environmental consequences and other nonmarket values. For example, invasives may cause changes in ecological services by disturbing the operation of the hydrological cycle, including flood control and water supply, waste assimilation, recycling of nutrients, conservation and regeneration of soils, pollination of crops, and seed dispersal. Such services have current-use value and option value (the potential value of such services in the future). In the South African fynbos, for example, the establishment of invasive tree species—which use more water than do native species—has decreased water supplies for nearby communities and increased fire hazards, justifying government expenditures equivalent to US$40 million per year for both manual and chemical control.[27]

> Customs and quarantine practices, developed in an earlier time, are inadequate safeguards against the rising tide of species that threaten native biodiversity.

Many people in today's globalized economy are driven especially by economic motivations. Those who are importing non-native species are usually doing so with a profit motive and often seek to avoid paying for possible associated negative impacts if those species become invasive. The fact that these negative impacts might take several decades to appear make it all the easier for the negative economic impacts to be ignored. Similarly, those who are ultimately responsible for such "accidental" introductions (for example, through infestation of packing materials or organisms carried in ballast water) seek to avoid paying the economic costs that would be required to prevent these "accidental," but predictable, invasions. In both cases, the potential costs are externalized to the larger society, and to future generations.

Responses

Customs and quarantine practices, developed in an earlier time to guard against diseases and pests of economic importance, are inadequate safeguards against the rising tide of species that threaten native biodiversity. Globally, about 165 million 6-meter-long, sealed containers are being shipped around the world at any given time. This number is far larger than custom officers can reasonably be expected to examine in detail. In the United States, some 1,300 quarantine officers are responsible for inspecting 410,000 planes and more than 50,000 ships, with each ship carrying hundreds of containers. While they intercept alien species nearly 50,000 times a year, it is highly likely that at least tens of thousands more enter the country uninspected each year. In Europe, inspection at the port of entry is also desperately overextended, and once a container enters the European Union, no further border inspections are done. This is a recipe for disaster.

Instead, a different set of strategies is now needed to deal with invasive species. These include prevention (certainly the most preferable), early eradication, special containment, or integrated management (often based on biological control). Mechanical, biological, and chemical means are available for controlling invasive species of plants and animals once they have arrived. Early warning, quarantine, and various other health measures are involved to halt the spread of pathogens.[28]

The international community has responded to the problem of invasive alien species through more than 40 conventions or programs, and many more are awaiting finalization or ratification.[29] The most comprehensive is the 1992 Convention on Biological Diversity, which calls on its 188 parties to "prevent the introduction of, control, or eradicate those alien species which threaten ecosystems, habitats, or species" (Article 8h).[30] A much older instrument, one that is virtually universally applied, is the 1952 International Plant Protection Convention, which applies primarily to plant pests, based on a system of phytosanitary certificates. Regional agreements further strengthen this convention. Other instruments deal with invasive alien species in specific regions (such as Antarctica), sectors (such as fishing in the Danube River), or vectors (such as invasive species in ballast water, through the International Maritime Organization). The fact that the problem continues to worsen indicates that the international response to date has been inadequate.

On the national level, some legal measures can offer very straightforward methods of preventing or managing invasions. For example, to deal with the problem of Asian beetle invasions, the United States now requires that all solid-wood packing material from China must be certified free of bark (under which insects may lurk) and heat-treated, fumigated, or treated with preservatives. China might reasonably issue a reciprocal regulation, as North American beetles are a hazard there.

The nursery industry is by far the largest intentional importer of new plant taxa. Issuing permits for imported species is a good way for the agencies responsible for managing such invasions to keep track of what is being traded and moved around the country. Some people believe that it is impossible to issue a regulation containing a list of permitted and prohibited species, at least partly because the ornamental horticulture industry is always seeking new species. But the Florida Nurserymen and Growers Association recently identified 24 marketed species on a black list drawn up by Florida's Exotic Pest Plant Council and decided to discourage trade in 11 of the species (the least promising sellers in any case).[31]

Sometimes nature itself can fight back against invasive alien species, at least when they reach plague proportions. For example, the zebra mussels that have invaded the North American Great Lakes with disastrous effects are now declining because a native sponge (*Eunapius fragilis*) is growing on the mussels, preventing them from opening their shells to feed or to breathe. The sponge has become abundant in some areas, while the zebra mussel population has fallen by up to 40 percent, although it is not yet clear whether the sponges will be effective in controlling the invasive mussels in the long term.[32]

Biological control—the intentional use of natural enemies to control an invasive species—is an important tool for managers. Some early efforts at biological control agents had disastrous effects, such as South American cane toads (*Bufo marinus*) in Australia, Indian common mynahs (*Acridotheres tristis*) in Hawaii, and Asian mongooses (*Herpestes javanicus*) in the Caribbean. Not only did these species not deal with the problem species upon which they were expected to prey, but they ended up causing havoc to native species and ecosystems. On the other hand, biological control programs are now much more carefully considered and in many cases are the most efficient, most effective, cheapest, and least damaging to the environment of any of the options for dealing with invasives that have already arrived.[33] Examples include the use of a weevil (*Cyrtobagous salviniae*) to control salvinia fern (*Salvinia molesta*), another weevil (*Neohydronomus affinis*) to control water lettuce (*Pistia stratiotes*), and a predatory beetle (*Hyperaspis pantherina*) to control orthezia scale (*Orthezia insignis*) that threatened the endemic national tree of Saint Helena (*Commidendrum robustum*).[34]

Those seeking to use viruses or other disease organisms to control an invasive species need to understand ecological links. When millions of rabbits died after the intentional introduction of the myxomatosis virus in the United Kingdom, for example, populations of their predators, including stoats, buzzards, and owls, declined sharply. The impact affected other species indirectly, leading to local extinction of the endangered large blue butterfly (*Maculina arion*) because of reduced grazing by rabbits on heathlands, which removed the habitat for an ant species that assists developing butterfly larvae.[35] But the use of the myxoma virus in conjunction with 1080 poison on the Phillip Island in the South Pacific successfully eradicated invasive rabbits, allowing the recovery of the island's vegetation (including the endemic *Hibiscus insularis*).[36]

At small scales of less than one hectare, it appears possible with current technology to eradicate invasive species of plants through use of herbicides, fire, physical removal, or a combination of these, but the costs of eradication rise quickly as the area covered increases. With the right approach and technology, invasive alien mammals can be eradicated from islands of thousands of hectares in size. Rat eradication from islands of larger than 2,000 hectares has been successful, and large mammals have been removed from much bigger ones than that, primarily by hunting and trapping.

Environmentally sensitive eradication also requires the restoration of the community or ecosystem following the removal of the invasive. For example, the eradication of Norway rats from Mokoia Island in New Zealand was followed by greatly increased densities of mice, also alien species. Similarly, the removal of Pacific rats (*Rattus exulans*) from Motupao Island, New Zealand, to protect a native snail led to increases of an exotic snail to the detriment of the natives. And on Motunau Island, New Zealand, the exotic box-thorn (*Lycium ferocissimum*) increased after the control of rabbits. On Santa Cruz Island, off the west coast of California, removing goats led to dramatic increases in the abundance of fennel (*Foeniculum vulgare*) and other alien species of weeds. Thus reversing the changes to native communities caused by non-native species will often require a sophisticated understanding of ecological relationships. It is now well recognized that eradication programs are only the first step in a long process of restoration.[37] Sometimes native species become dependent on invasive ones, causing dilemmas for managers. For example, giant kangaroo rats (*Dipodomys ingens*) in the American West continually modify their burrow precincts by digging tunnels, clipping plants, and other activities. This chronic disturbance to soil and vegetation sometimes promotes the establishment of invasive species of plants that were originally imported as ornamentals from the Mediterranean so that they constitute a very large proportion of the vegetation on giant kangaroo rat territories. They have significantly larger seeds than do native species so are favored by the grain-eating kangaroo rats.[38] Because the kangaroo rats depend on non-native plant species for food and the non-native plant species depend upon the kangaroo rats to disturb their habitat continually, the relationship is mutualistic. This strong relationship may also inhibit population growth of native grassland plants that occupy disturbed habitats but have difficulty competing with nonnative

weeds for resources. This mutualism presents an intractable conservation management dilemma, suggesting that it may be impossible to restore valley grasslands occupied by endangered kangaroo rats to conditions where native species dominate.

High-tech management measures are also being tried. For example, Australian scientists are planning to insert a gene known as "daughterless" into invasive male carp (*Cyprinus carpio*) in the Murray-Darling River, the country's longest, thereby ensuring that their offspring are male. The objective is to release them into the wild, sending wild carp populations into a decline and making room for the native species that are being threatened by the invasive European carp.[39] Using genetic modification can help eradicate an invasive alien species, but if the detrimental gene is released into nature and starts to flourish, many other species could be negatively affected. Thus the precautionary approach needs to be applied to control techniques as well as to introductions.

The problems of invasive alien species are so serious that actions must be taken even before we can be "certain" of all of their effects. However, mechanical removal, biocontrol, chemical control, shooting, or any other approach to controlling alien invasive species needs to be carefully considered prior to use to ensure that the implications have been fully and carefully considered, including impacts on human health, other species, and so forth. A public information program is also needed to ensure that the proposed measures are likely to be effective as well as socially and politically acceptable. Many animal-rights groups oppose the killing of any species of wildlife, for instance, even if they are causing harm to native species of plants and animals. The recent controversy surrounding the population of mute swans in the Chesapeake Bay is a good example.[40]

Conclusions

Ecosystems have been significantly influenced by people in virtually all parts of the world; some have even called these "engineered ecologies." Thus, a much more conscious

and better-informed management of ecosystems—one that deals with non-native species—is critical.

In just a few hundred years, major global forces have rendered natural barriers ineffective, allowing non-native species to travel vast distances to new habitats and become invasive alien species. The globalization and growth in the volume of trade and tourism, coupled with the emphasis on free trade, provide more opportunities than ever for species to be spread accidentally or deliberately. This inadvertent ending of millions of years of biological isolation has created major ongoing environmental problems that affect developed and developing countries, with profound economic and ecological implications.

Because of the potential for economic and ecological damage when an alien species becomes invasive, every alien species needs to be treated for management purposes as if it is potentially invasive, unless and until convincing evidence indicates that it is harmless in the new range. This view calls for urgent action by a wide range of governmental, intergovernmental, private sector, and civil institutions.

A comprehensive solution for dealing with invasive alien species has been developed by the Global Invasive Species Programme.[41] It includes 10 key elements:

- *An effective national capacity to deal with invasive alien species.* Building national capacity could include designing and establishing a "rapid response mechanism" to detect and respond immediately to the presence of potentially invasive species as soon as they appear, with sufficient funding and regulatory support; as well as implementing appropriate training and education programs to enhance individual capacity, including customs officials, field staff, managers, and policymakers. It could also include developing institutions at national or regional levels that bring together biodiversity specialists with agricultural quarantine specialists. Building basic border control and quarantine capacity and ensuring that agricultural quarantine, customs, and food inspection officers are aware of the elements of the Biosafety Protocol are other ways to deal with invasive alien species on a national level.

- *Fundamental and applied research at local, national, and global levels.* Research is required on taxonomy, invasion pathways, management measures, and effective monitoring. Further understanding on how and why species become established can lead to improved prediction on which species have the potential to become invasive; improved understanding of lag times between first introduction and establishment of invasive alien species; and better methods for excluding or removing alien species from traded goods, packaging material, ballast water, personal luggage, and other methods of transport.

> **The problems of invasive alien species are so serious that actions must be taken even before we can be "certain" of all their effects.**

- *Effective technical communications.* An accessible knowledge base, a planned system for review of proposed introductions, and an informed public are needed within countries and between countries. Already, numerous major sources of information on invasive species are accessible electronically and more could also be developed and promoted, along with other forms of media.

- *Appropriate economic policies.* While prevention, eradication, control, mitigation, and adaptation all yield economic benefits, they are likely to be undersupplied, because it is difficult for policymakers to identify specific beneficiaries who should pay for the benefits received. New or adapted economic instruments can help ensure that the costs of addressing invasive alien species are better reflected in market prices. Economic principles relevant to national strategies include

ensuring that those responsible for the introduction of economically harmful invasive species are liable for the costs they impose; ensuring that use rights to natural or environmental resources include an obligation to prevent the spread of potential invasive alien species; and requiring importers of such potential species to have liability insurance to cover the unanticipated costs of introductions.

- *Effective national, regional, and international legal and institutional frameworks.* Coordination and cooperation between the relevant institutions are necessary to address possible gaps, weaknesses, and inconsistencies and to promote greater mutual support among the many international instruments dealing with invasive alien species.

- *A system of environmental risk analysis.* Such a system could be based on existing environmental impact assessment procedures that have been developed in many countries. Risk analysis measures should be used to identify and evaluate the relevant risks of a proposed activity regarding alien species and determine the appropriate measures that should be adopted to manage the risks. This would also include developing criteria to measure and classify impacts of alien species on natural ecosystems, including detailed protocols for assessing the likelihood of invasion in specific habitats or ecosystems.

- *Public awareness and engagement.* If management of invasive species is to be successful, the general public must be involved. A vigorous public awareness program would involve the key stakeholders who are actively engaged in issues relevant to invasive alien species, including botanic gardens, nurseries, agricultural suppliers, and others. The public can also be involved as volunteers in eradication programs of certain non-native species, such as woody invasives of national parks for suggested actions that individuals can take.)

- *National strategies and plans.* The many elements of controlling inva-

sive alien species need to be well coordinated, ensuring that they are not simply passed on to the Ministry of Environment or a natural resource management department. A national strategy should promote cooperation among the many sectors whose activities have the greatest potential to introduce them, including military, forestry, agriculture, aquaculture, transport, tourism, health, and water-supply sectors. The government agencies with responsibility for human health, animal health, plant health, and other relevant fields need to ensure that they are all working toward the same broad objective of sustainable development in accordance to national and international legislation. Such national strategies and plans can also encourage collaboration between different scientific disciplines and approaches that can seek new approaches to dealing with problems caused by invasive alien species.

- *Invasive alien species issues built into global change initiatives.* Global change issues relevant to invasives begin with climate change but also include changes in nitrogen cycles, economic development, land use, and other fundamental changes that might enhance the possibilities of these species becoming established. Further, responses to global change issues, such as sequestering carbon, generating biomass energy, and recovering degraded lands, should be designed in ways that use native species and do not increase the risk of the spread of non-native invasives.

- *Promotion of international cooperation.* The problem of invasive alien species is fundamentally international, so international cooperation is essential to develop the necessary range of approaches, strategies, models, tools, and potential partners to ensure that the problems of such species are effectively addressed. Elements that would foster better international cooperation could include developing an international vocabulary, widely agreed upon and adopted; cross-sector collaboration

WHAT CAN AN INDIVIDUAL DO?

While the problem of invasive alien species seems daunting, an individual can make an important contribution to the problem, and if thousands of individuals work toward reducing the spread of invasive aliens, real progress can be made. Here are some steps that can be taken:

- Become informed about the issue.
- Grow native plants, keep native pets, and avoid releasing non-natives into the wild.
- Avoid carrying any living materials when traveling.
- Never release plants, fish, or other animals into a body of water unless they came out of that body of water.
- Clean boats before moving them from one body of water to another, and avoid using non-native species as bait.
- Support the work of organizations that are addressing the problem of invasive alien species.

among international organizations involved in agriculture, trade, tourism, health, and transport; and improved linkages among the international institutions dealing with phytosanitary, biosafety, and biodiversity issues and supporting these by strong linkages to coordinated national programs.

Because the diverse ecosystems of our planet have become connected through numerous trade routes, the problems caused by invasive alien species are certain to continue. As with maintaining and enhancing health, education, and security, perpetual investments will be required to manage the challenge they present. These 10 elements will ensure that the clear and present danger of invasive species is addressed in ways that build the capacity to address any future problems arising from expanding international trade.

NOTES

1. H. A. Mooney, J. A. McNeely, L. E. Neville, P. J. Schei, and J. K. Waage, eds., *Invasive Alien Species: Searching for Solutions* (Washington, DC: Island Press, 2004).

2. B. Groombridge, ed., *Global Biodiversity: Status of the Earth's Living Resources* (Cambridge, UK: World Conservation Monitoring Centre, 1992).

3. C. Hilton-Taylor, IUCN *Red List of Threatened Species* (Gland, Switzerland: IUCN–The World Conservation Union (IUCN). 2000).

4. R. May, "British Birds by Number," *Nature*, 6 April 2000, 559–60.

5. Ibid., and Groombridge, note 2 above.

6. J-Y. Meyer, "Tahiti's Native Flora Endangered by the Invasion of *Miconia calvescens*," *Journal of Geography* 23 (1997): 775–81.

7. D. Pimentel, L. Lach, R. Zuniga, and D. Morrison, "Environmental and Economic Costs of Nonindigenous Species in the United States," *BioScience* 50 (2000): 53–65.

8. World Trade Organization (WTO), *International Trade Statistics 2003* (Geneva: WTO, 2004).

9. G. M. Ruiz et al., "Global Spread of Microorganisms by Ships," *Nature*, 2 November 2000, 49–50.

10. J. E. Pasek, "Assessing Risk of Foreign Pest Introduction via the Solid Wood Packing Material Pathway," presentation made at the North American Plant Protection Organization Symposium on Pet Risk Analysis, Puerto Vallerta, Mexico, 18–21 March 2002.

11. U.S. Department of Agriculture, *Agricultural Research Service Research to Combat Invasive Species*, www.invasivespecies.gov/ toolkit/arsisresearch.doc (accessed 27 April 2004).

12. S. L. Chown and K. J. Gaston, "Island-Hopping Invaders Hitch a Ride with Tourists in South Georgia, *Nature*, 7 December 2000, 637.

13. K. R. McKaye et al., "African Tilapia in Lake Nicaragua," *BioScience* 45 (1995): 406–11.

14. L. Winter, "Cats Indoors!" *Earth Island Journal*, Summer 1999, 25–26.

15. G. Zorpette, "Parrots and Plunder," *Scientific American*, July 1997, 15–17.

16. Ruiz et al., note 9 above.

17. McKaye et al., note 13 above.

18. A. J. Ribbink, "African Lakes and Their Fishes: Conservation Scenarios and Suggestions," *Environmental Biology of Fishes* 19 (1987): 3–26; T. Goldschmit, F. Witte, and J. Wanink, "Cascading Effects of the Introduced Nile Perch on the Detritivorousphytoplanktivorous Species in the Sublittoral Areas of Lake Victoria," *Conservation Biology* 7 (1993): 686–700; and R. Ogutu-Ohwayo, "Nile Perch in Lake Victoria: The Balance between Benefits and Negative Impacts of Aliens," in O. T. Sandlund, P. J. Schei, and A. Viken, eds., *Invasive Species and Biodiversity Management* (Dordrecht, Netherlands: Kluwer Academic Publishers, 1999), 47–64.

19. M. L. McKinney and J. L. Lockwood, "Biotic Homogenization: A Few Winners Replacing Many Losers in the Next Mass Extinction," *Tree* 14 (1999): 450–53.

20. C. M. D'Antonio, T. L. Dudley, and M. Mack, "Disturbance and Biological Invasions: Direct Effects and Feedbacks," in L. Locker, ed., *Ecosystems of Disturbed Ground* (Amsterdam: Elziveer, 1999).

21. B. Groombridge and M. D. Jenkins, *World Atlas of Biodiversity: Earth's Living Resources in the 21st Century* (Berkeley, CA: University of California Press, 2002)

22. Pimentel, Lach, Zuniga, and Morrison, note 7 above.

23. N. L. Larson, P. J. Anderson, and W. Newton, "Alien Plant Invasion in Mixed-Grass Prairie: Effects of Vegetation Type and Anthropogenic Disturbance," *Ecological Applications* 11 (2001): 128–41; and J. M. Levine, "Species Diversity and Biological Invasions: Relating Local Process to

Community Pattern," *Science*, 5 May 2000, 852–54.

24. Ibid.

25. S. Noemdoe, "Putting People First in an Invasive Alien Clearing Programme: Working for Water Programme," in J. A. McNeely, ed., *The Great Reshuffling: Human Dimensions of Invasive Alien Species*, (Gland, Switzerland: IUCN, 2001), 121–26

26. D. Pimentel et al., "Economic and Environmental Threats of Alien Plant, Animal, and Microbe Invasions," *Agriculture, Ecosystems and Environment* 84 (2001): 1–20.

27. Ibid.

28. J. Kaiser, "Stemming the Tide of Invading Species," *Science*, 17 September 2000, 1836–841.

29. C. Shine, N. Williams, and L. Gündling, *A Guide to Designing Legal and Institutional Frameworks on Alien Invasive Species* (Bonn, Germany: IUCN, 2000).

30. L. Glowka, F. Burhenne-Guilmin, and H. Synge, *A Guide to the Convention on Biological Diversity* (Gland, Switzerland: IUCN, 1994). See also K. Raustiala and D. G. Victor, "Biodiversity Since Rio: The Future of the Convention on Biological Diversity," *Environment*, May 1996, 16–20, 37–45.

31. Kaiser, note 28 above.

32. A. Ricciardi, F. Sneider, D. Kelch, and H. Reiswig, "Lethal and Sub-Lethal Effects of Sponge Overgrowth on Introduced Dreissenide Mussels in the Great Lakes—St. Lawrence River System," *Canadian Journal of Fisheries and Aquatic Sciences* 52: 2695–703.

33. M. S. Hoddle, "Restoring Balance: Using Exotic Species to Control Invasive Exotic Species," *Conservation Biology* 18 (2004): 38–49; S. M. Louda and P. Stiling, "The Double-Edged Sword of Biological Control in Conservation and Restoration," *Conservation Biology* 18 (2004): 50–53; and R. Wittenberg and M. J. W. Cock, eds., *Invasive Alien Species: A Tool Kit of Best Prevention and Management Practices* (Wallingford, UK: Global Invasive Species Programme and Commonwealth Agricultural Bureau International, 2001).

34. Wittenberg and Cock, ibid.

35. P. Daszak, A. Cunningham, and A. D. Hyatt. "Emerging Infectious Diseases of Wildlife: Threats to Biodiversity and Human Health," *Science*, 21 January 2000, 443–49.

36. P. Coyne, "Rabbit Eradication on Phillip Island," in Wittenberg and Cock, note 33 above, page 176.

37. R. C. Klinger, P. Schuyler, and J. D. Sterner, "The Response of Herbaceous Vegetation and Endemic Plant Species to the Removal of Feral Sheep from Santa Cruz Island, California," in C. R. Veitch and M. N. Klout, eds., *Turning the Tide: the Eradication of Invasive Species* (Gland, Switzerland, and Cambridge, UK: IUCN, 2002), 14 1–54.

38. P. Schiffman, "Promotion of Exotic Weed Establishment by Endangered Giant Kangaroo Rats (*Dipodomys ingens*) in a California Grassland," *Biodiversity and Conservation* 3 (1994): 524–37.

39. R. Nowak, "Gene Warfare: One Small Tweak and a Whole Species Will Be Wiped Out," *New Scientist*, 11 May 2002, 6.

40. See B. Engle, "No Swansong in the Chesapeake Bay," *Environment*, December 2003, 7.

41. The Global Invasive Species Programme (GISP) was established in 1997 as a consortium of the Scientific Committee on Problems of the Environment (SCOPE), CABI, and IUCN, in partnership with the United Nation Environment Programme and with funding from the Global Environment Facility (GEF). See J. A. McNeely, H. A. Mooney, L. E. Neville, P. J. Schei, and J. K. Waage, eds., *Global Strategy on Invasive Alien Species* (Gland, Switzerland: IUCN, 2001).

Jeffrey A. McNeely is chief scientist at IUCN-The World Conservation Union in Gland, Switzerland. His research focuses on a broad range of topics relating to conservation and sustainable use of biodiversity, with a particular focus in recent years on the relationship between agriculture and wild biodiversity, the relationship between biodiversity and human health, and the impacts of war on biodiversity. McNeely has written or edited more than 30 books, from *Mammals of Thailand* (Association for the Conservation of Wildlife, 1975), his first, to *Ecoagriculture: Strategies to Feed the World and Save Wild Biodiversity* (Island Press, 2003). He has also published extensively on biodiversity, protected areas, and cultural aspects of conservation. He may be reached at jam@iucn.org.

From *Environment,* July/August 2004, pp. 17-31. Reprinted by permission of the Helen Dwight Reid Educational Foundation. Published by Heldref Publications, 1319, Eighteenth St., NW, Washington, DC 20036-1802. Copyright © 2004.

Markets for Biodiversity Services

POTENTIAL ROLES AND CHALLENGES

By Michael Jenkins, Sara J. Scherr, and Mira Inbar

Historically, it has been the responsibility of governments to ensure biodiversity protection and provision of ecosystem services. The main instruments to achieve such objectives have been

- direct resource ownership and management by government agencies;
- public regulation of private resource use;
- technical assistance programs to encourage improved private management; and
- targeted taxes and subsidies to modify private incentives.

But in recent decades, several factors have stimulated those concerned with biodiversity conservation services to begin exploring new market-based instruments. The model of public finance for forest and biodiversity conservation is facing a crisis as the main sources of finance have stagnated, despite the recognition that much larger areas require protection. At the same time, increasing recognition of the roles that ecosystem services play in poverty reduction and rural development is highlighting the importance of conservation in the 90 percent of land outside protected areas. It is thus urgent to find new means to finance the provision of ecosystem services, yet under current conditions private actors lack financial incentives to do so.

Crisis in Biodiversity Conservation Finance

Financing and management of natural protected areas has historically been perceived as the responsibility of the public sector. According to the United Nations Environment Programme, there are presently 102,102 protected areas worldwide, covering an area of 18.8 million square kilometers. Seventeen million square kilometers of these areas—11.5 percent of the Earth's terrestrial surface—are forests. Two-thirds of these have been assigned to one of the six protected-area management categories designated by the World Conservation Union (IUCN).

However, over the last few decades, severe cutbacks in the availability of public resources have undermined the effectiveness of such strategies. Protected areas in the tropics are increasingly dependent on international public or private donors for financing. Yet budgets for government protection and management of forest ecosystem services are declining, as are international sources from overseas development assistance. Land acquisition for protected areas and compensation for lost resource-based livelihoods are often prohibitively expensive. For example, it has been estimated that $1.3 billion would be required to fully compensate inhabitants in just nine central African parks.[1] The donation-driven model is often unsustainable, both economically and environmentally. Sovereignty is also an issue: About 30 percent of private forest concessions in Latin America and the Caribbean and 23 percent in Africa are already foreign owned. At the same time, public responsibility for nature protection is shifting with processes of devolution and decentralization, and new sources of financing for local governments to take on biodiversity and ecosystem service protection have not been forthcoming.

Table 1. Estimated financial flows for forest conservation (in millions, U.S. dollars)				
Sources of finance	SFM (early 1990s)	SFM (early 2000)	PAS (early 1990s)	PAS (early 2000)
Official development assistance	$2,000–$2,200	$1,000–$1,200	$700–$770	$350–$420
Public expenditure	NA	$1,600	NA	$598
Philanthropy[a]	$85.6	$150	NA	NA
Communities[b]	$365–$730	$1,300–$2,600	NA	NA
Private companies	NA	NA	NA	NA

[a]Underestimates self-financing and in-kind nongovernmental organization contributions.
[b]Self-financing and in-kind contributions from indigenous and other local communities.

NOTE: In 1990, there were an estimated 100 million hectares of community-managed forests worldwide. SFM is "sustainable forest management." PAS stands for "protected area system."

SOURCE: A. Molnar, S. J. Scherr, and A. Khare, *Current Status and Future Potential of Markets for Ecosystem Services of Tropical Forests: An Overview* (Washington, DC: Forest Trends, 2004).

Moreover, scientific studies increasingly indicate that biodiversity cannot be conserved by a small number of strictly protected areas.[2] Conservation must be conceived in a landscape or ecosystem strategy that links protected areas within a broader matrix of land uses that are compatible with and support biodiversity conservation in situ. To achieve such outcomes, it will be essential to engage private actors in conservation finance on a large scale. Yet the markets for products from natural areas and forests face at least three serious challenges: declining commodity prices for traditionally important products, such as timber; competition from illegal sources; and poorly functioning, overregulated markets. Thus, private forest owners and landowners need to find new revenue streams to justify retaining forests on the landscapes and to manage them well in the context of declining commodity prices and competition in natural forests from illegal sources of timber.

Rural Development, Poverty Reduction, and Biodiversity

The vast majority of biodiversity resources in the world are found in populated landscapes, and it can be argued that the biodiversity that underpins ecosystem services critical to human health and livelihoods should have high priority in conservation efforts. An estimated 240 million rural people live in the world's high-canopy forest landscapes. In Latin America, for example, 80 percent of all forests are located in areas of medium to high human popula-

tion density.[3] Population growth in the world's remaining "tropical wilderness areas" is twice the global average. More than a billion people live in the 25 biodiversity "hotspots" identified by Conservation International; in 16 of these hotspots, population growth is higher than the world average.[4] While species richness is lower in drylands and other ecosystems not represented among the "hot spots," the species that play functional ecosystem roles are all the more important and difficult to replace.

Poor rural communities are especially dependent upon natural biodiversity. Low-income rural people rely heavily on the direct consumption of wild foods, medicines, and fuels, especially for meeting micronutrient and protein needs, and during "hungry" periods. An estimated 350 million poor people rely on forests as safety nets or for supplemental income. Farmers earn as much as 10 to 25 percent of household income from nontimber forest products. Bushmeat is the main source of animal protein in West Africa. The poor often harvest, process, and sell wild plants and animals to buy food. Sixty million poor people depend on herding in semiarid rangelands that they share with large mammals and other wildlife. Thirty million low-income people earn their livelihoods primarily as fishers, twice the number of 30 years ago. The depletion of fisheries has serious impacts on food security. Wild plants are used in farming systems for fodder, fertilizer, packaging, fencing, and genetic materials. Farmers rely on soil microorganisms to maintain soil fertility and structure for crop production, and they also rely on wild species in natural ecological communities for crop pollination and

pest and predator control. Wild relatives of domesticated crop species provide the genetic diversity used in crop improvement. The rural poor rely directly on ecosystem services for clean and reliable local water supplies. Ecosystem degradation results in less water for people, crops, and livestock; lower crop, livestock, and tree yields; and higher risks of natural disasters.

> **More than a billion people live in the 25 biodiversity "hotspots" identified by Conservation International; in 16 of these hotspots, population growth is higher than the world average.**

Three-quarters of the world's people living on less than $1 per day are rural. Strategies to meet the United Nations Millennium Development Goals in rural areas—to reduce hunger and poverty and to conserve biodiversity—must find ways to do so in the same landscapes. Crop and planted pasture production—mostly in low-productivity systems—dominate at least half the world's temperate, subtropical, and tropical forest areas; a far larger area is used for grazing livestock.[5] Food insecurity threatens biodiversity when it leads to overexploitation of wild plants and animals. Low farm productivity leads to depletion of soil and water resources and increases the pressure to clear additional land that serves as wildlife habitat. Some 40 percent of cropland in developing countries is degraded. Of more than 17,000 major protected areas, 45 percent (accounting for one-fifth of total protected areas) are heavily used for agriculture, while many of the rest are islands in a sea of farms, pastures, and production forests that are managed in ways incompatible for long-term species and ecosystem survival.[6]

Despite this high level of dependence by the poor on biodiversity, the dominant model of conservation seeks to exclude people from natural habitats. In India, for example, 30 million people are targeted for resettlement from protected areas.[7] From the perspective of poverty reduction and rural development, it is thus urgent to identify alternative conservation systems that respect the rights of forest dwellers and owners and address conservation objectives in the 90 percent of forests outside public protected areas. Markets for ecosystem services potentially offer a more efficient and lower-cost approach to forest conservation.[8]

Need for Financial Incentives to Provide Ecosystem Services

There is growing recognition that regulatory and protected area approaches, while critical, are insufficient to adequately conserve biodiversity. A fundamental problem is financial, especially for resources that lie outside protected areas. For these to be conserved, they need to be more valuable than the alternative uses of the land. And for such resources to be well managed, good stewardship needs to be more profitable than bad stewardship. The failure of forest owners and producers to capture financial benefits from conserving ecosystem benefits leads to overexploitation of forest resources and undersupply of ecosystem services.

This reality is hard for many people to accept, because most ecosystem services are considered "public goods." The "polluter pays" principle has argued that the right of the public to these services trumps the private rights of the landowner or manager. Yet good management has a cost. While the individual who manages his or her resources to protect biodiversity produces public benefits, the costs incurred are private. Under current institutions, those who benefit from these services have no incentive to compensate suppliers for these services. In most of the world, forest ecosystem services are not traded and have

no "price." Thus, where the opportunity costs of forest land for agricultural enterprises, infrastructure, and human settlements are higher than the use or income value of timber and nontimber forest products (NTFPs), habitats will be cleared and wild species will be allowed to disappear. Because they receive little or no direct benefit from them, resource owners and producers ignore the real economic and noneconomic values of ecosystem services in making decisions about land use and management.

A lower-cost approach to securing conservation is to pay only for the biodiversity services themselves, by paying landowners to manage their assets so as to achieve biodiversity or species conservation.

Mechanisms are needed by which resource owners are rewarded for their role as stewards in providing biodiversity and ecosystem services. Anticipation of such income flows would enhance the value of natural assets and thus encourage their conservation. Compared to previous approaches to forest conservation, market-based mechanisms promise increased efficiency and effectiveness, at least in some situations. Experience with market-based instruments in other sectors has shown that such mechanisms, if carefully designed and implemented, can achieve environmental goals at significantly less cost than conventional "command-and-control" approaches, while creating positive incentives for continual innovation and improvement. Markets for ecosystem services could potentially contribute to rural development and poverty reduction by providing financial benefits from the sale of ecosystem services, improving human capital through associated training and education, and strengthening social capital through investment in local cooperative institutions.

New Market Solutions to Conserve Biodiversity

The market for biodiversity protection can be characterized as a nascent market. Many approaches are emerging to finan-

cially remunerate the owners and managers of land and resources for their good stewardship of biodiversity (see Table 2). Market mechanisms to pay for other ecosystem services—watershed services, carbon sequestration or storage, landscape beauty, and salinity control, for example—can be designed to conserve biodiversity as well. However, in general, biodiversity services are the most demanding to protect because of the need to conserve many different elements essential for diverse, interdependent species to thrive. Figure 1 illustrates potential market solutions and some of the complexities involved.

Land Markets for High-Biodiversity-Value Habitat

National governments (in the form of public parks and protected areas), NGO conservation organizations (for example, The Nature Conservancy), and individual conservationists have long paid for the purchase of high-biodiversity-value forest habitats. Direct acquisition can be expensive, as underlying land and use values are also included. Local sovereignty concerns arise when buyers are from outside the country—or even the local area—or where extending the area of noncommercial real estate reduces the local tax base. New commercial approaches are being developed to encourage the establishment of privately owned conservation areas, such as conservation communities (the purchase of a plot of land by a group of people mainly for recreation or conservation purposes), ecotourism-based land protection projects, and ecologically sound real estate projects being organized in Chile.[9] These build on growing consumer demand for housing and vacation in biodiverse environments.

Payments for Use or Management

A lower-cost approach to securing conservation is to pay only for the biodiversity services themselves, by paying landowners to manage their assets so as to achieve biodiversity or species conservation. It is likely that the largest-scale payments for land-use or management agreements belong to one of two categories. One encompasses government agroenvironmental payments made to farmers in North America and Europe for reforesting conservation

Table 2: Types of payments for biodiversity protection

Purchase of high-value habitat

Type	Mechanism
Private land acquisition	Purchase by private buyers or nongovernmental organizations explicitly for biodiversity conservation
Public land acquisition	Purchase by government agency explicitly for biodiversity conservation

Payment for access to species or habitat

Bioprospecting rights	Rights to collect, test, and use genetic material from a designated area
Research permits	Right to collect specimens, take measurements in area
Hunting, fishing, or gathering permits for wild species	Right to hunt, fish, and gather
Ecotourism use	Rights to enter area, observe wildlife, camp, or hike

Payment for biodiversity-conserving management

Conservation easements	Owner paid to use and manage defined piece of land only for conservation purposes; restrictions are usually in perpetuity and transferable upon sale of the land
Conservation land lease	Owner paid to use and manage defined piece of land for conservation purposes for defined period of time
Conservation concession	Public forest agency is paid to maintain a defined area under conservation uses only; comparable to a forest logging concession
Community concession in public protected areas	Individuals or communities are allocated use rights to a defined area of forest or grassland in return for commitment to protect the area from practices that harm biodiversity
Management contracts for habitat or species conservation on private farms, forests, or grazing lands	Contract that details biodiversity management activities and payments linked to the achievement of specified objectives

Tradable rights under cap-and-trade regulations

Tradable wetland mitigation credits	Credits from wetland conservation or restoration that can be used to offset obligations of developers to maintain a minimum area of natural wetlands in a defined region
Tradable development rights	Rights allocated to develop only a limited total area of natural habitat within a defined region
Tradable biodiversity credits	Credits representing areas of biodiversity protection or enhancement that can be purchased by developers to ensure they meet a minimum standard of biodiversity protection

Support biodiversity-conserving businesses

Biodiversity-friendly businesses	Business shares in enterprises that manage for biodiversity conservation
Biodiversity-friendly products	Eco-labeling

SOURCE: S. J. Scherr, A. White, and A. Khare, *Current Status and Future Potential of Markets for Ecosystem Services in Tropical Forests: An Overview* (Washington, DC: Forest Trends, 2003).

easements. The other category describes management contracts aiming to conserve aquatic and terrestrial wildlife habitat. In Switzerland, "ecological compensation areas," which use farming systems compatible with biodiversity conservation, have expanded to include more than 8 percent of total agricultural land. In the tropics, diverse approaches include nationwide public payments in Costa Rica for forest conservation and in Mexico for forested watershed protection.

Conservation agencies are organizing direct payments systems, such as conservation concessions being negotiated by Conservation International, and forest conservation easements negotiated by the *Cordão de Mata* ("linked forest") project with dairy farmers in Brazil's Atlantic Forest. The dairy farmers in the latter example receive, in exchange, technical assistance and investment resources to raise crop and livestock productivity. Some countries that use land taxes are using tax policies in innovative

ways to encourage the expansion of private and public protected areas.

Payment for Private Access to Species or Habitat

Private sector demand for biodiversity has tended to take the form of payments for access to particular species or habitats that function as "private goods" but in practice serve to cover some or all of the costs of providing broader ecosystem services. Pharmaceutical compa-

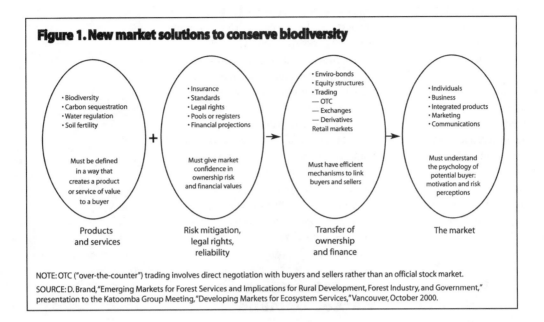

Figure 1. New market solutions to conserve biodiversity

- Biodiversity
- Carbon sequestration
- Water regulation
- Soil fertility

Must be defined in a way that creates a product or service of value to a buyer

Products and services

+

- Insurance
- Standards
- Legal rights
- Pools or registers
- Financial projections

Must give market confidence in ownership risk and financial values

Risk mitigation, legal rights, reliability

- Enviro-bonds
- Equity structures
- Trading
  — OTC
  — Exchanges
  — Derivatives
  Retail markets

Must have efficient mechanisms to link buyers and sellers

Transfer of ownership and finance

- Individuals
- Business
- Integrated products
- Marketing
- Communications

Must understand the psychology of potential buyer: motivation and risk perceptions

The market

NOTE: OTC ("over-the-counter") trading involves direct negotiation with buyers and sellers rather than an official stock market.

SOURCE: D. Brand, "Emerging Markets for Forest Services and Implications for Rural Development, Forest Industry, and Government," presentation to the Katoomba Group Meeting, "Developing Markets for Ecosystem Services," Vancouver, October 2000.

nies have contracted for bioprospecting rights in tropical forests. Ecotourism companies have paid forest owners for the right to bring tourists into their lands to observe wildlife, while private individuals are willing to pay forest owners for the right to hunt, fish, or gather nontimber forest products.

Tradable Rights and Credits within a Regulatory Framework

Multifactor markets for ecosystem services have been successfully established, notably for sulfur dioxide emissions, farm nutrient pollutants, and carbon emissions. These create rights or obligations within a broad regulatory framework and allow those with obligations to "buy" compliance from other landowners or users. Developing such markets for biodiversity is more complicated, because specific site conditions matter so much. The United States has operated a wetlands mitigation program since the early 1980s in which developers seeking to destroy a wetland must offset that by buying wetland banks conserved or developed elsewhere. A similar approach is used for "conservation banking," described in the box.

A variant of this approach is being designed for conserving forest biodiversity in Brazil by permitting flexible enforce-

ment of that country's "50 percent rule," which requires landholders in Amazon forest areas to maintain half of their land in forest. This rule is also applied in other regions in Brazil, where lesser proportional areas are set aside for forest use. Careful designation of comparable sites is required.

Another approach, biodiversity credits, is under development in Australia. In this system, legislation creates new property rights for private landholders who conserve biodiversity values on their land. These landholders can then sell resulting "credits" to a common pool. The law also creates obligations for land developers and others to purchase those credits. The approach requires that the "value" of the biodiversity unit can be translated into a dollar value.

Biodiversity-Conserving Businesses

Conservation values are beginning to inform consumer and investor decisions. Eco-labeling schemes are being developed that advertise or certify that products were produced in ways consistent with biodiversity conservation. The global trade in certified organic agriculture was worth $21 billion worldwide in 2000.[10] International organic standards are expanding to landscape-scale biodiversity impacts. The Rainforest Alliance and the Sustainable

A NEW FUND TO FINANCE FOREST ECOSYSTEM SERVICES

The Mexican government recently announced the creation of a new fund to pay indigenous and other communities for the forest ecosystem services produced by their land.[1] Indigenous and other communities own approximately 80 percent of all forests in Mexico—totaling some 44 million hectares—as collectively held, private land. The Mexican Forestry Fund has been under design since 2002, guided by a consultative group with government, nongovernmental organization, and industry representatives. The purpose of the US$20 million fund is to promote the conservation and sustainable management of natural forests, leverage additional financing, contribute to the competitiveness of the forest sector, and catalyze the development of mechanisms to finance forest ecosystem services. Operational manuals are being prepared, and priority conservation sites have already been identified. The fund proposes to pay $40 per hectare (ha) per year to owners of deciduous forests in critical mountain areas and $30 per ha per year to other forest types.

Notes

1. Comisión Nacional Forestal (CONAFOR), presentation given at the Mexican Forestry Expo, Guadalajara, Mexico, 8 August 2003.

Agriculture Network certify coffee, bananas, oranges, and other products grown in and around high-biodiversity-value areas. The Sustainable Agriculture Initiative is a coalition of multinational commercial food producers (Nestle, Dannon, Unilever, and others) who are seeking to ensure that all of the products they purchase along the supply chain come from producers who are protecting biodiversity. In 2002, more than 100 million hectares of forest were certified (a fourfold increase over 1996), although only 8 percent of the total certified area is in developing countries, and most of that is in temperate forests.

Current Market Demand

Available information suggests that biodiversity protection services are presently the largest market for ecosystem services. A team from McKinsey & Company, the World Resources Institute, and The Nature Conservancy estimated the annual international finance for the conservation market (conservation defined as protecting land from development) at $2 billion, with the forest component a large share of that.[11] Buyers are predominantly development banks and foundations in the United States and Europe.

A study by the International Institute for Environment and Development (IIED) of 72 cases of markets for forest biodiversity protection services in 33 countries found that the main buyers of biodiversity services (in declining order of prevalence) were private corporations, international NGOs and research institutes, donors, governments, and private individuals.[12] Communities, public agencies, and private individuals predominate as sellers. Most of these cases took place in Latin America and in Asia and the Pacific. Only four cases were found in Europe and Russia and one was found in the United States.

Three-quarters of the cases in the IIED study were international markets, and the rest were distributed among regional, national, and local buyers. International actors—as well as many on the national level—who demand biodiversity protection services tend to focus on the most biodiverse habitats (in terms of species

richness) or those perceived to be under the greatest threat globally (for example, places like the Amazon, where there are a high number of endemic species and where habitat area has greatly declined). Most of the private corporations were interested in eco-labeling schemes for crops or timber, investment in biodiversity-friendly companies, horticultural companies concerned with ecosystem services, or pharmaceutical bioprospecting. Such private payments are usually site-specific. Local actors more commonly focus on protecting species or habitats of particular economic, subsistence, or cultural value.

Projected Growth in Market Demand

The fastest-growing component of future market demand for biodiversity services is likely to be in eco-labeling of crop, livestock, timber, and fish products for export and for urban consumers. In 1999, the value of the organic foods market was US$14.2 billion. Its value is growing at 20–30 percent a year in the industrialized world, as the international organic movement is strengthening standards for biodiversity conservation.[13] Pressures continue to increase on major international trading and food processing companies to source from suppliers who are not degrading ecosystem services. Donor and international NGO conservation will continue to expand as NGOs begin to establish entire research departments aimed at developing new market-based instruments. Voluntary biodiversity offsets are also a promising source of future demand, as many large companies are seeking ways to maintain their "license to operate" in environmentally sensitive areas, and offsets are of increasing interest to them.

The costs of and political resistance to land acquisition are rising. Construction of biological corridors in and around production areas is an increasingly important conservation objective. At the same time, however, many of the most important sites for biodiversity conservation are in more densely populated areas with high opportunity costs for land. Thus we are likely to see a major shift from land acquisition to various types of

CONSERVATION BANKING IN THE UNITED STATES

Amendments to the United States Endangered Species Act in 1982 provided for an "incidental take" of enlisted species, if "a landowner provides a long-term commitment to species conservation through development of a Habitat Conservation Plan (HCP)." These amendments have opened the door to a series of market-based transactions, described as conservation banking, which permits land containing a natural resource (such as wetlands, forests, rivers, or watersheds) that is conserved and maintained for specified enlisted species to be used to offset impacts occurring elsewhere to the same natural resource.[1] A private landowner may request an "incidental take" permit and mitigate it by purchasing "species credits" from preestablished conservation banks. Credits are administered according to individuals, breeding pairs, acres, nesting sites, and family units. Conservation banking has maximized the value of underutilized commercial real estate and given private landowners incentive to conserve habitat.

California was the first state to authorize the use of conservation banking and has established 50 conservation banks since 1995. Other states, including Alabama, Colorado, and Indiana, have followed suit. In April 2002, the Indiana Department of Transportation, the Federal Highway Association Indiana Division, and four local government agencies finalized an HCP for the endangered Indiana bat as part of the improvement of transportation facilities around Indianapolis International Airport. These highway improvements will occur in an area of known Indiana bat habitat that is predicted to experience nearly $1.5 billion in economic development during the next ten years. Under the HCP, approximately 3,600 acres will be protected, including 373 acres of existing bat habitat.

Notes

1. A. Davis, "Conservation Banking," presentation to the Katoomba Group-Lucarno Workshop, Lucarno, Switzerland, November 2003.

direct payments for easements, land leases, and management contracts.

> The fastest-growing component of future market demand for biodiversity services is likely to be in eco-labeling of crop, livestock, timber, and fish products for export and for urban consumers.

A rough back-of-the-envelope estimate suggests that the current value of international, national, and local direct payments and trading markets for ecosystem services from tropical forests alone could be worth several hundred million dollars per year, while the value of certified forest and tropical tree crop products may reach as much as a billion dollars. While this is a large and significant amount, it represents a small fraction of the value of conventional tropical timber and other forest product markets. For example, by comparison, the total value of tropical timber exports is $8 billion (including only logs, sawnwood, veneer, and plywood), which is a small fraction of the total exports and domestic timber, pulpwood, and fuelwood markets in tropical countries. NTFP markets are far larger still.[14] The total value of international trade for NTFPs is $7.5 billion-$9 billion per year, with another $108 billion in processed medicines and medicinal plants.[15] Domestic markets for NTFPs are many times larger (for example, domestic consumption accounted for 94 percent of the global output of fresh tropical fruits 1995–2000.[16] Nonetheless, these rough figures are quite interesting when compared with the scale of public and donor forest conservation finance summarized in Table 1.

Scaling Up Payments for Biodiversity: Next Steps

Markets for ecosystem services are steadily growing and can be expected to grow even more rapidly in the next decade. Yet they predominate as pilot projects. What will it take to transform these markets to impact ecosystem conservation on the global scale? The four most strategic and catalytic areas for policy and action are to

- structure emerging markets to support community-driven conservation;
- mobilize and organize buyers for ecosystem services;
- connect global and national action on climate change to biodiversity conservation; and
- invest in the policy frameworks and institutions required for functioning ecosystem service payment systems.

Supporting Community-Driven Conservation

The benefits of investments in ecosystem services will be maximized over the long term if markets reward local participation and utilize local knowledge. In community forests and agroforestry landscapes, communities have already established sophisticated conservation strategies. Studies of indigenous timber enterprises document conservation investments on the order of $2 per hectare per year apart from other management activities and investments of community time and labor; this is equal to the average available budget per hectare for protected areas worldwide. Conservation policies must recognize the role that local people are playing in the conservation of forest ecosystems worldwide and support them (either with cash or in-kind support) to continue to be good environmental stewards.

To enable conservation-oriented management to remain or become economically viable, it is important that ecosystem service payments and markets are designed so that they strategically channel financial payments to rural communities. Such payments can be used to develop and invest in new production systems that increase productivity and rural incomes, and enhance biodiversity at a landscape scale—an approach referred to as "ecoagriculture."[17] Ecosystem service payments to poor rural communities that are providing stewardship services of national or international value can help to meet multiple Millennium Development Goals. For any semblance of a sustainable future to be realized, it is crucial that our long-term vision includes biodiversity and natural ecosystems as part of the "natural infrastructure" of a healthy economy and society.

Mobilizing and Organizing Buyers for Ecosystem Services

Turning beneficiaries into buyers is the driving force of ecosystem service markets. Because beneficiaries are often hesitant to pay for goods previously considered free, "willingness to pay" for ecosystem services must be organized on a greater scale. The private sector must be called upon to engage in responsible corporate behavior in conserving biodiversity. For example, Insight Investment, a major financial firm, has developed a biodiversity policy that uses conservation as a screen for investment. Voluntary payments by consumers, retail firms, and other actors can be encouraged through social advertising. This approach is growing rapidly now for eco-labeling programs (labeling of some personal care products and foods) and voluntary carbon emission offset programs involving investment in reforestation. Stockholder pressure is beginning to influence some firms to avoid investments and activities that harm biodiversity, and this is evolving to positive action. Civil society campaigns can also mobilize willingness to pay for biodiversity offsets and payments to local partners for conservation.

Connecting Climate Action with Biodiversity Conservation

Far more aggressive action must and will be taken to mitigate and adapt to climate change. Land use and land-use change currently contribute more than 20 percent of carbon emissions and other greenhouse gases. Action to reduce these emissions must be a central part of our response, and it is critical that action to sequester carbon through improved land uses accompanies strategies to reduce industrial emissions. There is thus an unprecedented opportunity at this time to structure our responses to climate change so that actions related to land use are also designed to protect and restore biodiversity. Moreover, such actions can be designed in ways that enhance and protect livelihoods, especially for those most vulnerable to the impacts of climate change. Indeed, it is imperative that they do so.

As a result of the deliberations at the Conference of the Parties of the United Nations Framework Convention on Climate Change last year, payments for forest carbon through the Clean Development Mechanism (CDM) of the Kyoto Protocol can be used to finance forest restoration and regeneration projects that conserve biodiversity while providing an alternative income source for local people.[18] But the scale of forest carbon under CDM is very small—too small to have a major impact on climate, biodiversity, or livelihoods. It is critical that we aim for a much larger program in the second commitment period, and it is crucial that nations affiliated with the Organisation for Economic Co-operation and Development (OECD) create initiatives to utilize carbon markets for biodiversity conservation in their own internal trading programs. It is imperative to develop a new principle of international agreements on climate response and carbon trading, one that builds a system that encourages overlap of the major international environmental agreements and the Millennium Development Goals. This could mobilize demand by creating an international framework for investing in good ecosystem service markets. It is also important that emerging private voluntary markets for carbon (that is, with actors who do not have a regulatory obligation) are encouraged to pursue such biodiversity goals as well. The Climate, Community and Biodiversity Alliance, for example, is seeking to develop guidelines and indicators for private investments in carbon projects that will achieve these multiple goals. The Forest Climate Alliance of The Katoomba Group is seeking to mobilize the international rural development community to advocate for such approaches.[19]

Investing in Policy Frameworks and Institutions for Biodiversity Markets

Ecosystem service markets are genuinely new—and biodiversity markets are the newest and most challenging. Every market requires basic rules and institutions in order to function, and this is equally true of biodiversity markets. The biodiversity

conservation community needs to act quickly and strategically to ensure that as these markets develop, they are effective, equitable, and operational and are used sensibly to complement other conservation approaches.

Policymakers and public agencies play a vital role in creating the legal and legislative frameworks necessary for market tools to operate effectively. This includes establishing regulatory rules, systems of rights over ecosystem services, and mechanisms to enforce contracts and settle ownership disputes. Ecosystem service markets pose profound equity implications, as new rules may fundamentally change the distribution of rights and responsibilities for essential ecosystem services. Forest producers and civil society will need to take a proactive role to ensure that rules support the public interest and create development opportunities.

New institutions will also be needed to provide the business services required in ecosystem service markets. For example, in order for beneficiaries of biodiversity services to become willing to pay for them, better methods of measuring and assessing biodiversity in working landscapes must be developed, as well as the institutional capacity to do so. New institutions must be created to encourage transactions and reduce transaction costs. Such institutions could include "bundling" biodiversity services provided by large numbers of local producers, as well as investment vehicles that have a diverse portfolio of projects to manage risks. Registers must be established and maintained, to record payments and trades. For example, The Katoomba Group is developing a Web-based "Marketplace" to slash the information and transaction costs for buyers, sellers, and intermediaries in ecosystem service markets.[20]

Conclusion

Conservation of biodiversity and of the services biodiversity provides to humans and to the ecological health of the planet requires financing on a scale many times larger than is feasible from public and philanthropic sources. It is essential to find

new mechanisms by which resource owners and managers can realize the economic values created by good stewardship of biodiversity. Moreover, private consumers, producers, and investors can financially reward that stewardship. New markets and payment systems, strategically shaped to deliver critical public benefits, are showing tremendous potential to move biodiversity conservation objectives to greater scale and significance.

NOTES

1. M. Cernea and K. Schmidt-Soltau, 2003. "Biodiversity Conservation versus Population Resettlement, Risks to Nature

and Risks to People," paper presented at CIFOR (Center for International Forestry Research) Rural Livelihoods, Forests and Biodiversity Conference, Bonn, Germany, 19–23 May 2003.

2. See S. Wood, K. Sebastian, and S. J. Scherr, *Pilot Analysis of Global Ecosystems: Agroecosystems* (Washington, DC: International Food Policy Research Institute and the World Resources Institute, 2000), 64; and E. W. Sanderson et al., "The Human Footprint and the Last of the Wild," *Bioscience* 52, no. 10 (2002): 891–904.

3. K. Chomitz, *Forest Cover and Population Density in Latin America*, research notes to the World Bank (Washington, DC: World Bank, 2003).

4. R. P. Cincotta and R. Engelman, *Nature's Place: Human Population and the Future of Biological Diversity* (Washington, DC: Population Action International, 2000).

5. Wood, Sebastian, and Scherr, note 2 above.

6. J. McNeely and S. J. Scherr, *Ecoagriculture: Strategies to Feed the World and Conserve Wild Biodiversity* (Washington, DC: Island Press, 2003).

7. A. Khare et al., *Joint Forest Management: Policy Practice and Prospects* (London: International Institute for Environment and Development, 2000).

8. S. J. Scherr, A. White, and D. Kaimowitz, *A New Agenda for Forest Conservation and Poverty Reduction: Making Markets Work for Low-Income Communities* (Washington, DC: Forest Trends, CIFOR, and IUCN-The World Conservation Union, 2004).

9. E. Corcuera, C. Sepulveda, and G. Geisse, "Conserving Land Privately: Spontaneous Markets for Land Conservation in Chile," in S. Pagiola et al., eds., *Selling Forest Environmental Services: Market-Based Mechanisms for Conservation and Development* (London: Earthscan Publications, 2002).

10. J. W. Clay, *Community-Based Natural Resource Management within the New Global Economy: Challenges and Opportunities*, a report prepared by the Ford Foundation (Washington, DC: World Wildlife Fund, 2002).

11. M. Arnold and M. Jenkins, "The Business Development Facility: A Strategy to Move Sustainable Forest Management and Conservation to Scale," proposal to the International Finance Corporation (IFC) Environmental Opportunity Facility from Forest Trends, Washington, DC, 2003.

12. N. Landell-Mills, and I. Porras. 2002. *Markets for Forest Environmental Services: Silver Bullet or Fool's Gold? Markets for Forest Environmental Services and the Poor, Emerging Issues* (London: International Institute for Environment and Development, 2002).

13. International Federation of Organic Agriculture Movements (IFOAM), "Cultivating Communities," 14th IFOAM Organic World Congress, Victoria, BC, 21–28 August 2002.

14. S. J. Scherr, A. White, and A. Khare, *Current Status and Future Potential of Markets for Ecosystem Services of Tropical Forests: A Report for the International Tropical Timber Organization* (Washington, DC: Forest Trends, 2003).

15. M. Simula, *Trade and Environment Issues in Forest Protection*, Environment Division working paper (Washington, DC: Inter-American Development Bank, 1999).

16. Food and Agricultural Organization of the United Nations (FAO), *FAOSTAT* database for 2000, accessible via `http://www.fao.org`.

17. For more information on ecoagriculture, see the Ecoagriculture Partners' Web site at `http://www.ecoagriculturepartners.org`.

18. S. J. Scherr and M. Inbar, *Clean Development Mechanism Forestry for Poverty Reduction and Biodiversity Conservation: Making the CDM Work for Rural Communities* (Washington, DC: Forest Trends, 2003).

19. For more information on this project, see `http://www.katoombagroup.org/Katoomba/forestcarbon`.

20. The Katoomba Group is a unique network of experts in forestry and finance companies, environmental policy and research organizations, governmental agencies and influential private, community, and nonprofit groups. It is dedicated to advancing markets for some of the ecosystem services provided by forests, such as watershed protection, biodiversity habitat, and carbon storage. For more information on the Katoomba Group, see `http://www.katoombagroup.org`. Forest Trends serves as the secretariat for the group. More information on Forest Trends can be found at `http://www.forest-trends.org`.

Michael Jenkins is the founding president of Forest Trends, a nonprofit organization based in Washington, D.C., and created in 1999. Its mission is to maintain and restore forest ecosystems by promoting incentives that diversify trade in the forest sector, moving beyond exclusive focus on lumber and fiber to a broader range of products and services. Previously he worked as a senior forestry advisor to the World Bank (1998), as associate director for the Global Security and Sustainability Program of the MacArthur Foundation (1988–1998), as an agroforester in Haiti with the U.S. Agency for International Development (1983–1986), and as technical advisor with Appropriate Technology International (1981–1982). He has also worked in forestry projects in Brazil and the Dominican Republic and was a Peace Corps volunteer in Paraguay. He speaks Spanish, French, Portuguese, Creole, and Guaraní, and can be contacted by telephone at (202) 298-3000 or via e-mail at mjenkins@forest-trends.org. Sara J. Scherr is an agricultural and natural resource economist who specializes in the economics and policy of land and forest management in tropical developing countries. She is presently director of the Ecosystem Services program at Forest Trends, and also director of Ecoagriculture Partners, the secretariat of which is based at Forest Trends. She previously worked as principal researcher at the International Center for Research in Agroforestry, in Nairobi, Kenya, as (senior) research fellow at the International Food Policy Research Institute in Washington, D.C., and as adjunct professor at the Agricultural and Resource Economics Department of the University of Maryland, College Park. Her current work focuses on policies to reduce poverty and restore ecosystems through markets for sustainably grown products and environmental services and on policies to promote ecoagriculture—the joint production of food and environmental services in agricultural landscapes. She also serves as a member of the Board of the World Agroforestry Centre, and as a member of the United Nations Millennium Project Task Force on Hunger. Scherr can be reached by telephone at (202) 298-3000 or via e-mail at sscherr@forest-trends.org. Mira Inbar is program associate with Forest Trends. She works in the Ecosystem Services program, supporting efforts to establish frameworks and instruments for emerging transactions in environmental services worldwide. Before joining Forest Trends, she worked with communities in the Urubamba River Valley of Peru to initiate a forest conservation plan. She has worked with the National Fishery Department in Western Samoa, the Marie Selby Botanical Gardens, and Environmental Defense. Inbar may be reached by telephone at (202) 298-3000 or via e-mail at minbar@forest-trends.org. This article is © The Aspen Institute and is published with permission.

From *Environment,* July/August 2004, pp. 32-42. Reprinted by permission of the Aspen Institute. Copyright © 2004.

On the Termination of Species

Ecologists' warnings of an ongoing mass extinction are being challenged by skeptics and largely ignored by politicians. In part that is because it is surprisingly hard to know the dimensions of the die-off, why it matters and how it can best be stopped.

By W. Wayt Gibbs

HILO, HAWAII—Among the scientists gathered here in August at the annual meeting of the Society for Conversation Biology, the despair was almost palpable. "I'm just glad I'm retiring soon and won't be around to see everything disappear," said P. Dee Boersma, former president of the society, during the opening night's dinner. Other veteran field biologists around the table murmured in sullen agreement.

At the next morning's keynote address, Robert M. May, a University of Oxford zoologist who presides over the Royal Society and until last year served as chief scientific adviser to the British government, did his best to disabuse any remaining optimists of their rosy outlook. According to his latest rough estimate, the extinction rate—the pace at which species vanish—accelerated during the past 100 years to roughly 1,000 times what it was before humans showed up. Various lines of argument, he explained, "suggest a speeding up by a further factor of 10 over the next century or so.... And that puts us squarely on the breaking edge of the sixth great wave of extinction in the history of life on Earth."

From there, May's lecture grew more depressing. Biologists and conservationists alike, he complained, are afflicted with a "total vertebrate chauvinism." Their bias toward mammals, birds and fish—when most of the diversity of life lies elsewhere—undermines scientists' ability to predict reliably the scope and consequences of biodiversity loss. It also raises troubling questions about the high-priority "hotspots" that environmental groups are scrambling to identify and preserve.

"Ultimately we have to ask ourselves why we care" about the planet's portfolio of species and its diminishment, May said. "This central question is a political and social question of values, one in which the voice of conservation scientists has no particular standing." Unfortunately, he concluded, of "the three kinds of argument we

Overview/Extinction Rates

- Eminent ecologists warn that humans are causing a mass extinction event of a severity not seen since the age of dinosaurs came to an end 65 million years ago. But paleontologists and statisticians have called such comparisons into doubt.
- It is hard to know how fast species are disappearing. Models based on the speed of tropical deforestation or on the growth of endangered species lists predict rising extinction rates. But biologists' bias toward plants and vertebrates, which represent a minority of life, undermine these predictions. Because 90 percent of species do not yet have names, let alone censuses, they are impossible to verify.
- In the face of uncertainty about the decline of biodiversity and its economic value, scientists are debating whether rare species should be the focus of conservation. Perhaps, some suggest, we should first try to save relatively pristine—and inexpensive—land where evolution can progress unaffected by human activity.

use to try to persuade politicians that all this is important... none is totally compelling."

Although May paints a truly dreadful picture, his is a common view for a field in which best-sellers carry titles such as *Requiem for Nature*. But is despair justified? *The Skeptical Environmentalist*, the new English translation of a recent book by Danish statistician Bjørn Lomborg, charges that reports of the death of biodiversity have been greatly exaggerated. In the face of such external skepticism, internal uncertainty and public apathy, some scientists are questioning the conservation movement's overriding emphasis on preserving rare species and the threatened hotspots in which they are concentrated. Perhaps, they suggest, we should focus instead on saving something equally at risk but even more valuable: evolution itself.

Doom...

MAY'S CLAIM that humans appear to be causing a cataclysm of extinctions more severe than any since the one that erased the dinosaurs 65 million years ago may shock those who haven't followed the biodiversity issue. But it prompted no gasps from the conservation biologists. They have heard variations of this dire forecast since at least 1979, when Normal Myers guessed in *The Sinking Ark* that 40,000 species lose their last member each year and that one million would be extinct by 2000. In the 1980s Thomas Lovejoy similarly predicted that 15 to 20 percent would die off by 2000; Paul Ehrlich figured half would be gone by now. "I'm reasonably certain that [the elimination of one fifth of species] didn't happen," says Kirk O. Winemiller, a fish biologist at Texas A&M University who just finished a review of the scientific literature on extinction rates.

More recent projections factor in a slightly slower demise because some doomed species have hung on longer than anticipated. Indeed, a few have even returned from the grave. "It was discovered only this summer that the Bavarian vole, continental Eurasia's one and only presumed extinct mammal [since 1500], is in fact still with us," says Ross D. E. MacPhee, curator of mammalogy at the American Museum of Natural History (AMNH) in New York City.

Still, in the 1999 edition of his often-quoted book *The Diversity of Life*, Harvard University biologist E. O. Wilson cites current estimates that between 1 and 10 percent of species are extinguished every decade, at least 27,000 a year. Michael J. Novacek, AMNH's provost of science, wrote in a review article this spring that "figures approaching 30 percent extermination of all species by the mid-21st century are not unrealistic." And in a 1998 survey of biologists, 70 percent said they believed that a mass extinction is in progress; a third of them expected to lose 20 to 50 percent of the world's species within 30 years.

"Although these assertions of massive extinctions of species have been repeated everywhere you look, they do not equate with the available evidence," Lomborg argues in *The Skeptical Environmentalist*. A professor of statistics and political science at the University of Århus, he alleges that environmentalists have ignored recent evidence that tropical deforestation is not taking the toll that was feared. "No well-investigated group of animals shows a pattern of loss that is consistent with greatly heightened

extinction rates," MacPhee concurs. The best models, Lomborg suggests, project an extinction rate of 0.15 percent of species per decade, "not a catastrophe but a problem—one of many that mankind still needs to solve."

... or Gloom?

"IT'S A TOUGH question to put numbers on," Wilson allows. May agrees but says "that isn't an argument for not asking the question" of whether a mass extinction event is upon us.

> "If you are looking for hard evidence of tens or hundreds or thousands of species disappearing each year, you aren't going to find it."
>
> —KIRK O. WINEMILLER, TEXAS A&M

To answer that question, we need to know three things: the natural (or "background") extinction rate, the current rate and whether the pace of extinction is steady or changing. The first step, Wilson explains, is to work out the mean life span of a species from the fossil record. "The background extinction rate is then the inverse of that. If species are born at random and all live exactly one million years—and it varies, but it's on that order—then that means one species in a million naturally goes extinct each year," he says.

In a 1995 article that is still cited in almost every scientific paper on this subject (even in Lomborg's book), May used a similar method to compute the background rate. He relied on estimates that put the mean species life span at five million to 10 million years, however; he thus came up with a rate that is five to 10 times lower than Wilson's. But according to paleontologist David M. Raup (then at the University of Chicago), who published some of the figures May and Wilson relied on, their calculations are seriously flawed by three false assumptions.

One is that species of plants, mammals, insects, marine invertebrates and other groups all exist for about the same time. In fact, the typical survival time appears to vary among groups by a factor of 10 or more, with mammal species among the least durable. Second, they assume that all organisms have an equal chance of making it into the fossil record. But paleontologists estimate that fewer than 4 percent of all species that ever lived are preserved as fossils. "And the species we do see are the widespread, very successful ones," Raup says. "The weak species confined to some hilltop or island all went extinct before they could be fossilized," adds John Alroy of the University of California at Santa Barbara.

The third problem is that May and Wilson use an average life span when they should use a median. Because "the vast majority of species are short-lived," Raup says, "the average is distorted by the very few that have very long life spans." All three oversimplifications underestimate the background rate—and make the current picture scarier in comparison.

Earlier this year U.C.S.B. biomathematician Helen M. Regan and several of her colleagues published the first attempt ever to correct for the strong biases and uncertainties in the data. They looked exclusively at mammals, the best-studied group. They estimated how many of the mammals now living, and how many of those recently extinguished, would show up as fossils. They also factored in the uncertainty for each number rather than relying on best guesses. In the end they concluded that "the current rate of mammalian extinction lies between 17 and 377 times the background extinction rate." The best estimate, they wrote, is a 36- to 78-fold increase.

Regan's method is still imperfect. Comparing the past 400 years with the previous 65 million unavoidably assumes, she says, "that the current extinction rate will be sustained over millions of years." Alroy recently came up with a way to measure the speed of extinctions that doesn't suffer from such assumptions. Over the past 200 years, he figures, the rate of loss among mammal species has been some 120 times higher than natural.

A Grim Guessing Game

ATTEMPTS TO FIGURE out the current extinction rate are fraught with even more uncertainties. The international conservation organization IUCN keeps "Red Lists" of organisms suspected to be extinct in the wild. But MacPhee complains that "the IUCN methodology for recognizing extinction is not sufficiently rigorous to be reliable." He and other extinction experts have formed the Committee on Recently Extinct Organisms, which combed the Red Lists to identify those species that were clearly unique and that had not been found despite a reasonable search. They certified 60 of the 87 mammals listed by IUCN as extinct but claim that only 33 of the 92 freshwater fish presumed extinct by IUCN are definitely gone forever.

For every species falsely presumed absent, however, there may be hundreds or thousands that vanish unknown to science. "We are uncertain to a factor of 10 about how many species we share the planet with," May points out. "My guess would be roughly seven million, but credible guesses range from five to 15 million," excluding microorganisms.

Taxonomists have named approximately 1.8 million species, but biologists know almost nothing about most of them, especially the insects, nematodes and crustaceans that dominate the animal kingdom. Some 40 percent of the 400,000 known beetle species have each been recorded at just one location—and with no idea of individual species' range, scientists have no way to confirm its extinction. Even invertebrates known to be extinct often go unrecorded: when the passenger pigeon was elim-

inated in 1914, it took two species of parasitic lice with it. They still do not appear on IUCN's list.

"It is extremely difficult to observe an extinction; it's like seeing an airplane crash," Wilson says. Not that scientists aren't trying. Articles on the "biotic holocaust," as Myers calls it, usually figure that the vast majority of extinctions have been in the tropical Americas. Freshwater fishes are especially vulnerable, with more than a quarter listed as threatened. "I work in Venezuela, which has substantially more freshwater fishes than all of North America. After 30 years of work, we've done a reasonable job of cataloguing fish diversity there," observes Winemiller of Texas A&M, "yet we can't point to one documented case of extinction."

A similar pattern emerges for other groups of organisms, he claims. "If you are looking for hard evidence of tens or hundreds or thousands of species disappearing each year, you aren't going to find it. That could be because the database is woefully inadequate," he acknowledges. "But one shouldn't dismiss the possibility that it's not going to be the disaster everyone fears."

The Logic of Loss

THE DISASTER SCENARIOS are based on several independent lines of evidence that seem to point to fast and rising extinction rates. The most widely accepted is the species-area relation. "Generally speaking, as the area of habitat falls, the number of species living in its drops proportionally by the third root to the sixth root," explains Wilson, who first deduced this equation more than 30 years ago. "A middle value is the fourth root, which means that when you eliminate 90 percent of the habitat, the number of species falls by half."

Extinction *Filters*

SURVIVAL OF THE FITTEST takes on a new meaning when humans develop a region. Among four Mediterranean climate regions, those developed more recently have lost larger fractions of their vascular plant species in modern times. Once the species least compatible with agriculture are filtered out by "artificial selection," extinction rates seem to fall.

REGION [in order of development]	EXTINCT [per 1,000]	THREATENED [percent]
Mediterranean	1.3	14.7
South African Cape	3.0	15.2
California	4.0	10.2
Western Australia	6.6	17.5

SOURCE: "Extinctions in Mediterranean Areas." Werner Greuter in Extinction Rates. Edited by J. H. Lawton and R. H. May. Oxford University Press, 1995

Why Biodiversity Doesn't [Yet] Pay

FOZ DO IGUAÇU, BRAZIL—At the International Congress of Entomologists last summer, Ebbe Nielsen, director of the Australian National Insect Collection in Canberra, reflected on the reasons why, despite the 1992 Convention on Biological Diversity signed here in Brazil by 178 countries, so little has happened since to secure the world's threatened species. "You and I can say extinction rates are too high and we have to stop it, but to convince the politicians we have to have convincing reasons," he said. "In developing countries, the economic pressures are so high, people use whatever they can find today to survive until tomorrow. As long as that's the case, there will be no support for biodiversity at all."

Not, that is, unless it can be made more profitable to leave a forest standing or a wetland wet than it is to convert the land to farm, pasture or parking lot. Unfortunately, time has not been kind to the several arguments environmentalists have made to assign economic value to each one of perhaps 10 million species.

A Hedge against Disease and Famine

"Narrowly utilitarian arguments say: The incredible genetic diversity contained in the population and species diversity that we are heirs to is ultimately the raw stuff of tomorrow's biotechnological revolution," observes Robert May of Oxford. "It is the source of new drugs." Or new foods, adds E. O. Wilson of Harvard, should something happen to the 30 crops that supply 90 percent of the calories to the human diet, or to the 14 animal species that make up 90 percent of our livestock.

"Some people who say that may even believe it," May continues. "I don't. Give us 20 or 30 years and we will design new drugs from the molecule up, as we are already beginning to do."

Hopes were raised 10 years ago by reports that Merck has paid $1.14 million to InBio, a Costa Rican conservation group, for novel chemicals extracted from rain-forest species. The contract would return royalties to InBio if any of the leads became drugs. But none have, and Merck terminated the agreement in 1999. Shaman Pharmaceuticals, founded in 1989 to commercialize traditional medicinal plants, got as far as late-stage clinical trials but then went bankrupt. And given, as Wilson himself notes in *The Diversity of Life,* that more than 90 percent of the known varieties of the basic food plants are on deposit in seed banks, national parks are hardly the cheapest form of insurance against crop failures.

Ecosystem Services

"Potentially the strongest argument," May says, "is a broadly utilitarian one: ecological systems deliver services we're only just beginning to think of trying to estimate. We do not understand how much you can simplify these sys-

tems and yet still have them function. As Aldo Leopold once said, the first rule of intelligent tinkering is to keep all the pieces."

The trouble with this argument, explains Columbia University economist Geoffrey Heal, is that "it does not make sense to ask about the value of replacing a life-support system." Economics can only assign values to things for which there are markets, he says. If all oil were to vanish, for example, we could switch to alternative fuels that cost $50 a barrel. But that does not determine the price of oil.

And although recent experiments suggest that removing a large fraction of species from a small area lowers its biomass and ability to soak up carbon dioxide, scientists cannot say yet whether the principle applies to whole ecosystems. "It may be that a grievously simplified world—the world of the cult movie *Blade Runner*—can be so run that we can survive in it," May concedes.

A Duty of Stewardship

Because science knows so little of the millions of species out there, let alone what complex roles each one plays in the ecosystems it inhabits, it may never be possible for economics to come to the aid of endangered species. A moral argument may thus be the best last hope—certainly it is appeals to leaders' sense of stewardship that have accomplished the most so far. But is it hazardous for scientists to make it?

They do, of course, in various forms. To Wilson, "a species is a masterpiece of evolution, a million-year-old entity encoded by five billion genetic letters, exquisitely adapted to the niche it inhabits." For that reason, conservation biologist David Ehrenfeld proposed in *The Arrogance of Humanism,* "long-standing existence in Nature is deemed to carry with it the unimpeachable right to continued existence."

Winning public recognition of such a right will take much education and persuasion. According to a poll last year, fewer than one quarter of Americans recognized the term "biological diversity." Three quarters expressed concern about species and habitat loss, but that is down from 87 percent in 1996. And May observes that the concept of biodiversity stewardship "is a developed-world luxury. If we were in abject poverty trying to put food in the mouth of the fifth child, the argument would have less resonance."

But if scientists "proselytize on behalf of biodiversity"—as Wilson, Lovejoy, Ehrlich and many others have done—they should realize that "such work carries perils," advises David Takacs of California State University at Monterey Bay. "Advocacy threatens to undermine the perception of value neutrality and objectivity that leads laypersons to listen to scientists in the first place." And yet if those who know rare species best and love them most cannot speak openly on their behalf, who will?

"From that rough first estimate and the rate of the destruction of the tropical forest, which is about 1 percent a year," Wilson continues, "we can predict that about one quarter of 1 percent of species either become extinct immediately or are doomed to much earlier extinction."

From a pool of roughly 10 million species, we should thus expect about 25,000 to evaporate annually.

Lomborg challenges that view on three grounds, however. Species-area relations were worked out by comparing the number of species on islands and do not necessarily apply to fragmented habitats on the mainland. "More than half of Costa Rica's native bird species occur in largely deforested countryside habitats, together with similar fractions of mammals and butterflies," Stanford University biologist Gretchen Daily noted recently in *Nature*. Although they may not thrive, a large fraction of forest species may survive on farmland and in woodlots—for how long, no one yet knows.

That would help explain Lomborg's second observation, which is that in both the eastern U.S. and Puerto Rico, clearance of more than 98 percent of the primary forests did not wipe out half of the bird species in them. Four centuries of logging "resulted in the extinction of only one forest bird" out of 200 in the U.S. and seven out of 60 native species in Puerto Rico, he asserts.

Such criticisms misunderstand the species-area theory, according to Stuart L. Pimm of Columbia University. "Habitat destruction acts like a cookie cutter stamping out poorly mixed dough," he wrote last year in *Nature*. "Species found only within the stamped-out area are themselves stamped out. Those found more widely are not."

Of the 200 bird types in the forests of the eastern U.S., Pimm states, all but 28 also lived elsewhere. Moreover, the forest was cleared gradually, and gradually it regrew as farmland was abandoned. So even at the low point, around 1872, woodland covered half the extent of the original forest. The species-area theory predicts that a 50 percent reduction should knock out 16 percent of the endemic species: in this case, four birds. And four species did go extinct. Lomborg discounts one of those four that may have been a subspecies and two others that perhaps succumbed to unrelated insults.

But even if the species-area equation holds, Lomborg responds, official statistics suggest that deforestation has been slowing and is now well below 1 percent a year. The U.N. Food and Agriculture Organization recently estimated that from 1990 to 2000 the world's forest cover dropped at an average annual rate of 0.2 percent (11.5 million hectares felled, minus 2.5 million hectares of new growth).

Annual forest loss was around half a percent in most of the tropics, however, and that is where the great majority of rare and threatened species live. So although "forecasters may get these figures wrong now and then, perhaps colored by a desire to sound the alarm, this is just a matter of timescale," replies Carlos A. Peres, a Brazilian ecologist at the University of East Anglia in England.

An Uncertain Future

ECOLOGISTS HAVE TRIED other means to project future extinction rates. May and his co-workers watched how vertebrate species moved through the threat categories in IUCN's database over a four-year period (two years for plants), projected those very small numbers far into the future and concluded that extinction rates will rise 12- to 55-fold over the next 300 years. Georgina M. Mace, director of science at the Zoological Society of London, came to a similar conclusion by combining models that plot survival odds for a few very well known species. Entomologist Nigel E. Stork of the Natural History Museum in London noted that a British bird is 10 times more likely than a British bug to be endangered. He then extrapolated such ratios to the rest of the world to predict 100,000 to 500,000 insect extinctions by 2300. Lomborg favors this latter model, from which he concludes that "the rate for all animals will remain below 0.208 percent per decade and probably be below 0.7 percent per 50 years."

It takes a heroic act of courage for any scientist to erect such long and broad projections on such a thin and lopsided base of data. Especially when, according to May, the data on endangered species "may tell us more about the vagaries of sampling efforts, of taxonomists' interests and of data entry than about the real changes in species' status."

Biologists have some good theoretical reasons to fear that even if mass extinction hasn't begun yet, collapse is imminent. At the conference in Hilo, Kevin Higgins of the University of Oregon presented a computer model that tracks artificial organisms in a population, simulating their genetic mutation rates, reproductive behavior and ecological interactions. He found that "in small populations, mutations tend to be mild enough that natural selection doesn't filter them out. That dramatically shortens the time to extinction." So as habitats shrink and populations are wiped out—at a rate of perhaps 16 million a year, Daily has estimated—"this could be a time bomb, an extinction event occurring under the surface," Higgins warns. But proving that that bomb is ticking in the wild will not be easy.

And what will happen to fig trees, the most widespread plant genus in the tropics, if it loses the single parasitic wasp variety that pollinates every one of its 900 species? Or to the 79 percent of canopy-level trees in the Samoan rain forests if hunters kill off the flying foxes on which they depend? Part of the reason so many conservationists are so fearful is that they expect the arches of entire ecosystems to fall once a few "keystone" species are removed.

Others distrust that metaphor. "Several recent studies seem to show that there is some redundancy in ecosystems," says Melodie A. McGeoch of the University of Pretoria in South Africa, although she cautions that what is redundant today may not be redundant tomorrow. "It really doesn't make sense to think the majority of species would go down with marginally higher pressures than if humans weren't on the scene," MacPhee adds. "Evolution should make them resilient."

If natural selection doesn't do so, artificial selection might, according to work by Werner Greuter of the Free University of Berlin, Thomas M. Brooks of Conservation International and others. Greuter compared the rate of recent plant extinctions in four ecologically similar regions and discovered that the longest-settled, most disturbed area—the Mediterranean—had the lowest rate. Plant extinction rates were higher in California and South Africa, and they were highest in Western Australia. The solution to this apparent paradox, they propose, is that species that cannot coexist with human land use tend to die out soon after agriculture begins. Those that are left are better equipped to dodge the darts we throw at them. Human-induced extinctions may thus fall over time.

"It turns out to be a lot easier to persuade a corporate CEO or a billionaire of the importance of the issue than it is to convince the American public."

—EDWARD O. WILSON, HARVARD UNIVERSITY

If true, that has several implications. Millennia ago our ancestors may have killed off many more species than we care to think about in Europe, Asia and other long-settled regions. On the other hand, we may have more time than we fear to prevent future catastrophes in areas where humans have been part of the ecosystem for a while—and less time than we hope to avoid them in what little wilderness remains pristine.

"The question is how to deal with uncertainty, because there really is no way to make that uncertainty go away," Winemiller argues. "We think the situation is extremely serious; we just don't think the species extinction issue is the peg that conservation movement should hang its hat on. Otherwise, if it turns out to be wrong, where does that leave us?"

Long-Term Savings

IT COULD LEAVE conservationists with less of a sense of urgency and with a handful of weak political and economic arguments [see box "Why Biodiversity Doesn't (Yet) Pay"]. It might also force them to realize that "many of the species in trouble today are in fact already members of the doomed, living dead," as David S. Woodruff wrote in the *Proceedings of the National Academy of Sciences* this past May. "Triage" is a dirty word to many environmentalists. "Unless we say no species loss is acceptable, then we have no line in the sand to defend, and we will be pushed back and back as losses build," Brooks argued at the Hilo meetings. But losses are inevitable, Wilson says, until the human population stops growing.

"I call that the bottleneck," Wilson elaborates, "because we have to pass through that scramble for remaining resources in order to get to an era, perhaps sometime in the 22nd century, of declining population. Our goal is to carry as much of the biodiversity through as possible." Biologists are divided, however, on whether the few charismatic species now recognized as endangered should determine what gets pulled through the bottleneck.

"The argument that when you protect birds and mammals, the other things come with them just doesn't stand up to close examination," May says. A smarter goal is "to try to conserve the greatest amount of evolutionary history." Far more valuable than a panda or rhino, he suggests, are relic life-forms such as the tuatara, a large iguanalike reptile that lives only on islets off the coast of New Zealand. Just two species of tuatara remain from a group that branched off from the main stem of the reptilian evolutionary tree so long ago that this couple make up a genus, an order and almost a subclass all by themselves.

But Woodruff, who is an ecologist at the University of California at San Diego, invokes an even broader principle. "Some of us advocate a shift from saving things, the products of evolution, to saving the underlying process, evolution itself," he writes. "This process will ultimately provide us with the most cost-effective solution to the general problem of conserving nature."

There are still a few large areas where natural selection alone determines which species succeed and which fail. "Why not save functioning ecosystems that haven't been despoiled yet?" Winemiller asks. "Places like the Guyana shield region of South America contain far more species than some of the so-called hotspots." To do so would mean purchasing tracts large enough to accommodate entire ecosystems as they roll north and south in response to the shifting climate. It would also mean prohibiting all human uses of land. It may not be impossible: utterly undeveloped wilderness is relatively cheap, and the population of potential buyers has recently exploded.

"It turns out to be a lot easier to persuade a corporate CEO or a billionaire of the importance of the issue than it is to convince the American public," Wilson says. "With a Ted Turner or a Gordon Moore or a Craig McCaw involved, you can accomplish almost as much as a government of a developed country would with a fairly generous appropriation."

"Maybe even more," agrees Richard E. Rice, chief economist for Conservation International. With money from Moore, McCaw, Turner and other donors, CI has outcompeted logging companies for forested land in Suriname and Guyana. In Bolivia, Rice reports, "we conserved an area the size of Rhode Island for half the price of a house in my neighborhood," and the Nature Conservancy was able to have a swath of rain forest as big as Yellowstone National Park set aside for a mere $1.5 million. In late July, Peru issued to an environmental group the country's first "conservation concession"—essentially a renewable lease for the right to *not* develop the land—for 130,000 hectares of forest. Peru has now opened some 60 million hectares of its public forests to such concessions, Rice says. And efforts are under way to negotiate similar deals in Guatemala and Cameroon.

"Even without massive support in public opinion or really effective government policy in the U.S., things are turning upward," Wilson says, with a look of cautious optimism on his face. Perhaps it is a bit early to despair after all.

MORE TO EXPLORE

Extinction Rates. Edited by John H. Lawton and Robert M. May. Oxford University Press, 1995.

The Currency and Tempo of Extinction. Helen M. Regan et al. in the *American Naturalist,* Vol. 157, No. 1, pages 1–10; January 2001.

Encyclopedia of Biodiversity. Edited by Simon Asher Levin. Academic Press, 2001.

The Skeptical Environmentalist. Bj&&0248rn Lomborg. Cambridge University Press, 2001.

W. Wayt Gibbs is senior writer.

UNIT 5
Resources: Land and Water

Unit Selections

Key Points to Consider

- Why is the number of farmers in the world decreasing? What kinds of social, economic, and cultural impacts will be produced by decreasing the number of family farmers and increasing the amount of land farmed by corporate farmers?

- What are the conflicting uses of water that prompt such environmental problems as those recently experienced in the Klamath River basin in Oregon? Are there ways to resolve conflicts over the use of water that have not yet been implemented?

- Explain the relationship between agricultural production and irrigation. Are irrigation systems inefficient and wasteful of water? If so, why and what could be done to eliminate such inefficiencies?

- How serious are claims that increasing demand for water places limits on agricultural production? What might be some of the contributing factors to a diminishing global water supply in either quantitative or qualitative terms?

 **Links: www.dushkin.com/online/**
These sites are annotated in the World Wide Web pages.

Global Climate Change
http://www.puc.state.oh.us/consumer/gcc/index.html

National Oceanic and Atmospheric Administration (NOAA)
http://www.noaa.gov

National Operational Hydrologic Remote Sensing Center (NOHRSC)
http://www.nohrsc.nws.gov

Virtual Seminar in Global Political Economy/Global Cities & Social Movements
http://csf.colorado.edu/gpe/gpe95b/resources.html

Terrestrial Sciences
http://www.cgd.ucar.edu/tss/

The worldwide situations regarding reduction of biodiversity, scarcity of energy resources, and pollution of the environment have received the greatest amount of attention among members of the environmentalist community. But there are a number of other resource issues that demonstrate the interrelated nature of all human activities and the environments in which they occur. One such issue is the declining quality of agricultural land. In the developing world, excessive rural populations have forced the overuse of lands and sparked a shift into marginal areas, and the total availability of new farmland is decreasing at an alarming rate of 2 percent per year. In the developed world, intensive mechanized agriculture has resulted in such a loss of topsoil that some agricultural experts are predicting a decline in food production. Other natural resources, such as minerals and timber, are declining in quantity and quality as well; in some cases they are no longer usable at present levels of technology. The overuse of groundwater reserves has resulted in potential shortages beside which the energy crisis pales in significance. And the very productivity of Earth's environmental systems—their ability to support human and other life—is being threatened by processes that derive at least in part from energy overuse and inefficiency and from pollution. Many environmentalists believe that both the public and private sectors, including individuals, are continuing to act in a totally irresponsible manner with regard to the natural resources upon which we all depend.

Uppermost in the minds of many who think of the environment in terms of an integrated whole, as evidenced by many of the selections in this unit, is the concept of the threshold or critical limit of human interference with natural systems of land and water. This concept suggests that the environmental systems we occupy have been pushed to the brink of tolerance in terms of stability and that destabilization of environmental systems has consequences that can only be hinted at, rather than predicted. Although the broader issue of system change and instability, along with the lesser issues such as the quantity of agricultural land, the quality of iron ore deposits, the sustained yield of forests, or the availability of fresh water seems to be quite diverse, all are closely tied to a pair of concepts—that of resource marginality and of the globalization of the economy that has made marginality a global rather than a regional problem. Many of these ideas are brought together in the lead article of this unit. In "Where Have All the Farmers Gone?" Brian Halweil of the World Watch Institute discusses the globalization of industry and trade that is creating a uniform approach to all forms of economic management, including management of agricultural resources. Halweil notes that increasing agribusiness and decreasing numbers of family farmers represent a loss of both biological and cultural diversity, as standardized farming and business practices that may work for some areas but not for others are applied uniformly.

In the second article in this unit, farming also figures importantly—but in entirely different ways. Journalist Bruce Barcott, writing in *Mother Jones,* discusses the conflicts between various water users, from farmers to fishermen, that have erupted in the drought-stricken American West. In "What's a River For?" Barcott focuses his attention on the Klamath River basin of northern California and southern Oregon where a struggle for an increasingly scarce resource has erupted between farmers who need water for irrigation, factories, and cities that need water for manufacturing and domestic purposes, ranchers whose livestock depend upon water for their very existence, and fishermen—both recreational and subsistence—who depend upon the river's salmon habitat for nourishment for both body and soul. The Klamath simply could not and cannot support all of these conflicting uses and, in 2002, a decision was made to release irrigation water from the river, resulting in one of the largest fish die-backs in American history. Use of water for irrigation impacts not only on fish populations, however. We now know that the loss of topsoil to accelerated erosion, the loss of living and farming space to reservoirs, and increasing the salinity in irrigation waters and soil are problems greater than the dryness that prompts irrigation to begin with. In the third and fourth articles in the unit, the struggle between different stakeholders in the land and water use wars also takes center stage. In "A Human Thirst," UN consultant Don Hinrichsen reports that human uses now consume more than half the world's freshwater, often leaving other animal species to go thirsty. As water is withdrawn from rivers and other sources to feed the demands of agriculture, industry, and over six billion humans, the aquatic ecosystems, the plants and animals they support, and the very surface of the land itself suffers. A similar set of problems is discussed by Mark Rosegrant, Ximing Cai, and Sarah Cline, all of the International Food Policy Research Institute, in their article "Will the World Run Dry? Global Water and Food Security." The authors argue that while water development is the underpinning of food production, jobs, economic growth, and environment sustainability throughout the world, little is being done on an international scale to manage this precious resource. In many areas of the world, water is already in short supply and this scarcity will worsen in the absence of investment commitments from governments, development agencies, and others.

The final article in this unit deals with another critical issue in resource management. In "Fire Fight," author Paul Trachtman describes the increasing clashes between officials of the U.S. federal government and environmentalists over ways to reduce the risk of catastrophic forest fires, particularly in the American West. The favored approach of federal land agencies such as the U.S. Forest Service has been to thin forest tracts by commercial logging. The argument here is that more than a century of fire protection has left so much unburned forest that it must be thinned in order to prevent the massive conflagrations that have characterized nearly each summer in the West for the past decade. The environmentalists counter with the argument that such a policy is simply a means of allowing commercial logging at low cost on public lands and favor a more "natural" solution of letting wildfires burn themselves out. The obvious problem with the latter strategy, counters the government, is that wildfires a century ago were in uninhabited areas that are now filled with towns, resort communities, and private homes. Neither side in the argument seems willing to listen to the other.

There are two possible solutions to all these problems posed by the use of increasingly marginal and scarce resources and by the continuing pollution of the global atmosphere. One is to halt the basic cause of the problems—increasing population and consumption. The other is to provide incentives and techniques for the conservation and management of existing resources and for the discovery of alternative resources to eliminate the demand for more marginal resources and the use of heavily polluting ones.

Where Have All the Farmers Gone?

The globalization of industry and trade is bringing more and more uniformity to the management of the world's land, and a spreading threat to the diversity of crops, ecosystems, and cultures. As Big Ag takes over, farmers who have a stake in their land—and who often are the most knowledgeable stewards of the land—are being forced into servitude or driven out.

by Brian Halweil

Since 1992, the U.S. Army Corps of Engineers has been developing plans to expand the network of locks and dams along the Mississippi River. The Mississippi is the primary conduit for shipping American soybeans into global commerce—about 35,000 tons a day. The Corps' plan would mean hauling in up to 1.2 million metric tons of concrete to lengthen ten of the locks from 180 meters to 360 meters each, as well as to bolster several major wing dams which narrow the river to keep the soybean barges moving and the sediment from settling. This construction would supplement the existing dredges which are already sucking 85 million cubic meters of sand and mud from the river's bank and bottom each year. Several different levels of "upgrade" for the river have been considered, but the most ambitious of them would purportedly reduce the cost of shipping soybeans by 4 to 8 cents per bushel. Some independent analysts think this is a pipe dream.

Around the same time the Mississippi plan was announced, the five governments of South America's La Plata Basin—Bolivia, Brazil, Paraguay, Argentina, and Uruguay—announced plans to dredge 13 million cubic meters of sand, mud, and rock from 233 sites along the Paraguay-Paraná River. That would be enough to fill a convoy of dump trucks 10,000 miles long. Here, the plan is to straighten natural river meanders in at least seven places, build dozens of locks, and construct a major port in the heart of the Pantanal—the world's largest wetland. The Paraguay-Paraná flows through the center of Brazil's burgeoning soybean heartland—second only to the

United States in production and exports. According to statements from the Brazilian State of Mato Grasso, this "Hidrovía" (water highway) will give a further boost to the region's soybean export capacity.

Lobbyists for both these projects argue that expanding the barge capacity of these rivers is necessary in order to improve competitiveness, grab world market share, and rescue farmers (either U.S. or Brazilian, depending on whom the lobbyists are addressing) from their worst financial crisis since the Great Depression. Chris Brescia, president of the Midwest River Coalition 2000, an alliance of commodity shippers that forms the primary lobbying force for the Mississippi plan, says, "The sooner we provide the waterway infrastructure, the sooner our family farmers will benefit." Some of his fellow lobbyists have even argued that these projects are essential to feeding the world (since the barges can then more easily speed the soybeans to the world's hungry masses) and to saving the environment (since the hungry masses will not have to clear rainforest to scratch out their own subsistence).

Probably very few people have had an opportunity to hear both pitches and compare them. But anyone who has may find something amiss with the argument that U.S. farmers will become more competitive versus their Brazilian counterparts, at the same time that Brazilian farmers will, for the same reasons, become more competitive with their U.S. counterparts. A more likely outcome is that farmers of these two nations will be pitted against each other in a costly race to maximize production, resulting in short-cut practices that essentially strip-mine their soil and throw long-term investments in the land to the wind. Farmers in Iowa will have stronger incentives to plow up land along stream banks, triggering faster erosion of topsoil. Their brethren in Brazil will find themselves needing to cut deeper into the savanna, also accelerating erosion. That will increase the flow of soybeans, all right—both north and south. But it will also further depress prices, so that even as the farmers are shipping more, they're getting less income per ton shipped. And in any case, increasing volume can't help the farmers survive in the long run, because sooner or later they will be swallowed by larger, corporate, farms that can make up for the smaller per-ton margins by producing even larger volumes.

So, how can the supporters of these river projects, who profess to be acting in the farmer's best interests, not notice the illogic of this form of competition? One explanation is that from the advocates' (as opposed to the farmers') standpoint, this competition isn't illogical at all—because the lobbyists aren't really representing farmers. They're working for the commodity processing, shipping, and trading firms who want the price of soybeans to fall, because these are the firms that buy the crops from the farmers. In fact, it is the same three agribusiness conglomerates—Archer Daniels Midland (ADM), Cargill, and Bunge—that are the top soybean processors and traders along both rivers.

Welcome to the global economy. The more brutally the U.S. and Brazilian farmers can batter each-other's prices (and standards of living) down, the greater the margin of profit these three giants gain. Meanwhile, another handful of companies controls the markets for genetically modified seeds, fertilizers, and herbicides used by the farmers—charging oligopolistically high prices both north and south of the equator.

In assessing what this proposed digging-up and reconfiguring of two of the world's great river basins really means, keep in mind that these projects will not be the activities of private businesses operating inside their own private property. These are proposed public works, to be undertaken at huge public expense. The motive is neither the plight of the family farmer nor any moral obligation to feed the world, but the opportunity to exploit poorly informed public sentiments about farmers' plights or hungry masses as a means of usurping public policies to benefit private interests. What gets thoroughly Big Muddied, in this usurping process, is that in addition to subjecting farmers to a gladiator-like attrition, these projects will likely bring a cascade of damaging economic, social, and ecological impacts to the very river basins being so expensively remodeled.

What's likely to happen if the lock and dam system along the Mississippi is expanded as proposed? The most obvious effect will be increased barge traffic, which will accelerate a less obvious cascade of events that has been underway for some time, according to Mike Davis of the Minnesota Department of Natural Resources. Much of the Mississippi River ecosystem involves aquatic rooted plants, like bullrush, arrowhead, and wild celery. Increased barge traffic will kick up more sediment, obscuring sunlight and reducing the depth to which plants can survive. Already, since the 1970s, the number of aquatic plant species found in some of the river has been cut from 23 to about half that, with just a handful thriving under the cloudier conditions. "Areas of the river have reached an ecological turning point," warns Davis. "This decline in plant diversity has triggered a drop in the invertebrate communities that live on these plants, as well as a drop in the fish, mollusk, and bird communities that depend on the diversity of insects and plants." On May 18, 2000, the U.S. Fish and Wildlife Service released a study saying that the Corps of Engineers project would threaten the 300 species of migratory birds and 12 species of fish in the Mississippi watershed, and could ultimately push some into extinction. "The least tern, the pallid sturgeon, and other species that evolved with the ebbs and flows, sandbars and depths, of the river are progressively eliminated or forced away as the diversity of the river's natural habitats is removed to maximize the barge habitat," says Davis.

The outlook for the Hidrovía project is similar. Mark Robbins, an ornithologist at the Natural History Museum at the University of Kansas, calls it "a key step in creating a Florida Everglades-like scenario of destruction in the Pantanal, and an American Great Plains-like scenario in the Cerrado in southern Brazil." The Paraguay-Paraná feeds the Pantanal wetlands, one of the most diverse habitats on the planet, with its populations of woodstorks, snailkites, limpkins, jabirus, and more than 650 other species of birds, as well as more than 400 species of fish and hundreds of other less-studied plants, mussels, and marshland organisms. As the river is dredged and the banks are built up to funnel the surrounding wetlands water into the navigation path, bird nesting habitat and fish spawning grounds will be eliminated, damaging the indigenous and other traditional societies that depend on these resources. Increased barge traffic will suppress river species here just as it will on the Mississippi. Meanwhile, herbicide-intensive soybean monocultures—on farms so enormous that they dwarf even the biggest operations in the U.S. Midwest—are rapidly replacing diverse grasslands in the fragile Cerrado. The heavy plowing and periodic absence of ground cover associated with such farming erodes 100 million tons of soil per year. Robbins notes that "compared to the Mississippi, this southern river system and surrounding grassland is several orders of magnitude more diverse and has suffered considerably less, so there is much more at stake."

Supporters of such massive disruption argue that it is justified because it is the most "efficient" way to do business. The perceived efficiency of such farming might be compared to the perceived efficiency of an energy system based on coal. Burning coal looks very efficient if you ignore its long-term impact on air quality and climate stability. Similarly, large farms look more efficient than small farms if you don't count some of their largest costs—the loss of the genetic diversity that underpins agriculture, the pollution caused by agro-chemicals, and the dislocation of rural cultures. The simultaneous demise of small, independent farmers and rise of multinational food giants is troubling not just for those who empathize with dislocated farmers, but for anyone who eats.

An Endangered Species

Nowadays most of us in the industrialized countries don't farm, so we may no longer really understand that way of life. I was born in the apple orchard and dairy country of Dutchess County, New York, but since age five have spent most of my life in New York City—while most of the farms back in Dutchess County have given way to spreading subdivisions. It's also hard for those of us who get our food from supermarket shelves or drive-thru windows to know how dependent we are on the viability of rural communities.

Whether in the industrial world, where farm communities are growing older and emptier, or in developing nations where population growth is pushing the number of farmers continually higher and each generation is inheriting smaller family plots, it is becoming harder and harder to make a living as a farmer. A combination of falling incomes, rising debt, and worsening rural poverty is forcing more people to either abandon farming as their primary activity or to leave the countryside altogether—a bewildering juncture, considering that farmers produce perhaps the only good that the human race cannot do without.

Since 1950, the number of people employed in agriculture has plummeted in all industrial nations, in some regions by more than 80 percent. Look at the numbers, and you might think farmers are being singled out by some kind of virus:

- In Japan, more than half of all farmers are over 65 years old; in the United States, farmers over 65 outnumber those under 35 by three to one. (Upon retirement or death, many will pass the farm on to children who live in the city and have no interest in farming themselves.)
- In New Zealand, officials estimate that up to 6,000 dairy farms will disappear during the next 10 to 15 years—dropping the total number by nearly 40 percent.
- In Poland, 1.8 million farms could disappear as the country is absorbed into the European Union—dropping the total number by 90 percent.
- In Sweden, the number of farms going out of business in the next decade is expected to reach about 50 percent.

- In the Philippines, Oxfam estimates that over the next few years the number of farm households in the corn–producing region of Mindanao could fall by some 500,000—a 50 percent loss.
- In the United States, where the vast majority of people were farmers at the time of the American Revolution, fewer people are now full-time farmers (less than 1 percent of the population) than are full-time prisoners.
- In the U.S. states of Nebraska and Iowa, between a fifth and a third of farmers are expected to be out of business within two years.

Of course, the declining numbers of farmers in industrial nations does not imply a decline in the importance of the farming sector. The world still has to eat (and 80 million more mouths to feed each year than the year before), so smaller numbers of farmers mean larger farms and greater concentration of ownership. Despite a precipitous plunge in the number of people employed in farming in North America, Europe, and East Asia, half the world's people still make their living from the land. In sub-Saharan Africa and South Asia, more than 70 percent do. In these regions, agriculture accounts, on average, for half of total economic activity.

Some might argue that the decline of farmers is harmless, even a blessing, particularly for less developed nations that have not yet experienced the modernization that moves peasants out of backwater rural areas into the more advanced economies of the cities. For most of the past two centuries, the shift toward fewer farmers has generally been assumed to be a kind of progress. The substitution of high-powered diesel tractors for slow-moving women and men with hoes, or of large mechanized industrial farms for clusters of small "old fashioned" farms, is typically seen as the way to a more abundant and affordable food supply. Our urban-centered society has even come to view rural life, especially in the form of small family-owned businesses, as backwards or boring, fit only for people who wear overalls and go to bed early—far from the sophistication and dynamism of the city.

Urban life does offer a wide array of opportunities, attractions, and hopes—some of them falsely created by urban-oriented commercial media—that many farm families decide to pursue willingly. But city life often turns out to be a disappointment, as displaced farmers find themselves lodged in crowded slums, where unemployment and ill-health are the norm and where they are worse off than they were back home. Much evidence suggests that farmers aren't so much being lured to the city as they are being driven off their farms by a variety of structural changes in the way the global food chain operates. Bob Long, a rancher in McPherson County, Nebraska, stated in a recent *New York Times* article that passing the farm onto his son would be nothing less than "child abuse."

As long as cities are under the pressure of population growth (a situation expected to continue at least for the next three or four decades), there will always be pressure for a large share of humanity to subsist in the countryside. Even in highly urbanized North America and Europe, roughly 25 percent of the population—275 million people—still reside in rural areas. Meanwhile, for the 3 billion Africans, Asians, and Latin Americans who remain in the countryside—and who will be there for the foreseeable future—the marginalization of farmers has set up a vicious cycle of low educational achievement, rising infant mortality, and deepening mental distress.

Hired Hands on Their Own Land

In the 18th and 19th centuries, farmers weren't so trapped. Most weren't wealthy, but they generally enjoyed stable incomes and strong community ties. Diversified farms yielded a range of raw and processed goods that the farmer could typically sell in a local market. Production costs tended to be much lower than now, as many of the needed inputs were home-grown: the farmer planted seed that he or she had saved from the previous year, the farm's cows or pigs provided fertilizer, and the diversity of crops—usually a large range of grains, tubers, vegetables, herbs, flowers, and fruits for home use as well as for sale—effectively functioned as pest control.

Things have changed, especially in the past half-century, according to Iowa State agricultural economist Mike Duffy. "The end of World War II was a watershed period," he says. "The widespread introduction of chemical fertilizers and synthetic pesticides, produced as part of the war effort, set in motion dramatic changes in how we farm—and a dramatic decline in the number of farmers." In the post-war period, along with increasing mechanization, there was an increasing tendency to "outsource" pieces of the work that the farmers had previously done themselves—from producing their own fertilizer to cleaning and packaging their harvest. That outsourcing, which may have seemed like a welcome convenience at the time, eventually boomeranged: at first it enabled the farmer to increase output, and thus profits, but when all the other farmers were doing it too, crop prices began to fall.

Before long, the processing and packaging businesses were adding more "value" to the purchased product than the farmer, and it was those businesses that became the

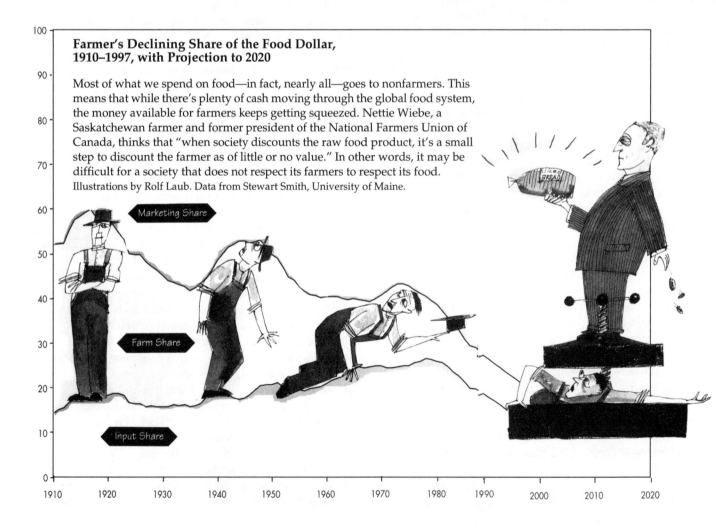

Farmer's Declining Share of the Food Dollar, 1910–1997, with Projection to 2020

Most of what we spend on food—in fact, nearly all—goes to nonfarmers. This means that while there's plenty of cash moving through the global food system, the money available for farmers keeps getting squeezed. Nettie Wiebe, a Saskatchewan farmer and former president of the National Farmers Union of Canada, thinks that "when society discounts the raw food product, it's a small step to discount the farmer as of little or no value." In other words, it may be difficult for a society that does not respect its farmers to respect its food. Illustrations by Rolf Laub. Data from Stewart Smith, University of Maine.

Marketing Share

Farm Share

Input Share

ConAgra: *Vertical Integration, Horizontal Concentration, Global Omnipresence*

Three conglomerates (ConAgra/DuPont, Cargill/Monsanto, and Novartis/ADM) dominate virtually every link in the North American (and increasingly, the global) food chain. Here's a simplified diagram of one conglomerate.

KEY: Vertical integration of production links, from seed to supermarket ◄► Concentration within a link

INPUTS
Distribution of farm chemicals, machinery, fertilizer, and seed

3 companies dominate North American farm machinery sector
6 companies control 63% of global pesticide market
4 companies control 69% of North American seed corn market
3 companies control 71% of Canadian nitrogen fertilizer capacity
ConAgra distributes all of these inputs, and is in a joint venture with DuPont to distribute DuPont's transgenic high-oil corn seed.

FARMS

The farm sector is rapidly consolidating in the industrial world, as farms "get big or get out." Many go under contract with **ConAgra** and other conglomerates; others just go under. In the past 50 years, the number of farmers has declined by 86% in Germany, 85% in France, 85% in Japan, 64% in the U.S., 59% in South Korea, and 59% in the U.K.

GRAIN COLLECTION

A proposed merger of Cargill and Continental Grain will control half of the global grain trade; **ConAgra** has about one-quarter.

GRAIN MILLING

ConAgra and 3 other companies account for 62% of the North American market.

PRODUCTION OF BEEF, PORK, TURKEY, CHICKEN, AND SEAFOOD

ConAgra ranks 3rd in cattle feeding and 5th in broiler production.

ConAgra Poultry, Tyson Foods, Perdue, and 3 other companies control 60% of U.S. chicken production

PROCESSING OF BEEF, PORK, TURKEY, CHICKEN, AND SEAFOOD

IBP, **ConAgra,** Cargill, and Farmland control 80% of U.S. beef packing

Smithfield, **ConAgra**, and 3 other companies control 75% of U.S. pork packing

SUPERMARKETS

ConAgra divisions own Wesson oil, Butterball turkeys, Swift Premium meats, Peter Pan peanut butter, Healthy Choice diet foods, Hunt's tomato sauce, and about 75 other major brands.

dominant players in the food industry. Instead of farmers outsourcing to contractors, it became a matter of large food processors buying raw materials from farmers, on the processors' terms. Today, most of the money is in the work the farmer no longer does—or even controls. In the United States, the share of the consumer's food dollar that trickles back to the farmer has plunged from nearly 40 cents in 1910 to just above 7 cents in 1997, while the shares going to input (machinery, agrochemicals, and seeds) and marketing (processing, shipping, brokerage, advertising, and retailing) firms have continued to expand. (See graph "Farmer's Declining Share of the Food Dollar") The typical U.S. wheat farmer, for instance, gets just 6 cents of the dollar spent on a loaf of bread—so when you buy that loaf, you're paying about as much for the wrapper as for the wheat.

Ironically, then, as U.S. farms became more mechanized and more "productive," a self-destructive feedback loop was set in motion: over-supply and declining crop prices cut into farmers' profits, fueling a demand for more technology aimed at making up for shrinking margins by increasing volume still more. Output increased dramatically, but expenses (for tractors, combines, fertilizer, and seed) also ballooned—while the commodity prices stagnated or declined. Even as they were looking more and more modernized, the farmers were becoming less and less the masters of their own domain.

On the typical Iowa farm, the farmer's profit margin has dropped from 35 percent in 1950 to 9 percent today. In order to generate the same income, this farm would need to be roughly four times as large today as in 1950—or the farmer would need to get a night job. And that's precisely what we've seen in most industrialized nations: fewer farmers on bigger tracts of land producing a greater share of the total food supply. The farmer with declining margins buys out his neighbor and expands or risks being cannibalized himself.

There is an alternative to this huge scaling up, which is to buck the trend and bring some of the input-supplying and post-harvest processing—and the related profits—back onto the farm. But more self-sufficient farming would be highly unpopular with the industries that now make lucrative profits from inputs and processing. And since these industries have much more political clout than the farmers do, there is little support for rescuing farmers from their increasingly servile condition—and the idea has been largely forgotten. Farmers continue to get the message that the only way to succeed is to get big.

The traditional explanation for this constant pressure to "get big or get out" has been that it improves the efficiency of the food system—bigger farms replace smaller farms, because the bigger farms operate at lower costs. In some respects, this is quite true. Scaling up may allow a farmer to spread a tractor's cost over greater acreage, for example. Greater size also means greater leverage in purchasing inputs or negotiating loan rates—increasingly important as satellite-guided combines and other equipment make farming more and more capital-intensive. But these economies of scale typically level off. Data for a wide range of crops produced in the United States show that the lowest production costs are generally achieved on farms that are much smaller than the typical farm now is. But large farms can tolerate lower margins, so while they may not *produce* at lower cost, they can afford to *sell* their crops at lower cost, if forced to do so—as indeed they are by the food processors who buy from them. In short, to the extent that a giant farm has a financial benefit over a small one, it's a benefit that goes only to the processor—not to the farmer, the farm community, or the environment.

This shift of the food dollar away from farmers is compounded by intense concentration in every link of the food chain—from seeds and herbicides to farm finance and retailing. In Canada, for example, just three companies control over 70 percent of fertilizer sales, five banks provide the vast majority of agricultural credit, two companies control over 70 percent of beef packing, and five companies dominate food retailing. The merger of Philip Morris and Nabisco will create an empire that collects nearly 10 cents of every dollar a U.S. consumer spends on food, according to a company spokesperson. Such high concentration can be deadly for the bottom line, allowing agribusiness firms to extract higher prices for the products farmers buy from them, while offering lower prices for the crop they buy from the farmers.

An even more worrisome form of concentration, according to Bill Heffernan, a rural sociologist at the University of Missouri, is the emergence of several clusters of firms that—through mergers, takeovers, and alliances with other links in the food chain—now possess "a seamless and fully vertically integrated control of the food system from gene to supermarket shelf." (See diagram "ConAgra") Consider the recent partnership between Monsanto and Cargill, which controls seeds, fertilizers, pesticides, farm finance, grain collection, grain processing, livestock feed processing, livestock production, and slaughtering, as well as some well-known processed food brands. From the standpoint of a company like Cargill, such alliances yield tremendous control over costs and can therefore be extremely profitable.

But suppose you're the farmer. Want to buy seed to grow corn? If Cargill is the only buyer of corn in a hundred mile radius, and Cargill is only buying a particular Monsanto corn variety for its mills or elevators or feedlots, then if you don't plant Monsanto's seed you won't have a market for your corn. Need a loan to buy the seed? Go to Cargill-owned Bank of Ellsworth, but be sure to let them know which seed you'll be buying. Also mention that you'll be buying Cargill's Saskferco brand fertilizer. OK, but once the corn is grown, you don't like the idea of having to sell to Cargill at the prices it dictates? Well, maybe you'll feed the corn to your pigs, then, and sell them to the highest bidder. No problem—Cargill's Excel Corporation buys pigs, too. OK, you're moving to the city, and renouncing the farm life! No more home-made grits for breakfast, you're buying corn flakes. Well, good news: Cargill Foods supplies corn flour to the top cereal makers. You'll notice, though, that all the big brands of corn flakes seem to have pretty much the same hefty price per ounce. After all, they're all made by the agricultural oligopoly.

As these vertical food conglomerates consolidate, Heffernan warns, "there is little room left in the global food system for independent farmers"—the farmers being increasingly left with "take it or leave it" contracts from the remaining conglomerates. In the last two decades, for example, the share of American agricultural output produced under contract has more than tripled, from 10 percent to 35 percent—and this doesn't include the contracts that farmers must sign to plant genetically engineered seed. Such centralized control of the food system, in which farmers are in effect reduced to hired hands on their own land, reminds Heffernan of the Soviet-style state farms, but with the Big Brother role now being played by agribusiness executives. It is also reminiscent of the "company store" which once dominated small American mining or factory towns, except that if you move out of town now, the store is still with you. The company store has gone global.

With the conglomerates who own the food dollar also owning the political clout, it's no surprise that agricultural policies—including subsidies, tax breaks, and environmental legislation at both the national and international levels—do not generally favor the farms. For example, the conglomerates command growing influence over both private and public agricultural research priorities, which might explain why the U.S. Department of Agriculture (USDA), an agency ostensibly beholden to farmers, would help to develop the seed-sterilizing Terminator technology—a biotechnology that offers farmers only greater dependence on seed companies. In some cases the influence is indirect, as manifested in government funding decisions, while in others it is more blatant. When Novartis provided $25 million to fund a research partnership with the plant biology department of the University of California at Berkeley, one of the conditions was that Novartis has the first right of refusal for any patentable inventions. Under those circumstances, of course, the UC officials—mindful of where their funding comes from—have strong incentives to give more attention to technologies like the Terminator seed, which shifts profit away from the farmer, than to technologies that directly benefit the farmer or the public at large.

Even policies that are touted to be in the best interest of farmers, like liberalized trade in agricultural products, are increasingly shaped by non-farmers. Food traders, processors, and distributors, for example, were some of the principal architects of recent revisions to the General Agreement on Trade and Tariffs (GATT)—the World Trade Organization's predecessor—that paved the way for greater trade flows in agricultural commodities. Before these revisions, many countries had mechanisms for assuring that their farmers wouldn't be driven out of their own domestic markets by predatory global traders. The traders, however, were able to do away with those protections.

The ability of agribusiness to slide around the planet, buying at the lowest possible price and selling at the highest, has tended to tighten the squeeze already put in place by economic marginalization, throwing every farmer on the planet into direct competition with every other farmer. A recent UN Food and Agriculture Organization assessment of the experience of 16 developing nations in implementing the latest phase of the GATT concluded that "a common reported concern was with a general trend towards the concentration of farms," a process that tends to further marginalize small producers and exacerbate rural poverty and unemployment. The sad irony, according to Thomas Reardon, of Michigan State University, is that while small farmers in all reaches of the world are increasingly affected by cheap, heavily subsidized imports of foods from outside of their traditional rural markets, they are nonetheless often excluded from opportunities to participate in food exports themselves. To keep down transaction costs and to keep processing standardized, exporters and other downstream players prefer to buy from a few large producers, as opposed to many small producers.

As the global food system becomes increasingly dominated by a handful of vertically integrated, international corporations, the servitude of the farmer points to a broader society-wide servitude that OPEC-like food cartels could impose, through their control over food prices and food quality. Agricultural economists have already noted that the widening gap between retail food prices and farm prices in the 1990s was due almost exclusively to exploitation of market power, and not to extra services provided by processors and retailers. It's questionable whether we should pay as much for a bread wrapper as we do for the nutrients it contains. But beyond this, there's a more fundamental question. Farmers are professionals, with extensive knowledge of their local soils, weather, native plants, sources of fertilizer or mulch, native pollinators, ecology, and community. If we are to have a world where the land is no longer managed by such professionals, but is instead managed by distant corporate bureaucracies interested in extracting maximum output at minimum cost, what kind of food will we have, and at what price?

Agrarian Services

No question, large industrial farms can produce lots of food. Indeed, they're designed to maximize quantity. But when the farmer becomes little more than the lowest-cost producer of raw materials, more than his own welfare will suffer. Though the farm sector has lost power and profit, it is still the one link in the agrifood chain accounting for the largest share of agriculture's public goods—including half the world's jobs, many of its most vital communities, and many of its most diverse landscapes. And in providing many of these goods, small farms clearly have the advantage.

Local economic and social stability: Over half a century ago, William Goldschmidt, an anthropologist working at the USDA, tried to assess how farm structure and size affect the health of rural communities. In California's San Joaquin Valley, a region then considered to be at the cutting edge of agricultural industrialization, he identified two small towns that were alike in all basic economic and geographic dimensions, including value of agricultural production, except in farm size. Comparing the two, he found an inverse correlation between the sizes of the farms and the well-being of the communities they were a part of.

The small-farm community, Dinuba, supported about 20 percent more people, and at a considerably higher level of living—including lower poverty rates, lower levels of economic and social class distinctions, and a lower crime rate—than the large-farm community of Arvin. The majority of Dinuba's residents were independent entrepreneurs, whereas fewer than 20 percent of Arvin's residents were—most of the others being agricultural laborers. Dinuba had twice as many business establishments as Arvin, and did 61 percent more retail business. It had more schools, parks, newspapers, civic organizations, and churches, as well as better physical infrastructure—paved streets, sidewalks, garbage disposal, sewage disposal and other public services. Dinuba also had more institutions for democratic decision making, and a much broader participation by its citizens. Political scientists have long recognized that a broad base of independent entrepreneurs and property owners is one of the keys to a healthy democracy.

The distinctions between Dinuba and Arvin suggest that industrial agriculture may be limited in what it can do for a community. Fewer (and less meaningful) jobs, less local spending, and a hemorrhagic flow of profits to absentee landowners and distant suppliers means that industrial farms can actually be a net drain on the local economy. That hypothesis has been corroborated by Dick Levins, an agricultural economist at the University of Minnesota. Levins studied the economic receipts from Swift County, Iowa, a typical Midwestern corn and soybean community, and found that although total farm sales are near an all-time high, farm income there has been dismally low—and that many of those who were once the financial stalwarts of the community are now deeply in debt. "Most of the U.S. Corn Belt, like Swift County, is a colony, owned and operated by people who don't live there and for the benefit of those who don't live there," says Levin. In fact, most of the land in Swift County is rented, much of it from absentee landlords.

This new calculus of farming may be eliminating the traditional role of small farms in anchoring rural economies—the kind of tradition, for example, that we saw in the emphasis given to the support of small farms by Japan, South Korea, and Taiwan following World War II. That emphasis, which brought radical land reforms and targeted investment in rural areas, is widely cited as having been a major stimulus to the dramatic economic boom those countries enjoyed.

Not surprisingly, when the economic prospects of small farms decline, the social fabric of rural communities begins to tear. In the United States, farming families are more than twice as likely as others to live in poverty. They have less education and lower rates of medical protection, along with higher rates of infant mortality, alcoholism, child abuse, spousal abuse, and mental stress. Across Europe, a similar pattern is evident. And in sub-Saharan Africa, sociologist Deborah Bryceson of the Netherlands-based African Studies Centre has studied the dislocation of small farmers and found that "as de-agrarianization proceeds, signs of social dysfunction associated with urban areas [including petty crime and breakdowns of family ties] are surfacing in villages."

People without meaningful work often become frustrated, but farmers may be a special case. "More so than other occupations, farming represents a way of life and defines who you are," says Mike Rosemann, a psychologist who runs a farmer counseling network in Iowa. "Losing the family farm, or the prospect of losing the family farm, can generate tremendous guilt and anxiety, as if one has failed to protect the heritage that his ancestors worked to hold onto." One measure of the despair has been a worldwide surge in the number of farmers committing suicide. In 1998, over 300 cotton farmers in Andhra Pradesh, India, took their lives by swallowing pesticides that they had gone into debt to purchase but that had nonetheless failed to save their crops. In Britain, farm workers are two-and-a-half times more likely to commit suicide than the rest of the population. In the United States, official statistics say farmers are now five times as likely to commit suicide as to die from farm accidents, which have been traditionally the most frequent cause of unnatural death for them. The true number may be even higher, as suicide hotlines report that they often receive calls from farmers who want to know which sorts of accidents (Falling into the blades of a combine? Getting shot while hunting?) are least likely to be investigated by insurance companies that don't pay claims for suicides.

Whether from despair or from anger, farmers seem increasingly ready to rise up, sometimes violently, against government, wealthy landholders, or agribusiness giants. In recent years we've witnessed the Zapatista revolution in Chiapas, the seizing of white-owned farms by landless blacks in Zimbabwe, and the attacks of European farmers on warehouses storing genetically engineered seed. In the book *Harvest of Rage*, journalist Joel Dyer links the 1995 Oklahoma City bombing that killed nearly 200 people—

In the Developing World, an Even Deeper Farm Crisis

"One would have to multiply the threats facing family farmers in the United States or Europe five, ten, or twenty times to get a sense of the handicaps of peasant farmers in less developed nations," says Deborah Bryceson, a senior research fellow at the African Studies Centre in the Netherlands. Those handicaps include insufficient access to credit and financing, lack of roads and other infrastructure in rural areas, insecure land tenure, and land shortages where population is dense.

Three forces stand out as particularly challenging to these peasant farmers:

Structural adjustment requirements, imposed on indebted nations by international lending institutions, have led to privatization of "public commodity procurement boards" that were responsible for providing public protections for rural economies. "The newly privatized entities are under no obligation to service marginal rural areas," says Rafael Mariano, chairman of a Filipino farmers' union. Under the new rules, state protections against such practices as dumping of cheap imported goods (with which local farmers can't compete) were abandoned at the same time that state provision of health care, education, and other social services was being reduced.

Trade liberalization policies associated with structural adjustment have reduced the ability of nations to protect their agricultural economies even if they want to. For example, the World Trade Organization's Agreement on Agriculture will forbid domestic price support mechanisms and tariffs on imported goods—some of the primary means by which a country can shield its own farmers from overproduction and foreign competition.

The growing emphasis on agricultural grades and standards—the standardizing of crops and products so they can be processed and marketed more "efficiently"—has tended to favor large producers, and to marginalize smaller ones. Food manufacturers and supermarkets have emerged as the dominant entities in the global agri-food chain, and with their focus on brand consistency, ingredient uniformity, and high volume, smaller producers often are unable to deliver—or aren't even invited to bid.

Despite these daunting conditions, many peasant farmers tend to hold on long after it has become clear that they can't compete. One reason, says Peter Rosset of the Institute for Food and Development Policy, is that "even when it gets really bad, they will cling to agriculture because of the fact that it at least offers some degree of food security—that you can feed yourself." But with the pressures now mounting, particularly as export crop production swallows more land, even that fallback is lost.

as well as the rise of radical right and antigovernment militias in the U.S. heartland—to a spreading despair and anger stemming from the ongoing farm crisis. Thomas Homer-Dixon, director of the Project on Environment, Population, and Security at the University of Toronto, regards farmer dislocation, and the resulting rural unemployment and poverty, as one of the major security threats for the coming decades. Such dislocation is responsible for roughly half of the growth of urban populations across the Third World, and such growth often occurs in volatile shantytowns that are already straining to meet the basic needs of their residents. "What was an extremely traumatic transition for Europe and North America from a rural society to an urban one is now proceeding at two to three times that speed in developing nations," says Homer-Dixon. And, these nations have considerably less industrialization to absorb the labor. Such an accelerated transition poses enormous adjustment challenges for India and China, where perhaps a billion and a half people still make their living from the land.

Ecological stability: In the Andean highlands, a single farm may include as many as 30 to 40 distinct varieties of potato (along with numerous other native plants), each having slightly different optimal soil, water, light, and temperature regimes, which the farmer—given enough time—can manage. (In comparison, in the United States, just four closely related varieties account for about 99 percent of all the potatoes produced.) But, according to Karl Zimmerer, a University of Wisconsin sociologist, declining farm incomes in the Andes force more and more growers into migrant labor forces for part of the year, with serious effects on farm ecology. As time becomes constrained, the farmer manages the system more homogenously—cutting back on the number of traditional varieties (a small home garden of favorite culinary varieties may be the last refuge of diversity), and scaling up production of a few commercial varieties. Much of the traditional crop diversity is lost.

Complex farm systems require a highly sophisticated and intimate knowledge of the land—something small-scale, full-time farmers are more able to provide. Two or three different crops that have different root depths, for example, can often be planted on the same piece of land, or crops requiring different drainage can be planted in close proximity on a tract that has variegated topography. But these kinds of cultivation can't be done with heavy tractors moving at high speed. Highly site-specific and management-intensive cultivation demands ingenuity and awareness of local ecology, and can't be achieved by heavy equipment and heavy applications of agrochemicals. That isn't to say that being small is always sufficient to ensure ecologically sound food production, because economic adversity can drive small farms, as well as big ones, to compromise sustainable food production by transmogrifying the craft of land stewardship into the crude labor of commodity production. But a large-scale, highly mechanized farm is simply not equipped to preserve landscape complexity. Instead, its normal modus is to use blunt management tools, like crops that have been genetically engineered to churn out insecticides, which obviate the need to scout the field to see if spraying is necessary at all.

In the U.S. Midwest, as farm size has increased, cropping systems have gotten more simplified. Since 1972, the number of counties with more than 55 percent of their acreage planted in corn and soybeans has nearly tripled, from 97 to 267. As farms scaled up, the great simplicity of managing the corn-soybean rotation—an 800 acre farm, for instance, may require no more than a couple of weeks planting in the spring and a few weeks harvesting in the fall—became its big selling point. The various arms of the agricultural economy in the region, from extension services to grain elevators to seed suppliers, began to solidify around this corn-soybean rotation, reinforcing the farmers' movement away from other crops. Fewer and fewer farmers kept livestock, as beef and hog production became "economical" only in other parts of the country where it was becoming more concentrated. Giving up livestock meant eliminating clover, pasture mixtures, and a key source of fertilizer in the Midwest, while creating tremendous manure concentrations in other places.

But the corn and soybean rotation—one monoculture followed by another—is extremely inefficient or "leaky" in its use of applied fertilizer, since low levels of biodiversity tend to leave a range of vacant niches in the field, including different root depths and different nutrient preferences. Moreover, the Midwest's shift to monoculture has subjected the country to a double hit of nitrogen pollution, since not only does the removal and concentration of livestock tend to dump inordinate amounts of feces in the places (such as Utah and North Carolina) where the livestock operations are now located, but the monocultures that remain in the Midwest have much poorer nitrogen retention than they would if their cropping were more complex. (The addition of just a winter rye crop to the corn-soy rotation has been shown to reduce nitrogen runoff by nearly 50 percent.) And maybe this disaster-in-the-making should really be regarded as a triple hit, because in addition to contaminating Midwestern water supplies, the runoff ends up in the Gulf of Mexico, where the nitrogen feeds massive algae blooms. When the algae die, they are decomposed by bacteria, whose respiration depletes the water's oxygen—suffocating fish, shellfish,

and all other life that doesn't escape. This process periodically leaves 20,000 square kilometers of water off the coast of Louisiana biologically dead. Thus the act of simplifying the ecology of a field in Iowa can contribute to severe pollution in Utah, North Carolina, Louisiana, *and* Iowa.

The world's agricultural biodiversity—the ultimate insurance policy against climate variations, pest outbreaks, and other unforeseen threats to food security—depends largely on the millions of small farmers who use this diversity in their local growing environments. But the marginalization of farmers who have developed or inherited complex farming systems over generations means more than just the loss of specific crop varieties and the knowledge of how they best grow. "We forever lose the best available knowledge and experience of place, including what to do with marginal lands not suited for industrial production," says Steve Gleissman, an agroecologist at the University of California at Santa Cruz. The 12 million hogs produced by Smithfield Foods Inc., the largest hog producer and processor in the world and a pioneer in vertical integration, are nearly genetically identical and raised under identical conditions—regardless of whether they are in a Smithfield feedlot in Virginia or Mexico.

As farmers become increasingly integrated into the agribusiness food chain, they have fewer and fewer controls over the totality of the production process—shifting more and more to the role of "technology applicators," as opposed to managers making informed and independent decisions. Recent USDA surveys of contract poultry farmers in the United States found that in seeking outside advice on their operations, these farmers now turn first to bankers and then to the corporations that hold their contracts. If the contracting corporation is also the same company that is selling the farm its seed and fertilizer, as is often the case, there's a strong likelihood that the company's procedures will be followed. That corporation, as a global enterprise with no compelling local ties, is also less likely to be concerned about the pollution and resource degradation created by those procedures, at least compared with a farmer who is rooted in that community. Grower contracts generally disavow any environmental liability.

And then there is the ecological fallout unique to large-scale, industrial agriculture. Colossal confined animal feeding operations (CAFOs)—those "other places" where livestock are concentrated when they are no longer present on Midwestern soy/corn farms—constitute perhaps the most egregious example of agriculture that has, like a garbage barge in a goldfish pond, overwhelmed the scale at which an ecosystem can cope. CAFOs are increasingly the norm in livestock production, because, like crop monocultures, they allow the production of huge populations of animals which can be slaughtered and marketed at rock-bottom costs. But the disconnection between the livestock and the land used to produce their feed means that such CAFOs generate gargantuan amounts of waste, which the surrounding soil cannot possibly absorb. (One farm in Utah will raise over five million hogs in a year, producing as much waste each day as the city of Los Angeles.) The waste is generally stored in large lagoons, which are prone to leak and even spill over during heavy storms. From North Carolina to South Korea, the overwhelming stench of these lagoons—a combination of hydrogen sulfide, ammonia, and methane gas that smells like rotten eggs—renders miles of surrounding land uninhabitable.

A different form of ecological disruption results from the conditions under which these animals are raised. Because massive numbers of closely confined livestock are highly susceptible to infection, and because a steady diet of antibiotics can modestly boost animal growth, overuse of antibiotics has become the norm in industrial animal production. In recent months, both the Centers for Disease Control and Prevention in the United States and the World Health Organization have identified such industrial feeding operations as principal causes of the growing antibiotic resistance in food-borne bacteria like *salmonella* and *campylobacter*. And as decisionmaking in the food chain grows ever more concentrated—confined behind fewer corporate doors—there may be other food safety issues that you won't even hear about, particularly in the burgeoning field of genetically modified organisms (GMOs). In reaction to growing public concern over GMOs, a coalition that ingenuously calls itself the "Alliance for Better Foods"—actually made up of large food retailers, food processors, biotech companies and corporate-financed farm organizations—has launched a $50 million public "educational" campaign, in addition to giving over $676,000 to U.S. lawmakers and political parties in 1999, to head off the mandatory labeling of such foods.

Perhaps most surprising, to people who have only casually followed the debate about small-farm values versus factory-farm "efficiency," is the fact that a wide body of evidence shows that small farms are actually more productive than large ones—by as much as 200 to 1,000 percent greater output per unit of area. How does this jive with the often-mentioned productivity advantages of large-scale mechanized operations? The answer is simply that those big-farm advantages are always calculated on the basis of how much of *one crop* the land will yield per acre. The greater productivity of a smaller, more complex farm, however, is calculated on the basis of how much food *overall* is produced per acre. The smaller farm can grow several crops utilizing different root depths, plant heights, or nutrients, on the same piece of land simultaneously. It is this "polyculture" that offers the small farm's productivity advantage.

To illustrate the difference between these two kinds of measurement, consider a large Midwestern corn farm. That farm may produce more corn per acre than a small farm in which the corn is grown as part of a polyculture that also includes beans, squash, potato, and "weeds" that serve as fodder. But in overall output, the polycrop—under close supervision by a knowledgeable farmer—produces much more food overall, whether you measure in weight, volume, bushels, calories, or dollars.

The inverse relationship between farm size and output can be attributed to the more efficient use of land, water, and other agricultural resources that small operations afford, including the efficiencies of intercropping various plants in the same field, planting multiple times during the year, targeting irrigation, and integrating crops and livestock. So in terms of converting inputs into outputs, society would be better off with small-scale farmers. And as population continues to grow in many nations, and the agricultural resources per person continue to shrink, a small farm structure for agriculture may be central to meeting future food needs.

Rebuilding Foodsheds

Look at the range of pressures squeezing farmers, and it's not hard to understand the growing desperation. The situation has become explosive, and if stabilizing the erosion of farm culture and ecology is now critical not just to farmers but to everyone who eats, there's still a very challenging question as to what strategy can work. The agribusiness giants are deeply entrenched now, and scattered protests could have as little effect on them as a mosquito bite on a tractor. The prospects for farmers gaining political strength on their own seem dim, as their numbers—at least in the industrial countries—continue to shrink.

A much greater hope for change may lie in a joining of forces between farmers and the much larger numbers of other segments of society that now see the dangers, to their own particular interests, of continued restructuring of the countryside. There are a couple of prominent models for such coalitions, in the constituencies that have joined forces to fight the Mississippi River Barge Capacity and Hidrovía Barge Capacity projects being pushed forward in the name of global soybean productivity.

The American group has brought together at least the following riverbedfellows:

- National environmental groups, including the Sierra Club and National Audubon Society, which are alarmed at the prospect of a public commons being damaged for the profit of a small commercial interest group;
- Farmers and farmer advocacy organizations, concerned about the inordinate power being wielded by the agribusiness oligopoly;
- Taxpayer groups outraged at the prospect of a corporate welfare payout that will drain more than $1 billion from public coffers;
- Hunters and fishermen worried about the loss of habitat;

157

- Biologists, ecologists, and birders concerned about the numerous threatened species of birds, fish, amphibians, and plants;
- Local-empowerment groups concerned about the impacts of economic globalization on communities;
- Agricultural economists concerned that the project will further entrench farmers in a dependence on the export of low-cost, bulk commodities, thereby missing valuable opportunities to keep money in the community through local milling, canning, baking, and processing.

A parallel coalition of environmental groups and farmer advocates has formed in the Southern hemisphere to resist the Hidrovía expansion. There too, the river campaign is part of a larger campaign to challenge the hegemony of industrial agriculture. For example, a coalition has formed around the Landless Workers Movement, a grassroots organization in Brazil that helps landless laborers to organize occupations of idle land belonging to wealthy landlords. This coalition includes 57 farm advocacy organizations based in 23 nations. It has also brought together environmental groups in Latin America concerned about the related ventures of logging and cattle ranching favored by large landlords; the mayors of rural towns who appreciate the boost that farmers can give to local economies; and organizations working on social welfare in Brazil's cities, who see land occupation as an alternative to shantytowns.

The Mississippi and Hidrovía projects, huge as they are, still constitute only two of the hundreds of agro-industrial developments being challenged around the world. But the coalitions that have formed around them represent the kind of focused response that seems most likely to slow the juggernaut, in part because the solutions these coalitions propose are not vague or quixotic expressions of idealism, but are site-specific and practical. In the case of the alliance forming around the Mississippi River project, the coalition's work has included questioning the assumptions of the Corps of Engineers analysis, lobbying for stronger antitrust examination of agribusiness monopolies, and calling for modification of existing U.S. farm subsidies, which go disproportionately to large farmers. Environmental groups are working to re-establish a balance between use of the Mississippi as a barge mover and as an intact watershed. Sympathetic agricultural extensionists are promoting alternatives to the standard corn-soybean rotation, including certified organic crop production, which can simultaneously bring down

Past and Future: Connecting the Dots

Given the direction and speed of prevailing trends, how far can the decline in farmers go? The lead editorial in the September 13, 1999 issue of *Feedstuffs*, an agribusiness trade journal, notes that "Based on the best estimates of analysts, economists and other sources interviewed for this publication, American agriculture must now quickly consolidate all farmers and livestock producers into about 50 production systems… each with its own brands," in order to maintain competitiveness. Ostensibly, other nations will have to do the same in order to keep up.

To put that in perspective, consider that in traditional agriculture, each farm is an independent production system. In this map of Ireland's farms circa 1930, each dot represents 100 farms, so the country as a whole had many thousands of independent production systems. But if the *Feedstuffs* prognosis were to come to pass, this map would be reduced to a single dot. And even an identically keyed map of the much larger United States would show the country's agriculture reduced to just one dot.

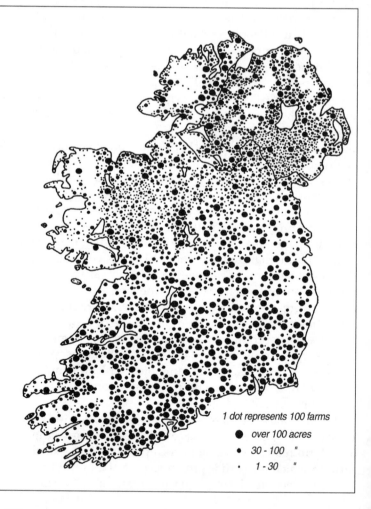

1 dot represents 100 farms

● over 100 acres

· 30 - 100 "

· 1 - 30 "

input costs and garner a premium for the final product, and reduce nitrogen pollution.

The United States and Brazil may have made costly mistakes in giving agribusiness such power to reshape the rivers and land to its own use. But the strategy of interlinked coalitions may be mobilizing in time to save much of the world's agricultural health before it is too late. Dave Brubaker, head of the Spira/GRACE Project on Industrial Animal Production at the Johns Hopkins University School of Public Health, sees these diverse coalitions as "the beginning of a revolution in the way we look at the food system, tying in food production with social welfare, human health, and the environment." Brubaker's project brings together public health officials focused on antibiotic overuse and water contamination resulting from hog waste; farmers and local communities who oppose the spread of new factory farms or want to close down existing ones; and a phalanx of natural allies with related campaigns, including animal rights activists, labor unions, religious groups, consumer rights activists, and environmental groups.

"As the circle of interested parties is drawn wider, the alliance ultimately shortens the distance between farmer and consumer," observes Mark Ritchie, president of the Institute for Agriculture and Trade Policy, a research and advocacy group often at the center of these partnerships. This closer proximity may prove critical to the ultimate sustainability of our food supply, since socially and ecologically sound buying habits are not just the passive *result* of changes in the way food is produced, but can actually be the most powerful *drivers of* these changes. The explosion of farmers' markets, community-supported agriculture, and other direct buying arrangements between farmers and consumers points to the growing numbers of nonfarmers who have already shifted their role in the food chain from that of choosing from the tens of thousands of food brands offered by a few dozen companies to bypassing such brands altogether. And, since many of the additives and processing steps that take up the bulk of the food dollar are simply the inevitable consequence of the ever-increasing time commercial food now spends in global transit and storage, this shortening of distance between grower and consumer will not only benefit the culture and ecology of farm communities. It will also give us access to much fresher, more flavorful, and more nutritious food. Luckily, as any food marketer can tell you, these characteristics aren't a hard sell.

Brian Halweil is a staff researcher at the Worldwatch Institute.

From *World Watch*, September/October 2000, pp. 12-28. © 2000 by the Worldwatch Institute (www.worldwatch.org). Reprinted by permission.

WHAT'S A RIVER FOR?

Thousands of dead salmon, acres of dying crops, pesticide-poisoned birds: How the Klamath River became the first casualty in the West's new water wars

By Bruce Barcott

ON THE MORNING of September 19, 2002, the Yurok fishermen who set their gill nets near the mouth of the Klamath River arrived to find the largest salmon run in years fully under way. The fish had returned from the ocean to the Klamath, on the Northern California coast, to begin their long trip upstream to spawn; there were thousands of them, as far as the eye could see. And they were dying. Fully-grown 30-pounders lay beached on shoreline rocks. Smaller fish floated in midriver eddies. Day after day they kept washing up; by the third day, biologists were estimating that 33,000 fish had been killed in one of the largest salmon die-offs in U.S. history.

The Yurok knew immediately what had happened. For months they, along with state experts and commercial fishermen, had been pleading with the federal government to stop diverting most of the river's water into the potato and alfalfa fields of Oregon's upper Klamath Basin. But the Bureau of Reclamation, the agency in charge of federal irrigation projects, refused to intervene. No one had proved, it argued, that the fish really needed the water.

When the die-off was discovered, federal authorities raced to send a flush of water downriver. But they didn't change their long-term policy—using the Klamath's water to support farmers over fish, fishermen, and some of the most environmentally critical wildlands in the nation.

From Montana to New Mexico, the Bush administration is favoring the water rights of farmers, ranchers, and developers over endangered species, Indian tribes—and the federal government itself.

FOR THE BUSH ADMINISTRATION, and for the advocacy groups that have joined the battle, the fight over the Klamath is more than a regional dispute. It's a bellwether signaling a key shift in the federal government's stance in a new generation of Western water wars. From Montana to New Mexico, conflicts over rivers, wetlands, and irrigation projects are pitting federal water rights against local and state governments and private interests. And in each case, the Bush administration is favoring farmers, ranchers, and developers over the rights of endangered species, Indian tribes, and the federal government itself. In the past year alone, administration officials have backed away from water policies designed to protect fish and birds along the Rio Grande and in California's Central Valley, given up their claim to protect thousands of acres of wetlands from being filled in for subdivisions and shopping malls, and moved toward ceding federal rights to water in several national parks and wildlife refuges.

"The administration sees water rights as property rights that come before other rights, including the right of ecosystems to exist," says Steve Malloch, executive director of the Western Water Alliance, a regional environmental advocacy group. "When faced with the choice between [environmental protection] and what they perceive to be inviolable property rights, they've made it clear what side they're going to come down on."

Nowhere have these issues played out more dramatically than on the 250-mile course the Klamath takes from Oregon's arid high country to the redwood forests of California's northern coast. When the federal government was forced by the courts in 2001 to withhold irrigation water and keep the Klamath flowing, national property-rights groups rallied to support struggling farmers. A year later, when the administration took water from the fish and gave it to farmers, the salmon kill made the river a cause célèbre for environmental groups. At least four separate lawsuits, with plaintiffs ranging from fishermen to property-rights advocates to environmentalists, are challenging government policies on the river. And a National Marine Fisheries Service biologist has filed for whistleblower protection, arguing that, under pressure from "a very high level," his agency changed a key scientific report to justify withholding water from the fish.

Decisions in some of the cases are expected soon, just as another season of trouble gets under way on the Klamath. After a winter of meager snowfalls, another drought is likely. Someone will have to go without water again. And that, in this part of the world, is the one thing no one can afford.

THE KLAMATH IS BORN A CRIPPLE. The river begins in Klamath Falls, Oregon, at the southern outlet of Upper Klamath Lake, a body of water so shallow that a tall man could nearly cross it without wetting his hat. Before the river becomes a river, the U.S. Bureau of Reclamation's "A" Canal diverts half its water into a labyrinthine system of aqueducts that irrigate more than 1,400 farms and ranches. The falls? There are no falls, never have been. The town's founders invented the name to attract homesteaders.

This was what you found when you arrived in Klamath Falls a century ago: swamps everywhere, and birds thick as flies. The Klamath Basin, a high arid plateau the size of Connecticut, didn't get much rain—about as much, on average, as the Texas Panhandle—but what it got it kept, draining the runoff from surrounding mountains into three shallow lakes and a vast system of wetlands. Ten million birds paused to feed here during their migration from Canada to Mexico. Egrets, terns, mallards, pelicans, eagles, tundra swans, and herons browsed amid thickets of 10-foot-tall bulrushes known as tule (too-lee). The Klamath Indians pulled tens of thousands of sucker fish from the lakes every year. This was the Everglades of the American West.

Where some saw a thriving wetland, the U.S. Bureau of Reclamation saw farmland drowning under water. Starting in 1906, the Bureau drained and replumbed nearly the entire basin, building a complex set of canals that shrank Lower Klamath Lake and Tule Lake to one-quarter their original size and sending the water to irrigate thousands of acres of crops and pasture. After the farmers had their share, some of the water would drain into the Tule Lake and Lower Klamath National Wildlife refuges, and some would be pumped back into the Klamath River. Over time, a map of the project came to resemble a guide to the London Underground.

A few weeks after low river levels caused one of the largest salmon kills in U.S. history, a federal biologist said his superiors had changed a key scientific assessment under pressure from "a very high level."

Thanks to irrigation, things hummed along for nearly a century. Homesteaders built farm towns along the Oregon-California border: Merrill, Malin, Tulelake. Teenage girls were crowned queen of the potato festival. Teenage boys hung out at the Tulelake bowling alley. A local historian proclaimed the Klamath Basin Project "one of the most successful government projects in American history."

But over those 100 years, any life that didn't grow in neat harrowed rows began to drain out of the basin.

The Klamath Indians noticed first. The shortnose sucker, a staple of the tribal diet, began growing scarce in Upper Klamath Lake. Farther downstream, the coho salmon that provided sustenance to the Yurok tribe—and to thousands of commercial fishermen along the coast—all but vanished. Environmentalists successfully fought to have both fish protected under the Endangered Species Act. The tribes, whose treaties guaranteed them the right to harvest sucker and salmon in perpetuity, filed lawsuits demanding protection for the fish. By 1995, the Interior Department's regional solicitor—the department's in-house lawyer—issued an opinion reflecting the new legal reality: Tribes, and the endangered species they sought to protect, would have first right to disputed federal waters. Irrigators would have to come second.

The Bureau of Reclamation, an agency created not to protect nature but to overcome it, was slow to implement the new priorities. But by the spring of 2001, it had no choice. A drought had struck the Klamath and a new lawsuit, from commercial fishermen along the coast, required that river levels be kept high enough for the fish. That meant withholding some of the farmers' irrigation water and sending it downriver—at least until the drought broke.

The drought did not break.

"WE WERE LOOKING GOOD AND GOING into 2001," says Dick Carleton. "We had our three tractors and potato equipment nearly paid off. Then we got word that the water wouldn't be coming." The 59-year-old farmer steers his red F-250 truck down a muddy track separating fields of alfalfa stubble. Carleton's farm sits three days' flow from the Klamath headwaters. He and his son Jim, 34, grow alfalfa and potatoes on 1,500 acres in Merrill, Oregon, 15 miles south of Klamath Falls. If you've eaten Campbell's cream of potato soup or snacked on Frito-Lay potato chips you may have tasted their crop. But you might not again. The Carletons are bankrupt.

"It's Chapter 12," Dick explains. "You don't lose everything, but you still owe your debts. Gives us a chance to regroup."

Without irrigation water, the Carletons lost their 2001 potato crop. Their bills piled up. A typical potato farmer can carry a $300,000 loan just on his equipment. Multiply Dick Carleton by 1,400 (the number of local growers who depend on Bureau irrigation) and you end up with a lot of angry farmers.

By the summer of 2001, those farmers were making national news. They paraded their tractors through Klamath Falls and carried signs that said "Klamath Basic Betrayed." National property-rights advocates flocked to the basin, accusing the government of engaging in "rural cleansing." Western members of Congress held hearings in Klamath Falls and vowed to rewrite the Endangered Species Act. The farmers started a vigil at the "A" Canal headgate. When somebody surreptitiously opened the canal—and local police refused to make arrests—federal marshals were called in.

"I wasn't a political person before 2001," says Dick Carleton. "But I spent a lot of time up at the headgate." For the farmers, things grew desperate. Some started selling off equipment for 10 cents on the dollar.

Many farmers lashed out at the Bush administration for locking up their water. But behind the scenes, top officials at the Interior Department, which oversees the Bureau of Reclamation, were scrambling to keep the farmers in business. "The department was in no position to say, 'We're not going to comply with the law,'" recalls Sue Ellen Wooldridge, deputy chief of staff for Interior Secretary Gale Norton. "We tried to see if there was any flex"—any wiggle room in the law that would allow delivery of irrigation water—"and there wasn't."

Officials knew that the only way to change the irrigation plan was a new scientific assessment of how much water the fish needed. In October 2001, Norton asked a committee of the National Research Council to review the available research on the Klamath. Four months later, the committee delivered a draft report: There wasn't enough evidence, it concluded, to know exactly how much water was enough. Environmentalists argue that the document left out key studies of the river; the council says it is still working on a final report.

But if the committee's findings were ambivalent, the Bureau of Reclamation's response was anything but. As long as science hadn't proved that low flows would harm the salmon, the agency announced, farmers would receive their full allocation of water. Wildlife experts—especially those at the National Marine Fisheries Service, which is charged with protecting the salmon—objected, but they were overruled. In a ceremony on March 29, 2002, with national TV cameras in attendance and farmers chanting, "Let the waters flow," Norton herself helped open the Klamath headgate.

The bureau's allocation to the farmers was so generous that by summer, irrigation ditches in Klamath County were overflowing with water, prompting Oregon state officials to complain about flooding. By September, dead salmon were washing up on the Yurok reservation. And the following month, Michael Kelly—the National Marine Fisheries Service's lead biologist in the Klamath case—filed a claim for protection as a whistleblower, charging that his superiors had changed a key scientific assessment, known as a biological opinion, in response to "political pressure." Environmentalists are now challenging the biological opinion—the foundation for the Klamath irrigation plan—in court. Kelly isn't speaking to the press while the claim is pending.

Throughout the controversy, the Interior Department has argued that its policy on the Klamath is driven by science; if the science changes, says Wooldridge, so will the irrigation plan. But the department's stance also reflects Interior Secretary Norton's own long-standing philosophy. The secretary cut her teeth working for the Rocky Mountain Legal Foundation, a property-rights group in Colorado where her boss was Reagan's controversial Interior Secretary James Watt, who once proposed selling off public lands, including some national parks. As Colorado attorney general in the 1990s, Norton frequently challenged the federal government's land and water rights in her state. Among her first actions at Interior was to appoint Colorado attorney Bennett Raley, who has represented cities and farmers in water disputes, as the assistant secretary overseeing the Bureau of Reclamation.

"There's a new wind blowing out of Washington," says Janet Neuman, an environmental law professor at Lewis & Clark College who, in the '90s, served on a federal water commission charged with evaluating Bureau of Reclamation policies. "At one point, the bureau was trying to move away from being perceived as being captive to [private interests]. Now they are pulling back from that."

In Colorado, for example, the administration has announced that it will not insist on its right to keep water flowing in the Gunnison River—a move that would endanger the unique ecosystems of Black Canyon of the Gunnison National Park, but could allow the water to be diverted toward Denver's booming suburbs. And on the Rio Grande, Norton is backing the city of Albuquerque in its attempts to withhold water from the river, even if that means destroying the habitat of an endangered fish called the silvery minnow.

Similar conflicts are likely to erupt throughout the West, experts believe, as drought and booming development exacerbate the pressure on already overtaxed water systems. "The Klamath is looked at as symptomatic," says Neuman. "There but for the grace of God go many, many other basins where there has been decades of over-commitment of the water resources. Endangered species listings and irrigation demands and unmet tribal demands—those circumstances exist in lots of places around the West, and all it takes is a particularly low water year to bring them to a head. It's just a matter of how low, and how soon."

"THERE—IN THAT TREE," says Bob Hunter. "Northern harrier." The marsh hawk perches in a bare cottonwood tree, scowling at a flock of bufflehead ducks bobbing on the marsh. A flock of white swans flaps overhead, their long necks slanted like 737s at takeoff.

Hunter is a lawyer for the Oregon environmental group Water Watch, and today he is guiding me through some of the most bizarre wetlands in the nation's wildlife refuge system. "What you're looking at is the only wildlife refuge in America that grows potatoes and hay," he says as we drive down a gravel levee road. To our left is the water of Tule Lake. To our right is an overwintering alfalfa field. Both are within the refuge's borders.

The peculiar setup goes back to the days of Teddy Roosevelt, who created both the National Wildlife Refuge System and the Bureau of Reclamation. The Klamath Basin became their battlefield. Reclamation engineers wanted it for farmland. Conservationists claimed it as a bird sanctuary. Over the years a compromise emerged that allowed local farmers to lease about one-third of the land in the refuges at Lower Klamath Lake and Tule Lake. To grow crops, they need irrigation water—the "excess" water that would otherwise keep the wetlands wet.

For years, environmentalists have argued for an end to farming in the refuge: "If you take those 32,000 acres of lease land out of the irrigation loop," says Hunter, "that's more than

15 percent of the Klamath Project put back into the river—at no cost to the government." In the '90s, the refuge's manager began doing just that, withholding water from the lease lands in drought years. But the Bush administration scrapped that policy last year.

If you imagine the Pacific flyway as an hourglass stretching from Alaska to South America, the Klamath marshes sit at its waist; 80 percent of all the migratory birds in the West stop here at some point on their journeys. "We don't really know if there's any other place in the United States that has the same significance for wildlife," says Wendell Wood, southern Oregon field representative for the Oregon Natural Resources Council. But since irrigation began a century ago, the number of birds in the basin has dropped by 90 percent, and those that remain—including nearly 1,000 bald eagles—find their breeding grounds drained and polluted. During the summer, many of the wetlands near Tule Lake turn into fields of cracked mud, even as the alfalfa fields next to them sprout a lush crop. At least 46 different insect- and weed-killing chemicals are regularly applied to the fields in the refuge, sometimes mixed in with irrigation water in a process called "chemigation"; the Bureau of Reclamation itself uses several toxic herbicides to keep irrigation canals weed-free. The refuge's longtime manager, Phil Norton, once said that when he first came to Tule Lake, he "couldn't believe they called it a wildlife refuge."

NOT FAR FROM THE REFUGE'S DUCK PONDS, on about 600 acres of what used to be Tule Lake, John Anderson grows alfalfa and mint. His father, Robert, won the land in a 1947 government lottery that offered World War II veterans the chance to be America's last homesteaders. "This was a thriving community back then," recalls Anderson, now 50. "Lot of neighbors, lot of kids around. Prices were better. Lot of folks got out of the business since then. At my church I'm one of the only farmers younger than 75."

Sure enough, a stroll down Tulelake's dusty streets reveals a town whose decline began long before the 2001 water crisis. The bowling alley is shuttered, and the hardware store closes all winter. The only signs of life come from a small grocery store, a county ag extension office, and Tulelake High School.

Years ago, subsidizing water-intensive crops like potatoes in a region that gets only 18 inches of rain each year seemed like an efficient use of federal resources. But, John Anderson says, farmers can tell that the winds have shifted. "Americans have changed their priorities," he says. "Now they want rivers, wetlands, clean water, wildlife. I can understand. But the American people should be willing to pay for it."

Anderson is among a growing number of farmers who support a seemingly simple solution to the water dilemma: What if the government paid some of the farmers to quit irrigating? A buyout of, say, 30 percent of the Klamath Projects' 200,000 acres of irrigated land, at roughly $2,500 an acre, would cost $150 million. The scheme would remove a lot of claims on the river and could leave enough water for both the fish and the remaining farmers. Variations of the plan have been floated by numerous conservation groups and some members of Oregon's congressional delegation.

When it comes to water in the West, it's appropriate to borrow from Faulkner: The past isn't dead here. It's not even past.

But when it comes to water in the West, it's appropriate to borrow from Faulkner: The past isn't dead here. It's not even past. Decades-old contracts and double crosses are recalled as if they happened last week. A century ago, farmers in California's Owens Valley took a water buyout and saw the lifeblood of their valley diverted to Los Angeles. Roman Polanski immortalized the scam in *Chinatown.* "Some of the Klamath pioneers," says Dan Keppen, head of the organization that represents the valley's irrigators, "moved here from the Owens Valley."

Keppen's group, the Klamath Water Users Association, is dead set against a buyout, which it says would simply usher in the end of farming in the basin. Last year the group successfully lobbied to block a $175 million congressional aid package for the Klamath because some of the money could have been used to buy out farms. National property-rights groups, who oppose turning private land into public property, have resisted the idea, and so has the Bush administration. As one official privately notes, Secretary Norton feels that the federal land portfolio "is quite large enough, thank you very much."

AS IT LEAVES the Klamath plateau, the river drops into Northern California's woolly Siskiyou country, home to Bigfoot sightings, marijuana patches, off-the-grid rednecks, and long-toothed hippies. State Highway 96 follows its course past a string of abandoned mines, through former mill towns that now get by on fly-fishing and rafting tours. As the Klamath Mountains segue into the Coast Range, moist Pacific air creeps up the river valley in cottony mists. Moss overcoats wrap around trunks of toyon, the California holly that inspired the name Hollywood. Strengthened by dozens of winter streams, the Klamath throws up rapids and surfable waves that draw kayakers from hundreds of miles away. At the village of Weitchpec (Witchpeck), the green Trinity River joins the muddy Klamath, and the combined channel veers away from the highway. A one-lane road on the Hoopa Indian Reservation continues to shadow the river until finally it, too, gives out and the Klamath rolls on through the woods, for the first time truly wild.

But the water wars don't stop when you leave the river. A few miles from the Klamath-Trinity confluence, I knocked at a house that had a sign posted in its yard. "DYING 4 WATER," it said, over a drawing of a salmon. Duane Sherman Sr., the 33-year-old former Hoopa tribal chairman, invited me in.

"I'm dividing up this deer we killed a week ago," Sherman said, offering me a bite of venison jerky. "You see, Indians are nothing but extended family. This deer feeds my grandmother, my sister, my aunt," he said, nodding to three women chatting at the kitchen table, "and my nephews too." Two boys watched cartoons in the next room. "The salmon's the same way. Those fish aren't just our livelihood. If we don't fish, we don't eat."

Most of the 33,000 fish lost in last September's salmon kill were headed upriver toward the Trinity, where the Hoopa have fishing rights. (Their Yurok neighbors have rights along the lower 40 miles of the Klamath, from the mouth to the Trinity confluence.) Which means there's a lot more deer than salmon on the Sherman family table these days.

For more than a decade the Hoopa, along with the Yurok, state and municipal governments, and landowners up and down the river, have been working to restore the Klamath fishery, once among the most productive on the West Coast. Last year was supposed to be the first season it all paid off. Instead, the salmon kill left local communities with what one study estimates will be about $20 million in losses, roughly as much as the farmers lost during the water crisis of 2001.

"I understand that the irrigators have a contract with the government," Susan Masten, the Yurok tribal chair, tells me over a plate of eggs at Sis' diner, near the mouth of the Klamath. "The government also has a contract with us. And it was the first contract."

This is Masten's second water war. The Yuroks' federal agreements, which date back to the mid-19th century, guarantee them access fish in perpetuity. But during the 1970s, tribal members battled the federal government and commercial fishermen for the right to set their nets. Federal agents with M-16s and riot gear occupied the reservation to keep tribal fishermen out of the river. A popular bumper sticker off the reservation read, "Can an Indian, Save a Salmon."

But over the years, the commercial fishermen have become the Yuroks' strongest allies. "After years of fighting with the tribes, we looked at each other and realized we had fewer and fewer fish to fight over," says Glen Spain, the Northwest regional director of the Pacific Coast Federation of Fishermen's Associations. "It's the same thing with the farmers. We're all workaday people trying to make a living. We understand the market forces driving the farmers down."

Like the farm towns farther upstream, the fishing communities on this stretch of coast are dotted with "For Sale" signs and shuttered family businesses. "From Fort Bragg, California, to Coos Bay, Oregon—the range of Klamath River salmon—we've lost 3,700 jobs and an $80 million-a-year economy," says Spain.

But the fishermen haven't been cast as victims in this crisis—not by the media, and not by the government. Their decline has come slowly, invisibly; they lack the ag industry's national lobbying muscle, or the property-rights movement's clout with the Bush administration. They are no more a political force than the ducks that land in the Tule Lake refuge.

SOME RIVERS ARE SO BIG you can't see them meet the sea. They simply fan out and merge with the tide. The amazing thing about the Klamath is that you can actually watch it pour into the Pacific. In Yurok country the river is a quarter-mile wide and quite; at its mouth it narrows into a frightening rush that cuts between a sandbar and a rocky cliff and then it's gone, lost in the foam. Standing there in the teeth of a January storm, watching a flock of gulls wheel in the cold wind, I thought of the classic closing line in *Chinatown:* "Forget it, Jake, it's Chinatown." Some things, it suggests, are simply too murky to understand.

But the Klamath water war comes down to a single, very simple equation: too many takers, not enough water.

"When I was a boy," John Anderson told me as we watched dusk settle over fields north of Tule Lake, "the ducks and geese came here by the millions. You could hear the flocks roar at night. We've given up a lot of the life that used to be here. Sometimes I question whether it's been a good trade-off."

It wasn't a wholly bad deal. A nation got fed, the West settled, families raised. But the math never did add up, and it's now becoming clear what got shortchanged in the bargain.

Additional reporting by Stephen Baxter.

Based in Seattle, **Bruce Barcott** is a contributing writer for *Outside*. In addition to this story about the battle over water rights on the Klamath River, Barcott covered the Pacific Northwest's salmon-fishing industry in "Aquaculture's Troubled Harvest" (November/December 2001).

A **Human** THIRST

Humans now appropriate more than half of all the freshwater in the world. Rising demands from agriculture, industry, and a growing population have left important habitats around the world high and dry.

by Don Hinrichsen

On March 20, 2000, a group of monkeys, driven mad with thirst, clashed with desperate villagers over drinking water in a small outpost in northern Kenya near the border with Sudan. The Pan African News Agency reported that eight monkeys were killed and 10 villagers injured in what was described as a "fierce two-hour melee." The fight erupted when relief workers arrived and began dispensing water from a tanker truck. Locals claimed that a prolonged drought had forced animals to roam out of their natural habitats to seek life-giving water in human settlements. The monkeys were later identified as generally harmless vervets.

The world's deepening freshwater crisis—currently affecting 2.3 billion people—has already pitted farmers against city dwellers, industry against agriculture, water-rich state against water-poor state, county against county, neighbor against neighbor. Inter-species rivalry over water, such as the incident in northern Kenya, stands to become more commonplace in the near future.

"The water needs of wildlife are often the first to be sacrificed and last to be considered," says Karin Krchnak, population and environment program manager at the National Wildlife Federation (NWF) in Washington, D.C. "We ignore the fact that working to ensure healthy freshwater ecosystems for wildlife would mean healthy waters for all." As more and more water is withdrawn from rivers, streams, lakes and aquifers to feed thirsty fields and the voracious needs of industry and escalating urban demands, there is often little left over for aquatic ecosystems and the wealth of plants and animals they support.

The mounting competition for freshwater resources is undermining development prospects in many areas of the world, while at the same time taking an increasing toll on natural systems, according to Krchnak, who co-authored an NWF report on population, wildlife, and water. In effect, humanity is waging an undeclared water war with nature.

"There will be no winners in this war, only losers," warns Krchnak. By undermining the water needs of wildlife we are not just undermining other species, we are threatening the human prospect as well.

Pulling Apart the Pipes

Currently, humans expropriate 54 percent of all available freshwater from rivers, lakes, streams, and shallow aquifers. During the 20th century water use increased at double the rate of population growth: while the global population tripled, water use per capita increased by six times. Projected levels of population growth in the next 25 years alone are expected to increase the human take of available freshwater to 70 percent, according to water expert Sandra Postel, Director of the Global Water Policy Project in Amherst, Massachusetts. And if per capita water consumption continues to rise at its current rate, by 2025 that share could significantly exceed 70 percent.

As a global average, most freshwater withdrawals—69 percent—are used for agriculture, while industry accounts for 23 percent and municipal use (drinking water, bathing and cleaning, and watering plants and grass) just 8 percent.

The past century of human development—the spread of large-scale agriculture, the rapid growth of industrial development, the construction of tens of thousands of large dams, and the growing sprawl of cities—has profoundly altered the Earth's hydrological cycle. Countless rivers, streams, floodplains, and wetlands have been dammed, diverted, polluted, and filled. These components of the hydrological cycle, which function as the Earth's plumbing system, are being disconnected and plundered, piece by piece. This fragmentation has been so extensive that freshwater ecosystems are perhaps the most severely endangered today.

Left High and Dry

Habitat destruction, water diversions, and pollution are contributing to sharp declines in freshwater biodiversity. One-fifth of all freshwater fish are threatened or extinct. On continents where studies have been done, more than half of amphibians are in decline. And more than 1,000 bird species—many of them aquatic—are threatened.

More than 40,000 large dams bisect waterways around the world, and more than 500,000 kilometers of river have been dredged and channelized for shipping. Deforestation, mining, grazing, industry, agriculture, and urbanization increase pollution and choke freshwater ecosystems with silt and other runoff.

Water diversion for irrigation, industry, and urban use has increased 35-fold in the past 300 years. In some cases, this increased demand has deprived entire ecosystems of water. Sprawl is an increasing concern, as the spread of urban areas is destroying important wetlands, and paved-over area is reducing the amount of water that is able to recharge aquifers.

Consider the plight of wetlands—swamps, marshes, fens, bogs, estuaries, and tidal flats. Globally, the world has lost half of its wetlands, with most of the destruction having taken place over the past half century. The loss of these productive ecosystems is doubly harmful to the environment: wetlands not only store water and transport nutrients, but also act as natural filters, soaking up and diluting pollutants such as nitrogen and phosphorus from agricultural runoff, heavy metals from mining and industrial spills, and raw sewage from human settlements.

In some areas of Europe, such as Germany and France, 80 percent of all wetlands have been destroyed. The United States has lost 50 percent of its wetlands since colonial times. More than 100 million hectares of U.S. wetlands (247 million acres) have been filled, dredged, or channeled—an area greater than the size of California, Nevada, and Oregon combined. In California alone, more than 90 percent of wetlands have been tilled under, paved over, or otherwise destroyed.

Destruction of habitat is the largest cause of biodiversity loss in almost every ecosystem, from wetlands and estuaries to prairies and forests. But biologists have found that the brunt of current plant and animal extinctions have fallen disproportionately on those species dependent on freshwater and related habitats. One fifth of the world's freshwater fish—2,000 of the 10,000 species identified so far—are endangered, vulnerable, or extinct. In North America, the continent most studied, 67 percent of all mussels, 51 percent of crayfish, 40 percent of amphibians, 37 percent of fish, and 75 percent of all freshwater mollusks are rare, imperiled, or already gone.

The global decline in amphibian populations may be the aquatic equivalent of the canary in the coal mine. Data are scarce for many species, but more than half of the amphibians studied in Western Europe, North America, and South America are in a rapid decline.

Around the world, more than 1,000 bird species are close to extinction, and many of these are particularly dependent on wetlands and other aquatic habitats. In Mexico's Sonora Desert, for instance, agriculture has siphoned off 97 percent of the region's water resources, reducing the migratory bird population by more than half, from 233,000 in 1970 to fewer than 100,000 today.

Pollution is also exacting a significant toll on freshwater and marine organisms. For instance, scientists studying beluga whales swimming in the contaminated St. Lawrence Seaway, which connects the Atlantic Ocean to North America's Great Lakes, found that the cetaceans have dangerously high levels of PCBs in their blubber. In fact the contamination is so severe that under Canadian law the whales actually qualify as toxic waste.

Waterways everywhere are used as sewers and waste receptacles. Exactly how much waste ends up in freshwater systems and coastal waters is not known. However, the UN Food and Agriculture Organization (FAO) estimates that every year roughly 450 cubic kilometers (99 million gallons) of wastewater (untreated or only partially treated) is discharged into rivers, lakes, and coastal areas. To dilute and transport this amount of waste requires at least 6,000 cubic kilometers (1.32 billion gallons) of clean water. The FAO estimates that if current trends continue, within 40 years the world's entire stable river flow would be needed just to dilute and transport humanity's wastes.

The Point of No Return?

The competition between people and wildlife for water is intensifying in many of the most biodiverse regions of the world. Of the 25 biodiversity hotspots designated by Conservation International, 10 are located in water-short regions. These regions—including Mexico, Central America, the Caribbean, the western United States, the Mediterranean Basic, southern Africa, and southwestern China—are home to an extremely high number of endemic and threatened species. Population pressures and overuse of resources, combined with critical water shortages, threaten to push these diverse and vital ecosystems over the brink. In a number of cases, the point of no return has already been reached.

China

China, home to 22 percent of the world's population, is already experiencing serious water shortages that threaten both people and wildlife. According to China's former environment minister, Qu Geping, China's freshwater supplies are capable of sustainably supporting no more than 650 million people—half its current population. To compensate for the tremendous shortfall, China is draining its rivers dry and mining ancient aquifers that take thousand of years to recharge.

As a result, the country has completely overwhelmed its freshwater ecosystems. Even in the water-rich Yangtze River Basin, water demands from farms, industry, and a giant population have polluted and degraded freshwater and riparian ecosystems. The Yangtze is one of the longest rivers in Asia, winding

6,300 kilometers on its way to the Yellow Sea. This massive watershed is home to around 400 million people, one-third of the total population of China. But the population density is high, averaging 200 people per square kilometer. As the river, sluggish with sediment and laced with agricultural, industrial, and municipal wastes, nears its wide delta, population densities soar to over 350 people per square kilometer.

The effects of the country's intense water demands, mostly for agriculture, can be seen in the dry lake beds on the Gianghan Plain. In 1950 this ecologically rich area supported over 1,000 lakes. Within three decades, new dams and irrigation canals had siphoned off so much water that only 300 lakes were left.

China's water demands have taken a huge toll on the country's wildlife. Studies carried out in the Yangtze's middle and lower reaches show that in natural lakes and wetlands still connected to the river, the number of fish species averages 100. In lakes and wetlands cut off and marooned from the river because of diversions and drainage, no more than 30 survive. Populations of three of the Yangtze's largest and more productive fisheries—the silver, bighead, and grass carp—have dropped by half since the 1950s.

Mammals and reptiles are in similar straits. The Yangtze's shrinking and polluted waters are home to the most endangered dolphin in the world—the Yangtze River dolphin, or Baiji. There are only around 100 of these very rare freshwater dolphins left in the wild, but biologists predict they will be gone in a decade. And if any survive, their fate will be sealed when the massive Three Gorges Dam is completed in 2013. The dam is expected to decrease water flows downstream, exacerbate the effects of pollution, and reduce the number of prey species that the dolphins eat. Likewise, the Yangtze's Chinese alligators, which live mostly in a small stretch near the river's swollen, silt-laden mouth, are not expected to survive the next 10 years. In recent years, the alligator population has dropped to between 800 and 1,000.

The Aral Sea

The most striking example of human water demands destroying an ecosystem is the nearly complete annihilation of the 64,500 square kilometer Aral Sea, located in Central Asia between Kazakhstan and Uzbekistan. Once the fourth largest inland sea in the world, it has contracted by half its size and lost three-quarters of its volume since the 1960s, when its two feeder rivers—the Amu Darya and the Syr Darya—were diverted to irrigate cotton fields and rice paddies.

The water diversions have also deprived the region's lakes and wetlands of their life source. At the Aral Sea's northern end in Kazakhstan, the lakes of the Syr Darya delta shrank from about 500 square kilometers to 40 square kilometers between 1960 and 1980. By 1995, more than 50 lakes in the Amu Darya delta had dried up and the surrounding wetlands had withered from 550,000 hectares to less than 20,000 hectares.

The unique *tugay* forests—dense thickets of small shrubs, grasses, sedges and reeds—that once covered 13,000 square kilometers around the fringes of the sea have been decimated. By

Alien Invaders

"Rapidly growing populations place heavy demand on freshwater resources and intensify pressures on wildlands," concludes a combined World Resources Institute and Worldwatch report called "Watersheds of the World." But increasingly, the introduction of exotic or alien species is playing a large role in wreaking havoc on freshwater habitats.

The spread of invasive species is a global phenomenon, and is increasingly fostered by the growth of aquaculture, shipping, and commerce. Whether introduced by accident or on purpose, these alien invaders are capable of altering habitats and extirpating native species en masse.

The invasion and insidious spread of the zebra mussel in the U.S. Great Lakes highlights the tremendous costs to ecosystems and species. A native of Eastern Europe, the zebra mussel arrived in the Great Lakes in 1988, released most likely through the discharge of ballast waters from a cargo ship. Once established, it spread rapidly throughout the region.

The mussels have crowded out native species that cannot compete with them for space and food. A study of the mussels in western Lake Erie found that all of the native clams at each of 17 sampling stations had been wiped out. Moreover, the last known population of the winged maple leaf clam, found in the St. Croix River in the upper Mississippi River basin, is now threatened by advancing ranks of the zebra mussel.

1999 less than 1,000 square kilometers of fragmented and isolated forest remained.

The habitat destruction has dramatically reduced the number of mammals that used to flourish around the Aral Sea: of 173 species found in 1960, only 38 remained in 1990. Though the ruined deltas still attract waterfowl and other wetland species, the number of migrant and nesting birds has declined from 500 species to fewer than 285 today.

Plant life has been hard hit by the increase in soil salinity, aridity, and heat. Forty years ago, botanists had identified 1,200 species of flowering plants, including 29 endemic species. Today, the endemics have vanished. The number of plant species that can survive the increasingly harsh climate is a fraction of the original number.

Most experts agree that the sea itself may very well disappear entirely within two decades. But the region's freshwater habitats and related communities of plants and animals have already been consigned to oblivion.

Lake Chad

Lake Chad, too, has shrunk—to one-tenth of its former size. In 1960, with a surface area of 25,000 square kilometers, it was the second-largest lake in Africa. When last surveyed, it was down to only 2,000 square kilometers. And here, too, massive water withdrawals from the watershed to feed irrigated agriculture have reduced the amount of water flowing into the lake to a trickle, especially during the dry season.

Lake Chad is wedged between four nations: populous Nigeria to the southwest, Niger on the northwest shore, Chad to the northeast, and Cameroon on a small section of the south shore. Nigeria has the largest population in Africa, with 130 million inhabitants. Population-growth rates in these countries average 3 percent a year, enough to double human numbers in one generation. And population growth rates in the regions around the lake are even higher than the national averages. People gravitate to this area because the lake and its rivers are the only sources of surface water for agricultural production in an otherwise dry and increasingly desertified region.

Although water has been flowing into the lake from its rivers over the past decade, the lake is still in serious ecological trouble. The lake's fisheries have more or less collapsed from over-exploitation and loss of aquatic habitats as its waters have been drained away. Though some 40 commercially valuable species remain, their populations are too small to be harvested in commercial quantities. Only one species—the mudfish—remains in viable populations.

As the lake has withered, it has been unable to provide suitable habitat for a host of other species. All large carnivores, such as lions and leopards, have been exterminated by hunting and habitat loss. Other large animals, such as rhinos and hippopotamuses, are found in greatly reduced numbers in isolated, small populations. Bird life still thrives around the lake, but the variety and numbers of breeding pairs have dropped significantly over the past 40 years.

A Blue Revolution

As these examples illustrate, the challenge for the world community is to launch a "blue revolution" that will help governments and communities manage water resources on a more sustainable basis for all users. "We not only have to regulate supplies of freshwater better, we need to reduce the demand side of the equation," says Swedish hydrologist Malin Falkenmark, a senior scientist with Sweden's Natural Science Re-

search Council. "We need to ask how much water is available and how best can we use it, not how much do we need and where do we get it." Increasingly, where we get it from is at the expense of aquatic ecosystems.

If blindly meeting demand precipitated, in large measure, the world's current water crisis, reducing demand and matching supplies with end uses will help get us back on track to a more equitable water future for everyone. While serious water initiatives were launched in the wake of the World Summit on Sustainable Development held in Johannesburg, South Africa, not one of them addressed the water needs of ecosystems.

There is an important lesson here: just as animals cannot thrive when disconnected from their habitats, neither can humanity live disconnected from the water cycle and the natural systems that have evolved to maintain it. It is not a matter of "either or" says NWF's Krchnak. "We have no real choices here. Either we as a species live within the limits of the water cycle and utilize it rationally, or we could end up in constant competition with each other and with nature over remaining supplies. Ultimately, if nature loses, we lose."

By allowing natural systems to die, we may be threatening our own future. After all, there is a growing consensus that natural ecosystems have immense, almost incalculable value. Robert Costanza, a resource economist at the University of Maryland, has estimated the global value of freshwater wetlands, including related riverine and lake systems, at close to $5 trillion a year. This figure is based on their value as flood regulators, waste treatment plants, and wildlife habitats, as well as for fisheries production and recreation.

The nightmarish scenarios envisioned for a water-starved not too distant future should be enough to compel action at all levels. The water needs of people and wildlife are inextricably bound together. Unfortunately, it will probably take more incidents like the one in northern Kenya before we learn to share water resources, balancing the needs of nature with the needs of humanity.

Don Hinrichsen *is a UN consultant. He is former editor-in-chief of* Ambio *and was a news correspondent in Europe for 15 years.*

From *World Watch*, January/February 2003. © 2003 by Worldwatch Institute, www.worldwatch.org.

WILL THE WORLD RUN DRY?

Global Water And Food Security

By Mark W. Rosegrant, Ximing Cai, and Sarah A. Cline

Demand for the world's increasingly scarce water supply is rising rapidly, challenging its availability for food production and putting global food security at risk. Agriculture, upon which a burgeoning population depends for food, is competing with industrial, household, and environmental uses for this scarce water supply. Even as demand for water by all users grows, groundwater is being depleted, other water ecosystems are becoming polluted and degraded, and developing new sources of water is becoming more costly.

These challenges are receiving significant international attention, notably with the third World Water Forum that was convened in Japan from 16 to 23 March 2003. This meeting was the third in a series of meetings held every three years, bringing together water experts, government leaders, representatives from nongovernmental organizations (NGOs), and other interested parties to examine the major dilemmas facing the water sector and to seek solutions to these problems. The third forum hosted more than 24,000 participants from more than 180 countries and held sessions focusing on many of the most crucial water issues facing the world today. Some of the key issues addressed include the need for safe, clean water for all individuals; good governance in water management, including an integrated water resources management approach; capacity building, including education and access to information; financing of water resources infrastructure; and increased participa-

tion of all stakeholders, including women and the poor. More than 100 commitments on water were made during the forum, based primarily on the key water issues.

But despite this attention, the challenge of meeting both water and food security remains formidable. Planning of how to meet the increasing needs of various water users depends upon an understanding of the current situation and potential impacts of policy decisions. To this end, a global model of water and food supply and demand was developed to examine long-term prospects for water and food security under alternative policies.

A Thirsty World

Water development underpins food security, people's livelihoods, industrial growth, and environmental sustainability throughout the world. In 1995 the world withdrew 3,906 cubic kilometers (km^3) of water for these purposes. By 2025 water withdrawal for most uses (domestic, industrial, and livestock) is projected to increase by at least 50 percent. This will severely limit irrigation water withdrawal, which will increase by only 4 percent, in turn constraining food production.[1]

About 250 million hectares are irrigated worldwide today, nearly five times more than at the beginning of the twentieth century. Irrigation has helped boost agricultural yields and outputs and stabilize food production and prices. But growth in pop-

ulation and income will only increase the demand for irrigation water to meet food production requirements. Although the achievements of irrigation have been impressive, in many regions poor irrigation management has markedly lowered groundwater tables, damaged soils, and reduced water quality.[2]

Water is also essential for drinking and household uses and for industrial production. Access to safe drinking water and sanitation is critical to maintain health, particularly for children. But more than 1 billion people across the globe lack enough safe water to meet minimum levels of health and income.[3] Although the domestic and industrial sectors use far less water than agriculture, the growth in water consumption in these sectors has been rapid.

Water is integrally linked to the health of the environment. Water is vital to the survival of ecosystems and the plants and animals that live in them, and in turn ecosystems help to regulate the quantity and quality of water. Wetlands retain water during high rainfall, release it during dry periods, and purify it of many contaminants. Forests reduce erosion and sedimentation of rivers and recharge groundwater. The importance of reserving water for environmental purposes has only recently been recognized.

Alternative Futures for Water

The future of water and food is highly uncertain. Some of this uncertainty is due to relatively uncontrollable factors

Figure 1. Total water withdrawal by region in 1995 and business-as-usual projections for 2025

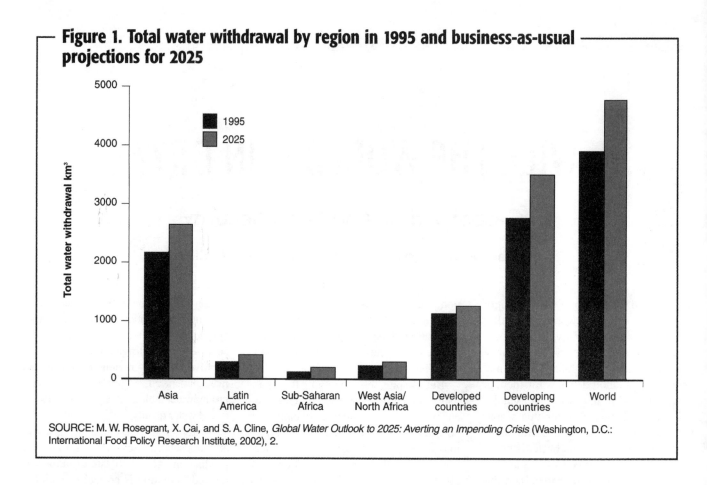

SOURCE: M. W. Rosegrant, X. Cai, and S. A. Cline, *Global Water Outlook to 2025: Averting an Impending Crisis* (Washington, D.C.: International Food Policy Research Institute, 2002), 2.

such as weather. But other critical factors can be influenced by the collective choices of the world's people. These factors include income and population growth; investment in water infrastructure; allocation of water to various uses; reform in water management; and technological changes in agriculture. Policy decisions—and the actions of billions of individuals—determine these fundamental, long-term drivers of water and food supply and demand. Three alternative futures for global water and food show the very different outcomes that policy choices produce.[4]

Business-As-Usual Scenario

In the business-as-usual scenario, current trends in water and food policy, management, and investment would remain as they are. International donors and national governments, complacent about agriculture and irrigation, would cut their investments in these sectors. Governments and water users would implement institutional and management reforms in a limited and piecemeal fashion. These conditions would leave the world ill-prepared to meet major challenges facing the water and food sectors.

> Policy decisions—and the actions of billions of individuals—determine fundamental, long-term drivers of water and food supply and demand.

Over the coming decades, the area of land devoted to cultivating food crops would grow slowly in most of the world because of urbanization, soil degradation, and slow growth in irrigation investment, as well as the fact that of arable land is already cultivated. Moreover, steady or declining real prices for cereals would make it unprofitable for farmers to expand harvested area. As a result, greater food production would depend primarily on increases in yield. Yet growth in crop yields would also diminish because of falling public investment in agricultural research and rural infrastructure. Moreover, many of the actions that produced yield gains in recent decades—such as increasing the density of crop planting, introducing strains that are more responsive to fertilizer, and improving management practices—cannot and would not easily be repeated.

In the water sector, the management of river basin and irrigation water would become more efficient, but slowly. Governments would continue to transfer management of irrigation systems to farmer organizations and water-user associations. Such transfers would increase water efficiency if they are built upon existing patterns of cooperation and backed by a supportive policy and legal environment. But these conditions are often lacking.

In some regions, farmers would adopt more efficient irrigation practices. Economic incentives to induce more efficient water management, however, would still face political opposition from those concerned about the impact of higher water prices on farmers' income and from entrenched interests that benefit from existing systems of allocating

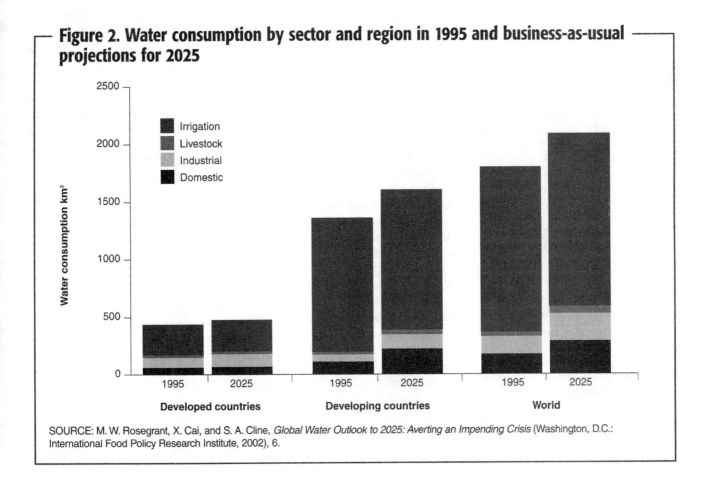

Figure 2. Water consumption by sector and region in 1995 and business-as-usual projections for 2025

SOURCE: M. W. Rosegrant, X. Cai, and S. A. Cline, *Global Water Outlook to 2025: Averting an Impending Crisis* (Washington, D.C.: International Food Policy Research Institute, 2002), 6.

water. Water management would also improve slowly in rainfed agriculture as a result of small advances in water harvesting, better on-farm management techniques, and the development of crop varieties with shorter growing seasons.

In the business-as-usual scenario, public investment in expanding irrigation and reservoir storage would decline as the financial, environmental, and social costs of building new irrigation systems escalate and the prices of cereals and other irrigated crops drop. Nevertheless, where benefits outweigh costs, many governments would construct dams, and reservoir water for irrigation would increase moderately.

With slow growth in irrigation from surface water, farmers would expand pumping from groundwater, which is subject to low prices and little regulation. Regions that currently pump groundwater faster than aquifers can recharge— such as the western United States, northern China, northern and western India,

and West Asia and North Africa—would continue to do so.[5]

The cost of supplying water to domestic and industrial users would rise dramatically. Better delivery and more efficient home water use would lead to some increase in the proportion of households connected to piped water. Many households, however, would remain unconnected. Small price increases for industrial water improvements in pollution control regulation and enforcement and new industrial technologies would cut industrial water-use intensity (water demand per $1,000 of gross domestic product). Yet industrial water prices would remain relatively low and pollution regulations would often be poorly enforced. Thus, significant potential gains would be lost.

Environmental and other interest groups would press to increase the amount of water allocated to preserving wetlands, diluting pollutants, maintaining riparian flora and other aquatic species, and supporting tourism and

recreation. Yet because of competition for water for other uses, the share of water devoted to environmental uses would not increase.

Almost all users would place heavy demands on the world's water supply under the business-as-usual scenario. Together, consumption of water for domestic, industrial, and livestock uses— that is, all nonirrigation uses—would increase dramatically, rising by 62 percent from 1995 to 2025. Because of rapid population growth and rising per capita water use, domestic consumption would increase by 71 percent, more than 90 percent of which would be in developing countries. Industrial water use would grow much faster in developing countries than in developed countries. The intensity of industrial water use would decrease worldwide, especially in developing countries (where initial intensity levels are very high), thanks to improvements in water-saving technology and demand policy. Nonetheless, the sheer size of the increase in the world's industrial production would still lead to an increase in total industrial water demand. Direct water consumption by livestock is very small compared with other sectors. But the rapid increase of livestock production, particularly in developing countries, means that livestock water demand is projected to increase 71 percent between 1995 and 2025. Although irrigation is by far the largest user of the world's water, use of irrigation water is projected to rise much more slowly than other sectors.

Water scarcity under the business-as-usual scenario would lead to slower growth of food production and substantial shifts in where the world's food is grown. Farmers would find themselves unable to raise crop yields as quickly as in the past in the face of a decline in relative water supply. Crop-harvested area is expected to grow even more slowly than crop yield in the coming decades, with all of the growth projected to occur in developing countries.

By substituting cereal and other food imports for irrigated agricultural production (so-called imports of virtual water), countries can effectively reduce their agricultural water use.[6] Under the business-as-usual scenario, developing countries would dramatically increase their reliance on food imports by 2025. The water (and land) savings from the projected large increases of food imports by the developing countries are particularly beneficial if they are the result of strong economic growth that generates the necessary foreign exchange to pay for the food imports. But even when rapidly growing food imports are primarily a result of rapid income growth, national policy makers concerned with heavy reliance on world markets often see them as a signal to set trade restrictions that can slow growth and food security in the longer term. More serious food security problems arise when high food imports are the result of slow agricultural and economic development that fails to keep pace with basic food demand driven by population and income growth. Under these conditions, countries may find it impossible to finance the required imports on a continuing basis, causing a further deterioration in the ability to bridge the gap between food consumption and the food required for basic livelihood.

Water Crisis Scenario

A moderate worsening of many of the current trends in water and food policy and in investment could build to a genuine water crisis. In the water crisis scenario, government budget problems would worsen. Governments would further cut their spending on irrigation systems and accelerate the turnover of irrigation systems to farmers and farmer groups but without the necessary reforms in water rights. Attempts to fund operations and maintenance in the main water system, still operated by public agencies, would cause water prices for irrigators to rise. Water users would fight price increases, and conflict would spill over to local management and cost-sharing arrangements. Spending on the operation and maintenance of secondary and tertiary systems would fall dramatically, and deteriorating infrastructure and poor management would lead to falling water-use efficiency. Likewise, attempts to organize river basin organizations to coordinate water management would fail because of inadequate funding and high levels of conflict among water stakeholders within the basin.

In the water crisis scenario, national governments and international donors would reduce their investments in crop breeding for rainfed agriculture in developing countries, especially for staple crops. Private agricultural research would fail to fill the investment gap for these commodities. This loss of research funding would lead to further declines in productivity growth in rainfed crop areas, particularly in more marginal areas. In search of improved incomes, people would turn to slash-and-burn agriculture, thereby deforesting the upper watersheds of many basins. Erosion and sediment loads in rivers would rise, in turn causing faster sedimentation of reservoir storage. People would increasingly encroach on wetlands for both land and water, and the integrity and health of aquatic ecosystems would be compromised. The amount of water reserved for environmental purposes would decline as unregulated and illegal withdrawals increase. The cost of building new dams would soar, discouraging new investment in many proposed dam sites. At other sites,

indigenous groups and NGOs would mount opposition over the environmental and human impacts of new dams. These protests and high costs would virtually halt new investment in medium and large dams and storage reservoirs. Net reservoir storage would decline in developing countries and remain constant in developed countries.

A moderate worsening of many of the current trends in water and food policy and in investment could build to a genuine water crisis.

In the attempt to get enough water to grow their crops, farmers would extract increasing amounts of groundwater for several years, driving down water tables. But because of the accelerated pumping, after 2010, key aquifers in northern China, northern and northwestern India, and West Asia and North Africa would begin to fail. With declining water tables, farmers would find the cost of extracting water too high, and a big drop in groundwater extraction from these regions would further reduce water availability for all uses.

As in the business-as-usual scenario, the rapid increase in urban populations would quickly raise demand for domestic water. However, governments would lack the funds to extend piped water and sewage disposal to newcomers. Governments would respond by privatizing urban water and sanitation services in a rushed and poorly planned fashion. The new private water and sanitation firms would be undercapitalized and able to do little to connect additional populations to piped water. An increasing number and percentage of the urban population must rely on high-priced water from vendors or spend many hours fetching often dirty water from standpipes and wells.

Total worldwide water consumption in 2025 under the water crisis scenario would be 13 percent higher than under the business-as-usual scenario, but much of this water would be wasted and of no benefit to anyone. Virtually all of the increase would go to irrigation, mainly because farmers would use water less efficiently and withdraw more water to compensate for water losses. The supply

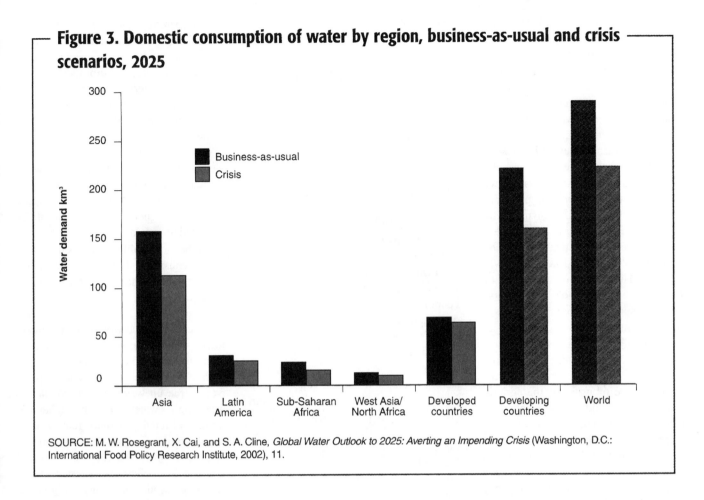

Figure 3. Domestic consumption of water by region, business-as-usual and crisis scenarios, 2025

SOURCE: M. W. Rosegrant, X. Cai, and S. A. Cline, *Global Water Outlook to 2025: Averting an Impending Crisis* (Washington, D.C.: International Food Policy Research Institute, 2002), 11.

of irrigation water would be less reliable, except in regions where so much water is diverted from environmental uses to irrigation that it balances the lower water-use efficiency.

For most regions, per capita demand for domestic water would be significantly lower than under the business-as-usual scenario, in both rural and urban areas. The result is that people would not have access to the water they would need for drinking and sanitation. Compared with outcomes under the business-as-usual scenario, the total domestic demand under the water crisis scenario would be 28 percent less in developing countries, 7 percent less in developed countries, and 23 percent less globally.

The water crisis scenario would also have significant impacts on other water users. Demand for industrial water would increase, owing to failed technological improvements and economic measures. With water diverted to make up for less efficient water use in other

sectors, the water crisis scenario would hit environmental uses particularly hard.

The water crisis scenario would have severe consequences for food production. Total cereal production, for example, would be 10 percent less than under the business-as-usual scenario—the result of declines in both cultivated area and yields. This reduction is the equivalent of an annual loss of the entire cereal crop of India, or the combined annual harvest of sub-Saharan Africa and West Asia and North Africa. The decline in food production would help push up food prices sharply under the water crisis scenario. These high prices would in turn dampen food demand.

The ultimate result of this scenario is growing food insecurity, especially in developing countries. Per capita cereal consumption in 2025 in the developing world would be 2 percent lower than 1995 levels. This scenario makes it clear that increasing water scarcity, combined with poor water policies and inadequate

investment in water, has the potential to generate sharp increases in cereal food prices over the coming decades. Price increases of this magnitude would take a significant bite out of the real income of poor consumers. Malnutrition would increase substantially, given that the poorest people in low-income developing countries spend more than half their income on food. Sharp price increases could also fuel inflation, place severe pressure on foreign exchange reserves, and have adverse impacts on macroeconomic stability and investment in developing countries.

Sustainable Water Scenario

A sustainable water scenario would dramatically increase the amount of water allocated to environmental uses, connect all urban households to piped water, and achieve higher per capita domestic water consumption while maintaining food production at the levels described in the business-as-usual scenario. It would achieve greater social equity and environmental protection through both careful reform in the water sector and sound government action.

Governments and international donors would increase their investments in crop research, technological change, and reform of water management to boost water productivity and the growth of crop yields in rainfed agriculture. Accumulating evidence shows that even drought-prone and high-temperature rainfed environments have the potential for dramatic increases in yield. Breeding strategies would directly target these rainfed areas. Improved policies and increased investment in rural infrastructure would help link remote farmers to markets and reduce the risks of rainfed farming.

To stimulate water conservation and free up agricultural water for environmental, domestic, and industrial uses, the effective price of water to the agricultural sector would be gradually increased. Agricultural water price increases would be implemented through incentive programs that provide farmers income for the water that they save, such as charge-subsidy schemes that pay farmers for reducing water use, and through the establishment, purchase, and trading of water-use rights. By 2025, agricultural water prices would be twice as high in developed countries and three times as high in developing countries compared with the business-as-usual scenario. The government would simultaneously transfer water rights and the responsibility for operation and management of irrigation systems to communities and water user associations in many countries and regions. The transfer of rights and systems would be facilitated with an improved legal and institutional environment for preventing and eliminating conflict and with technical and organizational training and support. As a result, farmers would increase their on-farm investments in irrigation and water management technology, and the efficiency of irrigation systems and basin water use would improve significantly.

River basin organizations would be established in many water-scarce basins to allocate water among stakeholder interests. Higher funding and reduced conflict over

water, thanks to better water management, would facilitate effective stakeholder participation in these organizations.

Farmers would be able to make more effective use of rainfall in crop production, thanks to breakthroughs in water harvesting systems and the adoption of advanced farming techniques, like precision agriculture, contour plowing, precision land leveling, and minimum-till and no-till technologies. These technologies would increase the share of rainfall that goes to infiltration and evapotranspiration.

Spurred by the rapidly escalating costs of building new dams and the increasingly apparent environmental and human resettlement costs, developing and developed countries would reassess their reservoir construction plans, with comprehensive analysis of the costs and benefits, including environmental and social effects, of proposed projects. As a result, many planned storage projects would be canceled, but others would proceed with support from civil society groups. Yet new storage capacity would be less necessary because rapid growth in rainfed crop yields would help reduce rates of reservoir sedimentation from erosion due to slash-and-burn cultivation.

Policy toward groundwater extraction would change significantly. Market-based approaches would assign rights to groundwater based on annual withdrawals as well as the renewable stock of groundwater. This step would be combined with stricter regulations and better enforcement of such tighter controls. Groundwater overdrafts would be phased out in countries and regions that previously pumped groundwater unsustainably.

Domestic and industrial water use would also be subject to reforms in pricing and regulation. Water prices for connected households would double, with targeted subsidies for low-income households. Revenues from price increases would be invested to reduce water losses in existing systems and to extend piped water to previously unconnected households. By 2025, all households would be connected. Industries would respond to higher prices, particularly in developing countries, by increasing in-plant recycling of water, which reduces water consumption.

With strong societal pressure for improved environmental quality, allocations for environmental uses of water would increase. Moreover, the reforms in agricultural and nonagricultural water sectors would reduce pressure on wetlands and other environmental uses of water. Greater investments and better water management would improve the efficiency of water use, leaving more water instream for environmental purposes. All reductions in domestic and urban water use, due to higher water prices, would be allocated to instream environmental uses.

In the sustainable water scenario, the world consumes less water but reaps greater benefits than under the business-as-usual scenario, especially in developing countries. In 2025, total worldwide water consumption would be 20 percent lower under the sustainable scenario than under the business-as-usual scenario. This reduction in consumption would free up water for environmental uses. Higher water prices and higher water-use efficiency would reduce consumption of irrigation water by 296 km^3 compared with the business-as-usual scenario. The reliability of irrigation water supply would be reduced slightly in the sustainable scenario—as compared with the business-as-usual scenario—because of a higher priority on environmental flows. Over time, however, more efficient water use in this scenario would counterbalance the transfer of water to the environment and would result in an improvement in the reliability of supply of irrigation water by 2025.

This scenario would improve the domestic water supply through universal access to piped water for rural and urban households. Other water sectors would also be affected under the sustainable water scenario. Industrial water demand would be reduced under the sustainable water scenario through technological improvements and effective economic incentives. The environment would be a major beneficiary of the sustainable water scenario, with large increases in the amount of water reserved for wetlands, instream flows, and other environmental purposes.

The sustainable water scenario can raise food production slightly over the business-as-usual scenario, while

Figure 4. Total and irrigation water consumption, by region, business-as-usual and sustainable scenarios, 2025

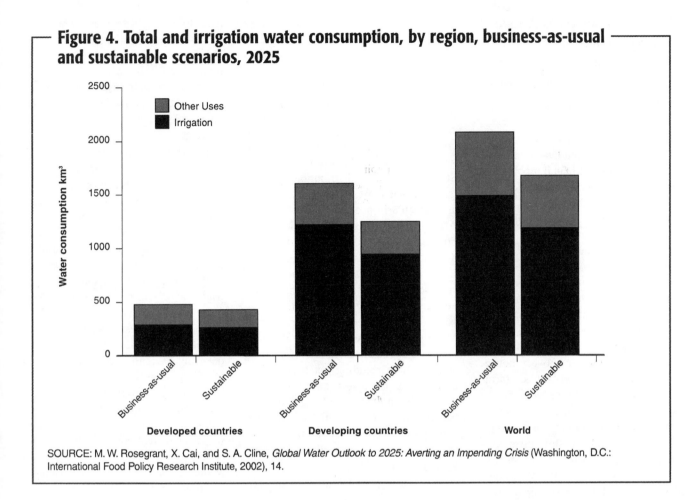

SOURCE: M. W. Rosegrant, X. Cai, and S. A. Cline, *Global Water Outlook to 2025: Averting an Impending Crisis* (Washington, D.C.: International Food Policy Research Institute, 2002), 14.

achieving much greater gains for domestic water use and the environment. The total harvested area under the sustainable water scenario in 2025 would be slightly lower than under the business-as-usual scenario, owing to less water for irrigation and slightly lower crop prices. With faster growth in rainfed yields making up for slower growth in harvested area and irrigated yields, total cereal production in 2025 would be 19 million tons more (a 1 percent difference) under the sustainable water scenario than under the business-as-usual scenario.

The sustainable scenario shows that with improved water policies, investments, and rainfed cereal crop management and technology, growth in food production can be maintained while universal access to piped water is achieved, and environmental flows are increased dramatically. Compared with the water crisis scenario, the increase in environmental flows under the sustainable water scenario would be about 1,490 km^3, equivalent to 5 times the annual flow of the Mississippi River, 20 times the annual flow of the Yellow River, and 4 times the annual flow of the Ganges River.

Implications for the Future

Water scarcity will get much worse if policy and investment commitments from national governments, international donors, and development banks weaken further. The water crisis scenario—predicated on the worsening of a number of already evident trends—would lead to a breakdown in domestic water service for hundreds of millions of people, a devastating loss of wetlands, serious reductions in food production, and skyrocketing food prices that would force declining per capita food consumption in much of the world.

Failure to adopt water-saving technology improvements and policy reforms could make demand for nonirrigation water grow even faster than projected, further worsening water scarcity.

In the sustainable water scenario, the world consumes less water but reaps greater benefits, especially in developing countries.

Water scarcity can lead to declining food demand and increasing food prices. As shown in the water crisis scenario, major cereal crop prices may be more than double the projections under the business-as-usual scenario, and at the same time food demand may be significantly reduced, especially in developing countries. Moreover, price increases can have an even larger impact on low-income consumers.

Water scarcity will get much worse if policy and investment commitments from national governments, international donors, and development banks weaken further.

Excessive diversion of water flows and overdraft of groundwater have already caused environmental problems in many regions around the world. The analysis shows that the problems, from a local to a worldwide scale, will likely be even more serious in the future. If current investment plans and recent trends in the water and food sectors continue, expanding the environmental uses of water would require reducing the consumption of irrigation water or domestic and municipal water or both. Thus, in the absence of policy and investment reform, competition over water between households and industries and between farmers and environmental uses will increase in many parts of the world.

With water becoming increasingly scarce, continued high flow diversions would become self-defeating. Excess extraction speeds the recession of ecological systems and lowers water quality, finally reducing the qualified water supply for human uses. This has already occurred in the Aral Sea Basin in Central Asia. Groundwater overdraft can likewise lead to the loss of an important water source for human uses, as is already happening in many regions.

However, the analysis also reveals cause for hope. The scenarios explored here point to three broad strategies that can address the challenge posed by water scarcity for food production:

- investment in infrastructure to increase the supply of water for irrigation, domestic, and industrial purposes;
- conservation of water and improvements in the efficiency of water use in existing systems, through reforms in water management and policy, including stronger incentives for water conservation; and
- improvements in crop productivity per unit of water and land through integrated water management and agricultural research

and policy efforts, including crop breeding and water management for rainfed agriculture.

Although the financial, environmental, and social costs of new water supply projects are high, in some regions, especially in developing countries, it is still crucial to selectively expand water supply, storage, and withdrawal capacities. Storage and water distribution systems (such as water lift projects and canals) are particularly needed for sub-Saharan Africa, some countries in South and Southeast Asia (such as Bangladesh, India, and Vietnam), and some countries in Latin America. These countries must consider not only the full social, economic, and environmental costs of development but also the costs of failure to develop new water sources. Projects must be designed to account for full costs and benefits, including not only irrigation benefits but also health, household water use, and catchment improvement benefits. It is also essential to improve compensation programs for those who are displaced or negatively affected by water projects.

Expanding water supplies can help alleviate water scarcity, but the results show that the most promising avenue is likely to be water management reforms, incentive policies, and investments in infrastructure and technology to enhance efficiency in existing uses. Feasible improvements in the efficiency of basin-scale irrigation water use can, on a global scale, compensate for irrigation reduction resulting from the phasing out of groundwater overdraft worldwide, increased committed environmental flows, higher prices for agricultural water use (which themselves encourage investments in improved efficiency), and low irrigated area development. In addition, improving irrigation water-use efficiency is an effective way to increase water productivity.

In severely water-scarce basins, however, relatively little room exists for improving water-use efficiency, and food production and farm incomes could fall significantly if water for irrigation is transferred to other uses. In these basins, governments will need to seek alternative means to compensate for the negative im-

pact of growing water scarcity on agriculture, such as investing in agriculture to obtain more rapid growth in crop yields, promoting the diversification of farming into less water-intensive crops, and diversifying the economy to reduce the economic role of agriculture over time.

Making big improvements in river-basin efficiency in specific river basins will require site-specific analysis and implementation. Basin efficiency depends on improvements both in water-saving technologies and in the institutions governing water allocation, water rights, and water quality. Industrial water recycling, such as recirculation of cooling water, can be a major source of water savings in many countries. Much potential also exists for improving the efficiency of domestic water use. Steps may include anything from detecting and repairing leaks in municipal systems to installing low-flow showerheads and low-water or waterless toilets. Treated wastewater can be used for a variety of nonpotable purposes including landscape and recreational irrigation, maintenance of urban stream flows and wetlands, wastewater-fed aquaculture, and toilet flushing. To encourage water-saving innovation, domestic and industrial water prices should be increased. Generalized subsidies should be replaced with subsidies targeted to the poor. Water providers should charge low prices for a basic entitlement of water, with increasing prices for greater amounts of water.

Improvements in the irrigation sector can be made at the technical, managerial, and institutional levels. Technical improvements include advanced irrigation systems, such as drip irrigation; sprinklers; conjunctive use of surface and groundwater; and precision agriculture, including computer monitoring of crop water demand. Managerial improvements include the adoption of demand-based irrigation scheduling systems and improved equipment maintenance. Institutional improvements involve the establishment of effective water user associations and water rights, the creation of a better legal environment for water allocation, and the introduction of higher water prices. Great care must be taken in designing a water pricing system for agriculture. Direct water price increases are likely to be punitive to farmers because water plays such a large role in their cost of production. Better alternatives would be pricing schemes that pay farmers for reducing water use, and water rights and water trading arrangements that provide farmers or water user associations with incentives to reduce wasteful water use.

Rainfed agriculture is a key to sustainable development of water and food. Rainfed agriculture still produces about 60 percent of total cereals, and its role remains very important in both the business-as-usual and the sustainable water scenarios. Improved water management and crop productivity in rainfed areas would relieve considerable pressure on irrigated agriculture and on water resources. Exploiting the full potential of rainfed agriculture, however, will require investing in water-harvesting technologies, crop breeding targeted to rainfed environments, agricultural extension services, and access to markets, credit, and input supplies in rainfed areas.

A large part of the world is facing severe water scarcity, but the impending water crisis can be averted. The precise mix of water policy and management reforms and investments, and the feasible institutional arrangements and policy instruments to be used, must be tailored to specific countries and basins. They will vary based on level of development, agroclimatic conditions, relative water scarcity, level of agricultural intensification, and degree of competition for water, But these solutions are not easy, and they take time, political commitment, and money. Fundamental reform of the water sector must start now.

NOTES

1. The source of these data is the 2002 IMPACT-WATER assessments and projections. IMPACT-WATER is a global modelling framework that combines an extension of the International Model for Policy Analysis of Agricultural Commodities and Trade (IMPACT) with a newly developed Water Simulation Model (WSM). For a more detailed description of the integrated model, see M. W. Rosegrant, X. Cai, and S. A. Cline, *World Water and Food to 2025: Dealing with Scarcity* (Washington, D.C.: International Food Policy Research Institute, 2002).

2. Food and Agricultural Organization of the United Nations (FAO), *FAOSTAT Database 2000* (Rome: FAO, 2000), accessible via `http://apps.fao.org/`.

3. World Health Organization (WHO) and United Nations Children's Fund (UNICEF), *Global Water Supply and Sanitation Assessment 2000 Report* (Geneva: UN, 2000).

4. For a more detailed analysis of these and other scenarios and a detailed discussion of methodology, see M. W. Rosegrant. X. Cai, and S. A. Cline, *Global Water Outlook to 2025: Averting an impending Crisis* (Washington, D.C.: International Food Policy Research Institute, 2002); and Rosegrant, Cai, and Cline, note 1 above.

5. West Asia and North Africa is one of the regional groupings used in the IMPACT-WATER model. The region consists of the following countries: Egypt, Turkey, Algeria, Cyprus, Iran, Iraq, Jordan, Kuwait, Lebanon, Libya, Morocco, Saudi Arabia, Syria, Tunisia, United Arab Emirates, and Yemen.

6. J. A. Allan, "Water Security Policies and Global Systems for Water Scarce Regions," in *Sustainability of Irrigated Agriculture—Transactions*. Vol. 1 E, special session: "The Future of Irrigation under Increased Demand from Competitive Uses of Water and Greater Needs for Food Supply—R.7" in the symposium on Management Information Systems in Irrigation and Drainage, Sixteenth Congress on Irrigation and Drainage, Cairn (New Delhi: International Commission on Irrigation and Drainage, 1996). 117–32.

Mark W. Rosegrant is senior research fellow at the international Food Policy Research Institute (IFPRI) and principal researcher at the International Water Management Institute (IWMI). Rosegrant has extensive experience in research and policy analysis in agriculture and economic development, with an emphasis on critical water issues as they impact world food security, rural livelihoods, and environmental sustainability. He alto developed IFPRI's International Model for Policy Analysis of Agricultural Commodities and Trade (IMPACT) and the IMPACT-WATER models. He continues to lead the team that maintains the models, which have been used to examine key policy options concerning food prices, food security, livestock and fisheries demand, agricultural research allocation, water resources, environment, and trade. He also currently coordinates a joint modeling team between IFPRI and IWMI, developing integrated global water and food models, and has written and edited numerous publications on agricultural economics, water resources, and food policy analysis. He can be reached via e-mail at m.rosegrant@cgiar.org. **Ximing Cai** is a research fellow at IFPRI and a researcher with IWMI, Cai's current research interests include water resource planning and management, operations research and their application to integrated water resources, and agricultural and economic systems. He can be reached via e-mail at x.cai@cgiar.org. **Sarah A. Cline** is a research analyst with IFPRI. Cline's work at IFPRI focuses on water resources policy and management, as well as global food supply, demand, and trade issues. She can be reached via e-mail at s.cline@cgiar.org. The authors retain copyright.

From *Environment,* September 2003, pp. 26-36. Published by Heldref Publications, 1319, Eighteenth St., NW, Washington, DC 20036-1802. Copyright © 2003. Reprinted by permission of the authors.

FIRE FIGHT

With forests burning, U.S. officials are clashing with environmentalists over how best to reduce the risk of catastrophic blazes.

BY PAUL TRACHTMAN

KATE KLEIN parks her U.S. Forest Service pickup truck along a muddy dirt road and climbs up a steep, rocky outcrop through a ghostly stand of burned ponderosa pines. Her boots sink into soot and ash. It is spring in the Apache-Sitgreaves National Forests in eastern Arizona and new grasses and seedlings should be turning the earth green. But from the top of the hill, she looks out over black trees as far as the eye can see, the remains of one of the largest wildfires in Arizona's history.

Klein, a 49-year-old district ranger with the Forest Service, had spent the better part of a decade trying to prevent a fire here (about 130 miles north of Tucson's June 2003 Aspen Fire, the first major blaze of the season) or at least minimize its effects. The 616,000 acres of the Black Mesa District under her care had long been a powder keg, she says, "a disaster waiting to happen," with too many trees per acre, too much deadwood littering the ground and everything made incendiary by years of drought. She came to believe that the only way to avoid catastrophic fires was to thin the forests through commercial logging, a process that would reduce what foresters call the "fuel load" and slow a fire's spread, giving firefighters a better chance of stopping it.

From 1996 to 1999, Klein and her staff studied the likely impact of logging on a 28,000-acre tract about six miles southwest of Heber-Overgaard, a mountain community of nearly 3,000 people. They had warned that a big fire could roar out of the forest and threaten Heber-Overgaard and nearby communities, places where more and more vacationers and retirees have built homes. "But

when we talked to these people about thinning," she recalls, "most of them opposed it, because they moved here for the forest."

If local resistance surprised her, it was nothing compared with the battles to come. In September 1999, having developed a plan to log a third of the tract, Klein's staff filed an 81-page report—required by U.S. regulations—outlining the possible environmental impacts. Environmentalists pounced. Lawyers for the Tucson-based Center for Biological Diversity, nicknamed nature's legal eagles, and two other nonprofit environmental groups said the study had insufficiently evaluated the effects on the environment and such wildlife as the Mexican spotted owl. They challenged the Forest Service computer model that suggested that the northern goshawk's habitat would actually be improved. They protested the harvest of large trees. The center barraged Klein with questions about logging trees infested with a parasitic plant called dwarf mistletoe: "What are the levels of infection in these stands? Have past harvests designed to stop dwarf mistletoe worked? Has the Forest Service monitored any such sales? Why is such a heavy-handed approach being used?" The environmental groups appealed to regional Forest Service officials to stop the project in November 1999. In February 2000, when the appeal was rejected, they notified the service that they intended to sue to block the project. Foresters continued to ready the forest for logging, marking trees to be cut.

Over the next two years, Klein's staff worked with lawyers on the legal case, responding to more questions and gathering more data. "If we don't write everything down, it's assumed we didn't consider it," she says. "Every time we lose a battle, we have to go back and do more analysis, computer models and evaluations. It's a downward spiral. We're forced to do so much writing that we spend less time in the woods knowing what we're making a decision about."

Until now, Klein had always thought of herself as an environmentalist. She had joined the Peace Corps and served in Honduras after receiving her forestry degree from Penn State in 1976. One of her first Forest Service assignments was at a New Mexico outpost, where she'd been proud to live in a house built by the pioneering forester and conservationist Aldo Leopold, author of the 1949 *A Sand County Almanac,* a bible of the environmental movement.

In mid-June of 2002, Klein prepared her final rebuttals to the complaints of the legal eagles. Meanwhile, the drought extended into its fourth year. "The week before the fire, there were three of us in the office working on our response," she says. "We worked all week and Friday night and Saturday, and we had just completed our report and sent it to the regional office on Monday. A fire broke out on Tuesday, a second fire started on Thursday, and four or five days later the whole area had burned up. Talk about frustration and hopelessness and anger and depression!"

The Tuesday fire had been set by an arsonist on the Fort Apache Indian Reservation, 22 miles from the Black Mesa Ranger Station in Heber-Overgaard. This fire was already burning out of control when, two days later, a hiker lost on the reservation started a fire to signal for help. Soon these two fires, the Rodeo and Chediski, would merge into an inferno.

FORESTS ACROSS THE WEST are primed for catastrophic fire, in part by a government policy put in place after the "Big Blowup," in 1910, a two-day firestorm that incinerated three million acres in Idaho and Montana and killed 85 people. The fire was so ferocious that people in Boston could see the smoke. The U.S. Forest Service, then five years old, decided to put out every fire in its domain, and within three decades the agency had formulated what it called the 10 a.m. policy, directing that fires be extinguished no later than the morning after their discovery. As fire-fighting methods improved through the years, the amount of burned forest and grassland declined from about 30 million acres annually in 1900 to about 5 million in the 1970s.

"If we had done all the thinning we wanted to over the years, we could have kept this fire from exploding," says the Forest Service's Kate Klein. In part, she blames environmentalists.

But the success of fire suppression, combined with public opposition to both commercial logging and preventive tree thinning on federal land, has turned Western forests into pyres, some experts say, with profound ecological effects. The vast ponderosa pine forests of the West evolved with frequent low-intensity ground fires. In some places, land that had as many as 30 or 40 large ponderosa pines scattered across an acre in the early 1900s, in grassy parklike stands, now have 1,000 to 2,000 smaller-diameter trees per acre. These fuel-dense forests are susceptible to destructive crown fires, which burn in the canopy and destroy most trees and seeds.

"It's as if we've spilled millions of gallons of gasoline in these forests," says David Bunnell, the recently retired manager of the Forest Service's Fire Use Program, in Boise, Idaho, which manages most wildland and prescribed fires and coordinates fire-fighting resources in the United States. During the past 15 years, the amount of acreage burned by wildfires has climbed, reversing a decades-long decline. In 2002, almost seven million acres burned—up from four million in 1987—and the federal government spent $1.6 billion and deployed 30,000 firefighters to suppress wildfires. Twenty-three firefighters were killed.

Decades ago, Aldo Leopold prophetically warned that working to keep fire out of the forest would throw nature out of balance and have untoward consequences. "A measure of success in this is all well enough," he wrote in the late 1940s, "but too much safety seems to yield only danger in the long run." Recently, the Forest Service has come around to Leopold's view, but many environmentalists continue to oppose agency plans to remove timber from forests.

Klein, who took over management of the Black Mesa District in 1991, places herself in Leopold's camp. "Over my years here, we've put out hundreds of lightning starts as quickly as we could," she says. The practice protected communities at the time, she adds, but also increased the risk of fire in the long run.

BY NIGHTFALL, June 18, firefighters dispatched to the Fort Apache Indian Reservation believed they might contain the arsonist's blaze. But the Rodeo Fire was burning too hot and too fast. On the morning of June 20, the other blaze—the Chediski Fire—was threatening to jump the Mogollon Rim and attack Heber-Overgaard and other communities. Klein's husband, Duke, a wildlife biologist, and their three children were evacuated from the family home in Heber-Overgaard along with everyone else as the flames closed in. For most of the day, she didn't know where they were.

Firefighters at the Black Mesa Ranger Station hoped to make a stand along a forest road on the rim, but they had only one bulldozer and fewer than 30 people. Klein called her boss and requested more firefighters. "He just said there aren't any; you're not going to get 'em," she recalls. Major fires had hit other states, and about 1,000 firefighters were already working above and below the rim.

The morning of June 22, the Chediski Fire raced 12 miles, jumped the rim and reached the Sitgreaves Forest tract that Klein had targeted for thinning. Returning from a briefing she'd given firefighters in nearby Honda that afternoon, Klein drove through "miles and miles of fire," she recalls, past burned-out houses and a blackened trailer park. "I got back to find it had overrun the town and was threatening the ranger station. It had run six or seven miles in a few hours. Its power awed me. Flames rose a couple of hundred feet in the air. It looked like the fire was boiling up there, and you'd see pieces of trees, branches going up. People were scared. I talked to the crews, and they had gotten into some very hairy situations trying to defend the station. In the evening, the fire died down a little, but around midnight we found out that a whole subdivision was threatened. So those guys went out and started fighting the fire again. They worked all night and kept at it until about noon the next day. We didn't have any replacements."

By the next day, the Rodeo Fire began to merge with the Chediski Fire, becoming one great conflagration, eventually stretching 50 miles across. It was what experts call a "plume-dominated fire," intense enough to generate its own weather, with towering thunderheads and rain that evaporated as it fell.

That night, Klein drove up a canyon and at 2 a.m. reached the head of the blaze, a harmless-looking ground fire just creeping along. But there was nobody she could dispatch to attack it. "I felt totally helpless." That morning, Monday the 24th, the fire made another run, which destroyed more houses. Then, on Tuesday, a team of fire-fighters arrived: soon there were more than 2,000 fire-fighters along Highway 260, which runs through Heber-Overgaard. Firefighters subdued part of the inferno with backfires—fires intentionally set to reduce fuel in the path of the oncoming blaze. The rest eventually burned itself out as it ran into patchier, less flammable piñon-juniper country.

Over 20 days, the Rodeo-Chediski Fire burned more than 460,000 acres. About 50,000 people were evacuated and 465 residences destroyed. Klein's house was spared, but many of her friends and neighbors were not so lucky; 15 percent of Heber-Overgaard was destroyed. Ultimately, more than 6,600 firefighters had fought the blaze, aided by 12 air tankers, 26 helicopters, 245 fire engines, 89 bulldozers and 95 water-supply trucks. Suppressing the fire cost about $43 million. It will cost another $42 million or so to do emergency rehabilitation in the forest, such as reseeding to prevent erosion and flooding, and long-term recovery work.

"The Forest Service is hijacking important concepts like fuels reduction to disguise traditional timber sales," says environmentalist Brian Segee. "They're turning the forest into a tree farm."

The tragedy still galls Klein. "If we had done all the thinning we wanted to over the years, we could have kept this fire from exploding, and we could have saved the towns it burned through." In a sense, she blames environmental activists. "All those arguments we heard about how 'your timber sale is going to destroy Mexican spotted owl habitat,' 'your timber sale is going to destroy the watershed.' And our timber sale wouldn't have had a fraction of the effect a severe wildfire has. It doesn't scorch the soil, it doesn't remove all the trees, it doesn't burn up all the forage. And then to hear their statements afterward! There was no humility, no acceptance of responsibility, no acknowledgement that we had indeed lost all this habitat that they were concerned about. All they could do was point their finger at us and say it was our fault."

For its part, the group that led the fight against Klein's tree-thinning proposal hasn't changed its thinking. Environmentalists at the Center for Biological Diversity believe that even if the project had gone ahead, it wouldn't have made a difference in halting such a large and destructive fire. "The Forest Service is hijacking important concepts like fuels reduction to disguise traditional timber sales," says Brian Segee, the center's Southwest public lands director. "I walked the ground and looked at the marking of trees, and they are turning the forest into a tree farm. When economics drives the decisions, it ultimately results in ecosystem degradation, and we just keep finding that when we don't resort to the courts, we're ignored."

NOT EVERY FORESTER has embraced the idea of fighting every fire. In 1972, in the Wilderness Area of Montana's Bitterroot National Forest, a handful of Forest Service heretics intentionally let a lightning strike burn—the first time the agency had done that. One of the maverick foresters, Bob Mutch, then a young researcher at the Forest Service Sciences Fire Laboratory, in Missoula, Montana, had had the idea that forest health might actually depend on fire. To be sure, a few foresters had previously argued that forests evolved with fire and were adapted to it, but they had been proverbial voices in the wilderness.

Mutch and the others are now retired, but in the midst of the destructive fire season of 2002—and only six weeks after the Rodeo-Chediski Fire scorched Arizona—they journeyed to the Bitterroot Mountains to assess the experiment they had begun three decades earlier. The Forest Service, whose orthodoxy they once challenged, now wanted their advice on preventing catastrophes from occurring in national forests.

In the Bitterroot Mountains, it's only a short way from Paradise to Hell's Half Acre. The ranger outpost at Paradise, where the veterans initially gathered, is a place of deep silence, sparkling water and tall ponderosa pines. The men were eager to look at "the scene of the crime," as they called it. They hardly looked like rebels. Among them was Bud Moore, in his mid-80s, who had grown up in a family of woodcutters and trappers in these mountains, and was hired as a Forest Service smoke chaser in 1935. There was Bill Worf, just a few years younger, who today is almost blind and last summer hiked the wilderness trail with black glasses and a white cane while someone ahead warned of fallen logs across the path. Orville Daniels, now 68, was the supervisor of the Bitterroot National Forest back in 1970. And there were Bob Mutch and Dave Aldrich, who now looked a bit like members of the Monkey Wrench Gang (as author Edward Abbey called a bunch of radical environmentalists in his 1975 novel of the same name). Aldrich, a muscular 63-year-old, had always looked at fire as the enemy until he joined the group. Mutch, 69, an intellectual and a researcher with a passion for ecology, had once been a smoke jumper, a Forest Service firefighter who parachutes from planes.

The only member of the group still employed at that time by the Forest Service was David Bunnell, 59. He was a firefighter before falling in with the Bitterroot bunch in the 1970s, and he remembers well his first encounter with them. "Renegades! Heretics!" he recalls thinking. "I'm surprised they weren't all fired."

As the group hiked a nine-mile trail from the Paradise guard station to a clearing called Cooper's Flat, every step took them through country they'd once watched burn. They pitched tents and talked late into the night over a campfire, reminiscing, and discussing what their experiment had told them about how best to manage America's national forests.

It was Bud Moore who had ignited their conspiracy. In 1969, he was transferred from Washington, D.C. to Missoula as regional director of what was then called Fire Control and Air Operations. As a Bitterroot native, he knew these woods deeply and sensed that fire was a part of their ecology. "When we were starting this program," he says, "we got tremendous support from the environmental community. The biggest resistance we had was in the Forest Service. We had that big culture of firefighters, and I was one of them."

Worf was one of them also. The idea that fire might belong in the wilderness didn't come easily to him. He'd spent years managing timber sales and fighting fires. In 1964, he landed on a task force in Washington, D.C. that was looking at how the Forest Service could implement the recently passed Wilderness Act, which defined wilderness as a place where "the earth and its community of life are untrammeled by man." Worf says, "They weren't talking about a pretty place to backpack!" He read Leopold, who had proposed that the United States set aside wilderness areas and watch nature without getting in its way. In 1969, Worf took a job as regional director of Recreation, Wilderness and Lands in Missoula, where he and Moore got together and agreed that managing wilderness meant leaving some natural fires alone.

"'We're thinking about a pilot project on fire use in wilderness,'" Daniels recalls Moore telling him in a phone call. "It just flashed through my mind, 'Of course this is what we should do.'" Mutch and Aldrich, who had recently joined Daniels' staff, began making inventories of trees and other vegetation, searching for clues to the history of fire in the forests. They cut into fire scars on ponderosa pine, revealing charred tree rings going back as far as the 1720s, showing that fires had burned there every 10 to 20 years. Those blazes evidently were ground fires that periodically cleared away flammable debris, stimulated regeneration of shrubs and grasses and, in general, did not kill large, healthy trees. "We were trying to re-create in our minds how fires had burned on these lands," Aldrich says, "and then write prescriptions for trying to bring fire back."

Their main concern was to keep wildfires from escaping beyond the wilderness, and they developed criteria for letting a fire burn and provisions for fighting the blaze if things went wrong. Aldrich remembers refining his ideas with Mutch late into many a night at Cooper's Flat. Finally, in August 1972, Daniels and Mutch flew to Washington and presented a plan to the agency's top brass to form what would become the Wilderness Prescribed Natural Fire Program. The plan was approved. Now all they needed was a fire.

They got their first one within days, but it petered out. It took a year of waiting before they got a big one. On August 10, 1973, lightning struck at Fitz Creek, which runs into White Cap Creek just above Paradise. As fire spread over the steep canyon slope along the White Cap, Daniels, Aldrich and Mutch stood by and watched. "Every day was a surprise," Aldrich recalls. "I learned more in a few days watching that fire than I did in the preceding 15 years fighting fire." He expected a much more intense fire. But up in the ponderosa pine forests, carpeted with thick layers of needles, the fire merely crept along. "I was able to step through the fire, or if it was burning intensely, I could run through it," he says. Blue grouse were picking away at the roasted pinecones. Mice and chipmunks scurried about. He saw a bull elk nonchalantly grazing about ten feet from the flames. Mutch noticed a black bear poking along the edges of the fire. Nowhere did they see any animals running scared.

But after five days, serenity gave way to shock. A "spot" of burning debris flew across White Cap Creek and ignited the north-facing slope, which was outside the area of the fire plan. Here, thick stands of highly flammable Douglas fir grew in the shade, surrounded by a heavy buildup of broken branches and other debris. "Dave and I were up at a lookout when we got the call that the fire was across the creek, and we turned around and saw this mushroom cloud," Mutch says. "In 30 minutes the fire had gone from the creek bottom 2,000 feet up to the ridge top, with 100-foot flame lengths, throwing spots everywhere. We just stared at it and said to each other, 'Oh my God, what have we done?'"

Daniels was called out of a public meeting in Missoula and raced back to the Bitterroot. Mutch was dispatched to brief a team of firefighters, some of them his old buddies, who were called in to stop the blaze. The firefighters wanted to put out the fire on both sides of the creek. But Daniels defended his turf. He designated the escaped fire the Snake Creek Fire and insisted that the firefighters leave his Fitz Creek Fire alone.

"They just absolutely thought we were crazy," he says, "but they put out the escaped fire after a week, at the cost of half a million dollars, and we allowed our fire to burn clear into the middle of September, and never had any more trouble with it."

THE UNCHECKED FITZ CREEK FIRE marked a profound change in Forest Service philosophy. Since 1972, says the Forest Service's Bunnell, federal agencies have made more than 4,000 decisions to stay the firefighter's hand, resulting in more than a million acres of public lands "treated" by natural wildland fires. In the Bitterroot Wilderness alone, Daniels and his successors have let more than 500 wildland fires burn freely, with impressive results. The Fitz Creek Fire veterans were amazed by what they saw in 2002. "It was the first time I've ever seen a forest working the way a natural forest should work," Daniels says. "You could see the results of all the old and new fires blended together in a mosaic; everything from old stands of decadent and dead trees where woodpeckers love to nest, to thick patches of young trees providing a home for the snowshoe hare, which in turn is prey to the lynx we're trying to recover. It's probably the way the forest looked before anyone began to influence it."

In 2000, a drought year, when Montana had its worst fire season in nearly a century, the Bitterroot Wilderness turned out to be fire resistant. A lot of fires got started, burning some 60,000 acres, but not one firefighter was needed to put them out. As the new fires kept running into places that had previously been allowed to burn, they stalled and expired for lack of fuels on the ground. "We've gained a lot of knowledge about natural fires in these ecosystems," says Jerry Williams, the Forest Service's director of Fire and Aviation Management, "and a lot of it came about watching wildland fires that we've let burn freely in the Bitterroot Wilderness over the past 30 years."

So far, though, the hands-off approach to fires has been mostly limited to wilderness areas. Other national forest areas are generally so dense and so loaded with debris and fuels that letting a lightning strike burn freely would lead to catastrophe. Foresters say that such areas would benefit from natural fires, but only after undergoing "mechanical treatment"—thinning trees and removing deadwood and other fuels. But when foresters propose such treatments, some environmentalists, who believe they, too, have the forest's best interests in mind, oppose the efforts.

Most Forest Service professionals advocate selling timber from national forests to help thin aging stands and also to defray the costs of noncommercial thinning. But some environmental groups argue that commercial logging does more to destroy the environment than to restore it, and some, like the Sierra Club, have called for an end to all commercial logging in national forests.

The environmental community's distrust of the Forest Service has deep roots, and veteran foresters acknowl-

edge past mistakes. Retired forester Bill Worf concedes that his generation was slow to accept the spirit, if not the letter, of environmental protection laws, and he even admits to a bit of creative obfuscation in times gone by. "You'd decide what you want to do, and then you would write an environmental impact statement that would support it. And that takes a lot of paper because you'd have to hide a lot of stuff." Environmentalists were particularly angered over the years by clear-cutting, or removing all trees from an area. Indeed, at the same time Daniels was approving fire use in the Bitterroot Wilderness, other parts of the Montana forest were the focus of a national battle over the practice. Mutch, the former forester, remembers that loggers "simply harvested what was there, then went in with bulldozers to put in terraces, and planted ponderosa pine seedlings." The result hardly replaced the complex forest that had been there. "It looked like rice paddy terraces in Southeast Asia," he went on. "It was very harsh treatment of the land. And people said, 'Hell no, that's timber mining!'"

The chief of the Forest Service, Dale Bosworth, says that clear-cutting is a thing of the past: "Most of what we harvest now is for stewardship purposes, habitat improvement for wildlife, restoration of watersheds and fuels reduction. All this shrill screaming about timber harvests is just a distraction from the real issue, which is about getting these fire-adapted ecosystems back into a healthy condition so they will be more resistant to catastrophic wildfires."

CONTROVERSY OVER fire management in national forests was boosted last year with the president's Healthy Forests Initiative, which followed the half-million-acre Biscuit Fire, in Oregon. The proposal, which is still being hammered into final form by the Forest Service and Congress, would let forest managers make some decisions about thinning and timber sales with less of the environmental impact analysis and documentation now required by law, and it would also limit the internal Forest Service appeals process, which some environmental groups have used to challenge decisions. A recently released audit by the General Accounting Office of Congress reports that 59 percent of the Forest Service's hazardous fuels reduction projects that were required to have environmental impact statements were appealed during fiscal years 2001 and 2002.

Political debate over the initiative has largely followed party lines, although a new Republican-sponsored Healthy Forests Restoration Act passed the House on May 21, 2003, with the support of 42 Democrats. For their part, Republican supporters say that the bill reflects the current thinking of Western governors and most foresters.

The National Resources Defense Council, a nonprofit environmental action organization, says the initiative is part of a Bush administration plan "to roll back 30 years of environmental progress." The proposal, the council says in a fund-raising flyer, "gives timber companies the right to cut down your last wild forests." Other environmental groups have called it a plan for "lawless logging" and "corporate giveaways."

Still, there are sprouts of compromise coming up though the ashes of last summer's major fires. Local citizens groups across the West are working with the Forest Service and other agencies, focusing on protecting communities near or within national forests. Everyone agrees that the "wildland-urban interface," where peoples' homes and other structures abut forest lands, is the place to start. "It's the first place you have to defend," says Klein. "But you can't devise all your strategies around the wildland-urban interface." That wouldn't protect watersheds, wildlife, old-growth stands, endangered species habitats, recreation areas and other parts of the fire-prone forest ecosystems out in the backcountry, she says.

The sometimes angry debate over healthy forests legislation rings hollow to Klein and many other foresters in the field. "We've almost gotten ourselves into a situation where nothing but a fire will fix it!" she says. "I think most of us working on the ground are disturbed with where we are, and we don't see an easy way out." She foresees a time when fire is allowed to play a larger role in forests, but not before communities are protected, forests thinned, the load of dead fuels reduced and political considerations tempered by ecological ones. Meanwhile, there will be more infernos, she says: "I think we have to accept that catastrophic wildfires are going to be part of getting back to a natural regime."

PAUL TRACHTMAN, a former science editor at SMITHSONIAN, lives in New Mexico. He wrote about artist James Turrell in May 2003.

UNIT 6

The Hazards of Growth: Pollution and Climate Change

Unit Selections

Key Points to Consider

- Why are agricultural pesticides that have been banned in the United States and other developed countries still used in the lesser-developed regions of the world? Do the benefits of increased food production outweigh the costs of using agricultural pesticides?

- How can monitoring of chemical concentrations in human tissues give clues as to environmental quality? Should such monitoring become a regular part of the health care system?

- How do dangerous contaminants enter organisms (such as plants) through the water system used for irrigation? Should companies that pollute sources of irrigation water be held accountable for potential damages to public health?

- How can increasing levels of carbon dioxide in ocean waters contribute to global warming? Is there a relationship between observed present-day phenomenae and climatic changes that have taken place in the past?

 **Links: www.dushkin.com/online/**
These sites are annotated in the World Wide Web pages.

Persistent Organic Pollutants (POP)
http://www.chem.unep.ch/pops/

School of Labor and Industrial Relations (SLIR): Hot Links
http://www.lir.msu.edu/hotlinks/

Space Research Institute
http://arc.iki.rssi.ru/eng/index.htm

Worldwatch Institute
http://www.worldwatch.org

Of all the massive technological changes that have combined to create our modern industrial society, perhaps none has been as significant for the environment as the chemical revolution. The largest single threat to environmental stability is the proliferation of chemical compounds for a nearly infinite variety of purposes, including the universal use of organic chemicals (fossil fuels) as the prime source of the world's energy systems. The problem is not just that thousands of new chemical compounds are being discovered or created each year, but that their long-term environmental effects are often not known until an environmental disaster involving humans or other living organisms occurs. The problem is exacerbated by the time lag that exists between the recognition of potentially harmful chemical contamination and the cleanup activities that are ultimately required.

A critical part of the process of dealing with chemical pollutants is the identification of toxic and hazardous materials, a problem that is intensified by the myriad ways in which a vast number of such materials, natural and man-made, can enter environmental systems. Governmental legislation and controls are important in correcting the damages produced by toxic and hazardous materials such as DDT, PCBs, or CFCs; in limiting fossil fuel burning; or in preventing the spread of living organic hazards such as pests and disease-causing agents. Unfortunately, as evidenced by most of the articles in this unit, we are losing the battle against harmful substances regardless of legislation, and chemical pollution of the environment is probably getting worse rather than better.

The first article in this unit deals with one of the most serious of all international pollution problems: that of soil and water pollution from the widespread application of agricultural pesticides. In "Agricultural Pesticides in Developing Countries" Sylvia Karlsson of Yale University's Center for Environmental Law and Policy notes that it was the use of agricultural pesticides in North America and Europe, along with their undesired and unanticipated side effects, that produced the first alerts (largely through Rachel Carson's *Silent Spring*) that modern society can have major impacts on Earth's ecological systems. While many of the most dangerous chemicals have been banned in countries like the United States, their benefits in terms of increased food production outweighs their perceived dangers in countries in the lesser-developed world. As a consequence, increases rather than decreases in biological pollution from pesticides have been noted in Africa and Asia, posing massive challenges for governments that now, whether they like it or not, operate in a large, interconnected system. In "Human Biomonitoring of Environmental Chemicals" scientists Ken Sexton, Larry Needham, and James Pirkle argue that the true measure of environmental pollution from chemical sources is the concentration of chemicals in human tissues. Unfortunately this most accurate measure of the true magnitude of chemical pollution is also the most expensive. In the future, suggest the authors, a full screening of biological markers of chemical concentrations in the human body may become a routine part of the annual physical examination. This monitoring, of course, will be part of the health care system in the world's richest countries—and largely absent in the remaining 70% of the world's population.

The final two articles in this selection deal with pollution at both ends of the spatial scale: from local to global. In "A Little Rocket Fuel with Your Salad?" science writer Gene Ayres begins with a brief description of the California reaction to the appearance of an insect pest that might damage crops in the state's important agribusiness areas—full coverage spraying with a toxic chemical, malathion. This introduction provides a segue into his real story, the dumping into groundwater and wastewater systems of perchlorates, a primary component of rocket fuel, by both industrial and government entities. Perchlorates are persistent chemicals—that is they stay within the environmental system for a long time—that eventually end up, in concentrated form, in crops irrigated with water that has been contaminated. Consumption of food items containing perchlorates is damaging to organic systems, including humans, and often causes significant organic damage. The section's last article also deals with the unwanted injection of harmful substances into the environment—but at the other end of the scale. Where Ayres' article dealt with local contamination, "Global Warming as a Weapon of Mass Destruction" by journalism professor Bruce Johansen, treats the carbonization of the world's oceans and the potentially-dire consequences for global climates. Unlike most studies that focus on atmospheric carbon dioxide and its impact

on the natural tendency of the atmosphere to retard the passage of the earth's radiant heat back into space (this is known as the "greenhouse effect"), Johansen examines the rising carbon dioxide levels in ocean water. These increasing concentrations threaten the health of marine organisms, including phytoplankton—the aquatic system's equivalent of green plants in terrestrial systems. Like green plants, phytoplankton absorb carbon dioxide and produce oxygen, helping to maintain a balance between the two gases in the atmosphere. Johansen's essential message is that the alterations in ocean organisms impact the atmosphere which, in turn, impacts the entire Earth and contributes further to enhancement of the greenhouse effect. The potential end result of this chain of events is atmospheric change that may eventually threaten life on earth.

The pollution problem might appear nearly impossible to solve. But solutions do exist: massive cleanup campaigns to remove existing harmful chemicals from the environment and to severely restrict their future use; strict regulation of the production, distribution, use, and disposal of potentially hazardous chemicals; the development of sound biological techniques to replace existing uses of chemicals for such purposes as pest control; the adoption of energy and material resource conservation policies; and the use of more conservative and protective agricultural and construction practices. We now possess the knowledge and the tools to ensure that environmental cleanup is carried through. It will not be an easy task, and it will be terribly expensive. It will also demand a new way of thinking about humankind's role in the environmental systems upon which all life forms depend. If we do not complete the task, however, the support capacity of the environment may be damaged or diminished beyond our capacities to repair it. The consequences would be fatal for all who inhabit this planet.

AGRICULTURAL PESTICIDES IN DEVELOPING COUNTRIES

A Multilevel Governance Challenge

By Sylvia I. Karlsson

The use of agricultural pesticides in North America and Europe and their non-desired side effects were among practices and subsequent effects that set alarm bells ringing—bells that, in the 1960s and 1970s, awakened the public and, eventually, government agencies to the impact our modern human life can have on the Earth's life-sustaining ecological systems. For the past 30-40 years in developing countries, agricultural pesticide use has set off a continuously ringing alarm, alerting us to a heavy toll on human health and the environment. It is an issue that clearly demonstrates links—between the local and the global and between developed and developing countries—often associated with environmental and societal problems. This creates a complex picture with many challenges to address in terms of governance.[1]

Following the trajectory of modern agriculture in developed countries, developing countries have, over the past half century, increasingly adopted a pest management approach that centers on the use of chemical pesticides. The pesticide world market's total value surpasses US$26 billion. Developing countries' share of this is approximately one-third.[2] It is reasonable to expect that developing countries will continue to experience similar—if not identical—patterns of negative human-health and environmental side effects that motivated industrialized countries to develop the pesticide regulatory apparatus in the 1970s. Looking at potential effects on human health, a variety of factors is likely to make local populations in developing countries—in particular farmers and agricultural workers but in general all food consumers—more vulnerable to the toxicological effects of pesticides.[3] Examples of such factors include

- low literacy and education levels;
- weak or absent legislative frameworks;
- climatic factors (which make the use of protective clothing while spraying pesticides uncomfortable);
- inappropriate or faulty spraying technology; and
- lower nutritional status (less physiological defense to deal with toxic substances).

The first four of these factors increase the likelihood of higher exposure; the last factor increases the toxic effects from that exposure on the human body. In addition, it has been found that when most organochlorine pesticides are banned or restricted, developing countries have often turned to substances that exert higher toxicity.[4]

According to a World Health Organization/United Nations Environment Programme (WHO/UNEP) working group report from 1990, unintentional acute poisonings with severe symptoms exceed 1 million cases each year, out of which 20,000 are fatal.[5] Additionally, there are estimated to be 2 million intentional poisonings (mainly suicide attempts) resulting in 200,000 deaths per year.[6] The vast majority of both unintentional and intentional intoxications occur in developing countries. However, good intoxication data are sparse in developing countries and these global estimates are often contested: Some stakeholders argue that they are too high, others insist that they are too low.[7] The data for estimating the toll of less acute effects, including long-term effects such as cancer, are even more sparse.[8]

Looking at potential environmental impacts of pesticide use in developing countries, many of the same factors that increase the vulnerability to health impacts are likely to exacerbate the release of the chemicals into the environment. Furthermore, in these regions there are unique ecosystems and species of ecological and economic importance and a range of managed systems (agricultural, silvicultural, and aquacultural) that are only marginally present in developed countries.[9] Because most species in developing countries are not subject to any tests before an industrialized country approves a particular pesticide, they may exhibit previously unknown sensitivities to pesticide exposure. Studies of pesticide impact on local environments in developing countries are few and far between: Data collection and research in these regions is extremely limited.[10] However, observed effects include contaminated ground, river, and coastal waters; fish kills; and impacts on cattle.[11] Environmental effects from organochlorine pesticides, a few of which are still in use, may also extend well beyond the region of use.[12] In the last decade, a hypothesis regarding the transport of such substances has received stronger support: It is believed that, due to their chemical characteristics, organochlorine pesticides are partially transported via the atmosphere from warmer tropical regions toward colder regions (the Arctic and Antarctica). According to this theory, the polar regions are thought to function as sinks for these substances as they condense and accumulate in ecosystems and food chains—ultimately affecting human health as well.[13]

Governance Challenges, from Local to Global

Pesticide use in developing countries is seen to produce benefits to society. At the same time, however, it has the potential to produce negative consequences on different scales, ranging from immediate health effects for those who spray them to environmental effects in remote regions. The reasons, or driving forces, behind these negative effects can be found in policies and actions covering levels of governance from the local to the global. The responses by stakeholders to reduce the negative effects are also found at each governance level. It is thus an issue that well illustrates an interconnected, globalized world and a multilevel governance challenge.[14]

To search for appropriate responses to such a globalized problem, it is necessary to examine the human activity—the use of pesticides in agriculture—that acts as the direct driving force for various undesired effects. In addition, it is helpful to analyze governance at the local, national, and global levels. It is equally important to determine how the diversity of stakeholders at these levels understand and structure the "pesticide problem" and address the problem through strategies of risk reduction.

Compared with other nations, Costa Rica produces the highest amount of coffee per hectare. Along with a number of other factors (many having to do with the banana trade), this has contributed to relatively high levels of pesticide use.

Such research—examining many aspects of pesticide use and governance in developing countries—was carried out in association with Linköping University, Sweden, in 1997-1999.

A close look at the stakeholders included in the study reveals three major categories: organizations (government and intergovernmental organizations (IGOs)), civil society (nongovernmental organizations (NGOs), private companies, and academia), and individuals (farmers and workers).[15]

At the global level of governance, some key stakeholders involved in pesticide use in developing countries (and which were examined in the 1997-1999 study) include IGOs such as the Food and Agriculture Organization of the United Nations (FAO) in Rome and the United Nations Environment Programme (UNEP) in Nairobi and Geneva (a more complete list appears below in the discussion of the activities of such organizations).

On the national level, two countries, Kenya and Costa Rica, served as important case studies of the issue. Both countries have enacted relatively ambitious pesticide legislation and other governance measures compared with other countries on their respective continents. Taken together, the countries' heavy dependence on—and the involvement of large businesses in—export agriculture, the type of export crops grown, and a number of other factors have made Kenya and Costa Rica substantial users of agricultural pesticides. The trend in this regard has been increasing for both nations in the past two decades.[16] In terms of product category, fungicides dominate import figures by volume in Costa Rica, followed by insecticides and fumigants.[17] In Kenya, fungicides account for half the market, insecticides 20 percent, and herbicides 18 percent.[18]

The high and increasing use of pesticides in Costa Rica has been attributed to such factors as the switch to horticultural crops; the expansion of land under banana production, a particularly difficult problem in banana plantations with the leaf spot disease *Mycosphaerella fijiensis*; the highest production of coffee per hectare (ha) in the world; and the fact that some transactions with pesticides have been exempted from taxes.[19] The volume of pesticide use is usually closely linked to the type of crop. In Costa Rica the average amount of active ingredient applied per ha was, according to one study, 0.25 kilograms (kg) for pasture, 3.5 kg for sugarcane, 6.5 kg for coffee, 10 kg for rice, 20 kg for vegetables/fruits, and 45 kg for bananas.[20]

Kenyan agricultural exports have been traditionally dominated by coffee and tea, but nontraditional crops such as pineapples, vegetables, and ornamentals have expanded quickly in recent years.[21] This expansion has contributed to increasing pesticide use. Historically, pesticide use per ha on the large agricultural farms has been much higher than on small farms. However, a substantial percentage of smallholder farmers in Kenya use pesticides, mainly on their export crops.[22]

When comparing approaches to pesticide use at the local level in different countries, it is helpful to look at the same crop. In the case of Kenya and Costa Rica, coffee is a good choice, and two prominent coffee growing regions are the Meru District in Kenya and the Naranjo *cantón* (district) in Costa Rica. Kenya's Meru District, which is located about 300 kilometers northeast of Nairobi (in the Eastern Province, covering the slopes of Mount Kenya), is one of the most fertile areas of the country with a range of subsistence and cash crops. Naranjo *cantón*, in Costa Rica's Alajuela province, is located in the hills northwest of San José. Coffee is the dominant crop, but there is also sugarcane and livestock production.[23] The land distribution is similar in both districts, with many small-scale landholders and a few medium- or large-scale landowners. However, the differences are substantial: Naranjo has significantly larger-sized farms, a higher level of modernization, and much higher literacy rates and general education levels than does Meru. Although the same variety of coffee, *Coffee arabica*, is grown in Meru and Naranjo, the answers farmers in the two districts gave to inquiries about pest problems share only some similarities. (Farmers in Meru were interviewed in early 1999; Naranjo

farmers were interviewed later that year.) In both areas, farmers said that the dominating pest problems had been fungi. However, Meru was also beset by a number of insect pests. In Naranjo, farmers said that insect pests were largely absent, but they mentioned that a number of nematodes (commonly known as roundworms) and different fungi have affected coffee production. In Meru, at least a quarter of the farmers responding to interviews were not spraying anything on their coffee. This was a new situation that had occurred for a few years prior to the interviews and were due primarily to falling coffee prices. Farmers in Meru have been spraying their coffee trees since the crop was introduced in the area in the 1950s. The interviews in Naranjo found that virtually all of the farmers there sprayed pesticides. Coffee was introduced in Costa Rica in the early 1900s—long before pesticides were available—but since the arrival of these chemicals in the 1950s, they have been applied ubiquitously.[24]

Which Problems and Whose Problems?

The stakeholders mentioned above raised five problem categories from pesticide use in developing countries: economic, production, human health, environment, and trade.[25] At the global level, pesticides are primarily seen as a health issue, especially for the poor, uneducated farmers and workers in developing countries who apply them. The pesticide-related trade problems for developing countries emerge in some agencies involved with regulations for pesticide residues in food crops. Persistent categories of pesticides—organochlorines—have emerged as a transboundary environmental issue in multilateral discussions and agreements, but there has been little mention of potential local and regional environmental effects in developing countries themselves. While the lack of data and research makes it impossible to give solid numbers and fig-

ures on health and environmental impacts of pesticides on a large scale in developing countries, the discussion above showed that there is enough isolated data coupled with higher risk factors in developing countries for IGOs to be taking the problem seriously.

At the national level in Kenya and Costa Rica, stakeholders involved in the study tended to either associate the entire assortment of pesticides with health and environmental effects or claimed there was no evidence for substantive negative effects. The government in Kenya tilted toward the latter position; various environmental NGOs in the country referred to substantial problems.[26] Costa Rican stakeholders made more references to significant health effects, and although environmental effects were usually discussed in general terms, the situation in banana plantations was often lifted out as potentially more serious. The trade problem has definitely remained high on the agenda for both countries, but stakeholders largely referred to it as a problem of the past, before appropriate policies had been put in place.[27]

There is little data on pesticide-related health effects in Kenya to substantiate opinions expressed regarding their magnitude. There was (at the time of the study) no working system to report intoxications to the authorities. In fact, one of the main areas of concern regarding chemical use generally was the absence of documentation on the risks in the country.[28] Figures surfaced in a national debate in the early 1990s claiming that 7 percent of the people in the agricultural sector—about 350,000 people—suffered pesticide poisoning each year.[29] Other sources reported that, in 1985, three major hospitals treated an average of two cases of pesticide poisoning every week.[30] Kenya's Ministry of Health estimated in 1996 that 700 deaths per year were caused by pesticide-related poisonings.[31] However, there were virtually no data at all on subacute poisonings from

Table 1. Toxicity classification of imported pesticides, Kenya and Costa Rica, 1993

World Health Organization (WHO) toxicity class	Percent total Kenya	Percent total Costa Rica
1a-Extremely hazardous	-	10
1b-Highly hazardous	3	8
Highly hazardous volatile fumigants	11	-
II-Moderately hazardous	22	24
Ill-Slightly hazardous	24	8
Unlikely to present acute hazard	25	40
Not classified by WHO	3	-
Unidentified	10	10

NOTE: It is very difficult to compare and interpret these kinds of import figures from different countries: Some of the products are imported as technical-grade material (concentrated) and others as ready-to-use formulations. For example, after formulation in Kenya the proportion of highly hazardous pesticides will be significantly higher, because a substantial portion of those substances are imported as technical-grade material. The figures show that nearly one-quarter of the pesticides used are moderately hazardous. More than 10 percent are highly hazardous.

SOURCE: H. Partow, *Pesticide Use and Management in Kenya* (Geneva: Institut Universitaire D'Etudes du Développement (Graduate Institute of Dévelopement Studies), 1995); and F. Chaverri, and J. Blanco, *Importación, Formulación y Use de Plaguicidas en Costa Rica*. Periodo 1992-1993 (Importation, Formulation, and Uso of Pesticides in Costa Rica between 1992-1993) (San José, Costa Rica: Programa de Plaguicidas: Desarollo, Salud y Ambiente, Escuela de Ciencias Ambientales Universidad Nacional (Pesticide Program: Development, Health, and Environment, School of Environmental Sciences, National University, 1995).

pesticides.[32] A few studies investigated residue levels of organochlorine pesticides in food and human tissue, but none of these looked at possible symptoms from such chronic, low-level exposure.[33]

Farmers and workers interviewed in Kenya's Meru District said the biggest problem with pesticides was their cost: None mentioned the possibility of long-term health effects.

Costa Rica had better data on pesticide-related health effects. In 1991, a new law there made it obligatory to report intoxications from pesticides to the Ministerio de Salud (Ministry of Health).[34] In the period 1980-1986, 3,347 cases of intoxications were treated in hospitals.[35] The figures for the years 1990-1996 indicate a significant increase of intoxications until 1995, followed by a slight decrease in 1996—when the reported number was 792.[36] Out of all the reported intoxications in 1996, farmers and workers associated with banana cultivation suffered the highest number (64 percent), followed by those working with ornamental plants (3.7 percent).[37] In terms of chemical classes, insecticides and nematicides were attributed to more than one-half of the intoxication cases, followed by herbicides.[38] One study identified young workers (under 30) and women as groups particularly affected by occupational poisonings.[39] The same study concluded that on a yearly basis, 1.5 percent of the agricultural workforce is medically treated for occupational poisonings.[40] Intoxications from cholinesterase-inhibiting pesticides (organophosphates and carbamates) and the herbicide paraquat together accounted for a majority of poisonings identified in hospitalization and fatality records in the 1980s in Costa Rica.[41] Unique studies—in a developing country context—have been done in Costa Rica on subacute health effects such as cancer. The organochlorines that were used in agriculture until the 1980s have shown a rather strong association with breast cancer in areas where rice—a crop that has been intensely sprayed with such products—is grown.[42] Other associations between pesticides and cancer include paraquat and lead arsenate, which have been linked to skin-related cancers; formaldehyde and leukemia; and dibromochloropropane (DBCP), which has been associated with lung cancer and melanoma.[43] Banana workers in particular showed increased incidence of some cancer types, specifically melanoma in men and cervical cancer in women.[44] Table 1 above shows pesticide import figures for Kenya and Costa Rica in the early 1990s, classified in categories according to acute toxicity. (Such classification gives an indication for assessing potential health risks.) Nearly one-quarter of the pesticides used are moderately hazardous, and more than 10 percent are highly hazardous.

Unfortunately, the lack of data on environmental effects is the rule and not the exception. The few studies that were found in Kenya only look at pesticide levels in the environment and do not indicate significant problems.[45] In Costa Rica, the Ministry of Environment and Energy had commissioned a report on the envi-

Table 2. Pesticides most frequently mentioned by local-area coffee farmers

Product Name	Pesticide category	Active ingredient	WHO hazard classification[1]
Meru			
Copper (various)	Fungicide	Copper	III
Lebaycid®	Insecticide	Fenthion	II
Sumithion®	Insecticide	Fenitrothion	II
Gramoxone®	Herbicide	Paraquat	II
Ambush®	Insecticide	Permethrin	II
Karate®	Insecticide	Lamba-cyhalothrin	II
Naranjo			
Atemi®	Fungicide	Cyproconazole	III
Silvacur®	Fungicide	Triadimenol	III
Gramoxone®	Herbicide	Paraquat	II
Roundup®	Herbicide	Glyphosphate	-[2]
Counter®	Nemtaicide	Terbufos	1a
Furudan®	Nematicide	Carbofuran	1b

[1]The World Health Organization (WHO) classification places products' active ingredients into the following categories: 1a-extremely hazardous; 1b-highly hazardous; highly hazardous volatile fumigants; II-moderately hazardous; III-slightly hazardous; and unlikely to present acute hazard.

[2]Unlikely to present hazard under normal use.

NOTE: Coffee farmers in Meru, Kenya, and Naranjo, Costa Rica, responded to questions about which pesticides they apply with a few product names: The table above lists the most commonly mentioned. In total, farmers in Meru identified more than 20 different products; in Naranjo more than 30 were mentioned. Fungicides, the products sprayed most frequently, belong to the less acutely toxic categories (II and III). The nematicides—used only in Costa Rica—stand out as having highly or even extremely toxic ingredients. The three pesticides most commonly involved with intoxications in Costa Rica during the years 1994-1996 were paraquat (bipyridylium), carbofuran (carbamate), and terbufos (organophosphate).

SOURCE: S. Karlsson, *Multilayered Governance. Pesticide Use in the South: Environmental Concerns in a Globalised World* (Linköping, Sweden: Linköping University, 2000); and R. Castro Córdoba, N. Morera González, and C. Jarquín Núñez, *Sistema de Vigilancia Epidemiologica de Intoxicaciones con Plaguicidas, la Experiencia de Costa Rica, 1994-1996* (Epidemiological Monitoring System of Pesticide Intoxications, Costa Rica's Experience, 1994-1996) (San José, Costa Rica: Departamento de Registro y Control de Sustancias y Medicina del Trabajo del Ministerio de Salud (Ministry of Health Department of Registration and Control of Occupational Substances and Medicine, 1998).

nmental effects of pesticide use on nana plantations in three regions the Atlantic zone. The report, wever, could also mainly refer to sticide residues detected in the enronment that have unknown efts: The only concrete direct gative effects reported were some cidences of fish killed in rivers and clining fish populations.[46]

At the local level, the chief complaint farmers in Meru expressed as an economic problem: They ere unable to purchase pesticides e to low coffee prices. Interview vealed that farmers in neither eru nor Naranjo saw production oblems from pesticides. Responnts said they had not seen pests develop resistance, and with few exceptions, all pests could be adequately controlled. Health effects that workers associated with using pesticides varied substantially. Some said they suffered problems such as dizziness, nausea, headache, skin problems, eye problems, or fever; others said they never had any problems.[47] There was hardly any mention in Meru of the possibility of more long-term health effects. However, it was not uncommon for respondents in Naranjo to mention such effects, including cancer. The general environment as a possible victim of negative effects was practically absent in the interviews in Meru, while respondents in Naranjo were slightly more aware of its vulnerability. Table 2 above shows some of the most commonly used pesticides in the districts. Many are moderately hazardous; some have notorious records as intoxicants in developing countries.

The upshot of all this is that there were, at the time of the study, substantial divergences in the understanding of the pesticide-linked problems among different stakeholder groups and at varying governance levels. This lack of common understanding is not surprising considering the diversity of priorities among stakeholders and prevalent lack of knowledge on so many aspects of the potential impacts.

Table 3. Strategies to reduce health and environmental risks, by governance level

Level	Country/district	Use	Type	Mode
Global		Integrated pest management Organic farming[a]	Phase-out Information exchange Risk assessment Toxicity classification	Codes of conduct Guidelines
National	Kenya	Integrated pest management Organic farming[a]	Registration Banning	Regulation and training
	Costa Rica	Integrated pest management Organic farming	Registration Banning	Regulation and training
Local	Meru, Kenya	Resistant coffee variety Getting others to spray	——	Safe-use training
	Naranjo, Costa Rica	Integrated pest management	——	Safe-use training

(a) Organic farming exists as a strategy on this level but is not strongly stressed.

NOTE: Strategies to reduce health and environmental risks are categorized under "use," "type," or "mode;" that is, they reduce pesticide use, address certain types of pesticides use related to their individual characteristics, or they address modes of the chemicals' application.

SOURCE: S. Karlsson, *Multilayered Governance. Pesticides in the South: Environmental Concerns in a Globalised World* (Linköping, Sweden: Linköping University, 2000).

Which Risk-Reduction Efforts, Where?

Three basic risk-reduction strategies emerge for addressing the risks with pesticides: reducing their use, using less-toxic types of substances, and using them in a more precautionary fashion or mode?[48]

These approaches can be linked to three factors of pesticides that contribute to their health and environmental risks—the volume of use, the type of pesticide used, and the mode in which they are used. Table 3 on page 6 shows which kind of risk-reduction efforts existed at each governance level and to which strategy they belong. At the global level, the predominant risk-reduction efforts focused on the type of pesticides, applying a chemical-by-chemical approach in many activities. Several IGOs have assisted developing countries to regulate and control the use of pesticides. This guidance has come in the form of facilitating the development of national legislation, helping to build capacity in chemical management, and providing scientific data on individual pesticides. Several international agreements have been developed to address problems associated with pesticides and other chemicals by targeting individual substances—by phasing them out, for example.[49] Some IGOs have also made significant efforts to support a better mode of use. For example, FAO developed the International Code of Conduct on the Distribution and Use of Pesticides, and there have been several efforts promoting its universal observance.[50] In addition, a number of technical guidelines have been made on how pesticides should be used.[51] Finally, there have been some efforts to reduce the overall use of pesticides. For instance, some IGO programs have encouraged the implementation of integrated pest management (IPM) techniques rather than supporting increased use of pesticides in agricultural projects.[52] However, there has been very low support for organic farming.[53] Table 4 on pages 7 and 8 gives more details on some of the relevant organizations and their activities as well as agreements on the international level.[54]

Most risk-reduction efforts by governments and industry at the national level in Kenya and Costa Rica belong to the type and mode categories, although IPM efforts have been increasingly encouraged. To address risks arising from the types of pesticides used, both countries have developed extensive laws and regulations on pesticides. The core of these have been registration processes.[55] Within these processes, pesticide companies submit each product for approval before it can be used in the country.[56] Both countries have banned a number of primarily organochlorine pesticides over the last two decades.[57] In both countries, training programs for the safe use of pesticides has involved government agencies as well as the pesticide industry.[58] The pesticide laws in both countries cover, for example, conditions for storage of pesticides by retailers, the training of pesticide retailer staff, and the application of pesticides by farmers (prescribing that they must be used safely).[59] In Kenya, there have been small efforts to establish IPM as a national policy, but there has been very little implementation in this regard. However, it appears that IPM has received more official support inCosta Rica.[60]

Table 4. Global-level actors, activities, and agreements on developing country pesticide risks

Actor/organization	Activities and international agreements
Intergovernmental Forum on Chemical Safety (IFCS) A noninstitutional arrangement where governments meet with intergovernmental and nongovernmental organizations every three years. Smaller meetings are held every year. Geneva, Switzerland (small secretariat) Established in 1994 www.who.int/ifcs	• Provides advice to governments, international organizations, intergovernmental bodies, and nongovernmental organizations on chemical risk assessment and environmentally sound management of chemicals. • Sets priorities and promotes coordination mechanisms at the national and international level.
Food and Agricultural Organization of the United Nations (FAO) Specialized agency Rome, Italy Established in 1945 www.fao.org	• Supports integrated pest management (IPM) projects with Farmer Field Schools in Asia and Africa (for example) at national and regional levels for various crop systems. • Assists countries with development of national pesticide legislation. • Produces technical guidelines on various aspects of pesticide risks. • Runs projects to clean up obsolete stocks of pesticides (wastes) in developing countries. • Facilitated the negotiation of the International Code of Conduct on the Distribution and Use of Pesticides (the FAO Code of Conduct) and monitors its observance. www.fao.org/ag/agp/agpp/pesticid/
Global IPM Facility Cosponsored by FAO, UNEP, United Nations Development Programme (UNDP), and the World Bank. FAO Headquarters, Rome, Italy Established in 1995 www.fao.org/globalipmfacility/home.htm	• Assists governments and nongovernmental organizations to initiate, develop, and expand IPM. • Strengthens IPM programs through, for example, initiation of pilot projects around the world. Programs include policy development and capacity building. • Applies a farmer-led, participatory approach to IPM.
United Nations Environment Programme (UNEP) Nairobi, Kenya Established in 1972	• Hosts the Interim Secretariat for the Stockholm Convention on Persistent Organic Pollutants (POPs), which initially targets 12 chemicals (out of which 9 are pesticides) for reduction and eventual elimination. It also sets up a system for identifying further chemicals for action. Signed by 151 countries, it will enter into force 17 May 2004. www.pops.int
UNEP Chemicals Geneva, Switzerland www.unep.org www.chem.unep.ch	• Hosts (with FAO) the Interim Secretariat for the Rotterdam Convention on Prior Informed Consent for Certain Hazardous Chemicals and Pesticides in International Trade, which prevents export of harmful pesticides (and industrial chemicals) unless the importing country agrees to accept them. Signed by 73 countries, it was entered into force 24 February 2004. www.pic.int • Produces the Legal File, which contains information on regulatory actions on hazardous chemicals in 13 countries and 5 international organizations. www.chem.unep.ch/irptc/legint.html

table continues on next page

Table 4, continued

Actor/organization	Activities and international agreements
International Programme on Chemical Safety (IPCS) A joint program of the World Health Organization (WHO), FAO, and the International Labour Organization (ILO). WHO Headquarters, Geneva, Switzerland Established in 1980 www.who.int/pcs/	• Produces and disseminates evaluations of the risk to human health and the environment from exposure to chemicals (including pesticides) and produces guideline values for exposure. • Carries out projects with governments to support their capacity in chemical safety. • Since the early 1990s, has carried out activities to develop a project for collecting data on pesticide poisoning. Several countries are testing the harmonized approach of data collection.
Pesticide Action Network (PAN) A network of more than 600 participating nongovernmental organizations, institutions, and individuals in more than 60 countries. Five regional centers (San Francisco, California; Santiago, Chile; London, United Kingdom; Dakar, Senegal; and Penang, Malaysia) www.pan-international.org/	• Works to replace the use of hazardous pesticides with ecologically sound alternatives. • Working from five autonomous regional centers, the network's programs include research, policy development, and media and advocacy campaigns.
CropLife International (formerly (pre-2000) Global Crop Protection Federation (GCPF)) Represents the crop-protection product manufacturers and their regional associations. Brussels, Belgium www.gcpf.org	• Supports the FAO Code of Conduct. • Initiated three pilot projects in 1991 to promote the safe use of pesticides (in Guatemala, Kenya, and Thailand). Now continuing to support safe-use programs through its National Associations. • Endorses and supports IPM. • Has carried out several projects addressing obsolete pesticide stocks. Supports a nongovernmental organization-initiated program to eliminate stocks of POPs in Africa.

SOURCE: Sylvia I. Karlsson, 2004.

On the margin, some NGOs in both countries—and even the Costa Rican government—have encouraged organic farming.[61]

On the local level in the two coffee-growing districts, safe-use training has been the only explicit approach to reduce the risks of pesticides. More than 20,000 farmers in Meru were trained in a project sponsored by the pesticide industry in the first half of the 1990s, but this number is still a small part of Meru's agricultural population: The project neglected groups such as women and casual and permanent farm workers, and it is likely that such farm workers are the group most heavily exposed to pesticides. In Naranjo, training on protective measures has been present for a number of years, although not in the form of an explicit safe-use project: It was incorporated in the general training from the national agricultural exten-sion system. Despite these training efforts, the adoption of the safe-use message among interview respondents was low in both districts. In Meru, a few said they used improvised or partial protective clothing, but most respondents neither owned nor had access to these. Farmers and workers interviewed in Meru said they were too expensive. In Naranjo, access was generally not a problem, but farmers said they were uncomfortable: Few used the protective wear, or if they did they wore it only in the early morning before it got too hot. When asked, most farmers said they believed there were no alternatives to pesticides. The few references to nonchemical alternatives to combat pests were primarily made by farmers in Naranjo and included pruning, hand weeding, and general soil conservation measures. In Meru, the extension system had provided a coffee variety that was resistant to the two most prominent fungal pests, but only some farmers at the time of the interviews had planted them. Interviewees had not applied any of these measures to address either health-risk or environmental problems. Accumulated experience and research in Costa Rica had, at the end of the 1990s, begun to show the negative consequences of herbicide use both on the productivity of the coffee plants and erosion levels. Consequently, the extension service had begun to discourage farmers from using herbicides. Many coffee farms were de facto organic: They could afford neither pesticides nor chemical fertilizers. But because there was no market infrastructure for organic coffee, they could not receive a higher price for their crop. In Naranjo, some farmers were aware that it was possible to receive a significant premium price for growing certified organic coffee. However, even the promoters of organic coffee farming did not encourage existing coffee farms to switch to organic farming: Organic certification requires a farm to be without agrochemical inputs for three years, and the harvest slumps in the meantime. (Generally, long-abandoned coffee farms are preferred instead; they can most successfully be developed into an organic farming system.) There were no efforts at the local level to reduce risks by avoiding certain types of pesticides. Farmers bought and applied the products that were recommended by companies, cooperatives, or the extension system.

Which Institutions, at What Level?

Despite large uncertainties in the precise nature and levels of risks, the description above of the scope of the health and environmental problems, globally and on a smaller scale in Kenya and Costa Rica, shows that there is by no means an effective governance system for pesticides in developing countries. As evidenced by the risk-reduction measures described above, considerable efforts of governance at different levels are not only insufficient but are also characterized by fragmentation and incoherence. One way to identify the reasons for this situation as well as potential options for change is to focus on institutions—here defined as formal and informal rules of human interactions—and their role in governance at various levels.[62] Institutions are at the core of governance: They influence who has access to what information, shape the incentives for various courses of action, and affect who has the capacity to act. In a multilevel governance context, the following question arises: Is it possible to identify some criteria to determine the types of institutions that should preferably be established, enforced, or changed at particular levels? Some suggestions for elements of such criteria can be found in theories of the management

of collectively owned natural resources. To identify current gaps in the governance system, it is useful to explore two sets of criteria. The first relate to the potential effectiveness of institutions if they were to match up the level of effects, driving forces and capacity; the second set relates to how the possibility to change institutions may vary across levels. Tables 5a and b illustrate these criteria.

Matching Institutions

It is particularly instructive to determine criteria that can address how well institutions and governance "match" the level where most negative effects occur, the driving forces behind the problems originate, and where there is capacity to take action.

• *Matching effects.* If institutions are not in place at the level to correspond with, or "fit" the geographical scope of the negative effects, their effectiveness can be limited.[63] The potential for persistent organic pollutants (POPs)—which include a number of organochlorine pesticides—to act as transboundary pollutants made countries like Canada and Sweden push for an international agreement to ban them. The development of the Stockholm Convention can thus be seen as an effort to match the global scope of the problem with global governance. The efforts to establish a harmonized registration system for pesticides in Central America can be seen in a similar light. Banning a pesticide in Costa Rica while it is allowed in a neighboring country invites smuggling, black market sales, and potential for cross-border pollution via rivers.[64] However, pesticides also exert very local and context-dependent effects on health and environment—which by aggregation can be seen as global problems—and these are not at all well matched with institutions. Consider, for example, the common use of paraquat in both Meru and Naranjo despite its well-known (on the international level) intoxication record, and the use of

category la and lb (extremely and highly hazardous, respectively) pesticides by small farmers in Naranjo (see Table 2).

The banana trade is a huge market in Costa Rica for pesticides. Unfortunately, data show that banana workers are more likely to suffer pesticide intoxications than other workers.

• *Matching driving forces.* If institutions are not sufficiently well established and implemented at the level where the driving forces for the problems originate, the governance measures at other levels will merely target symptoms. Moreover, from an ethical perspective, it would be preferable if those who are explicitly responsible for the problems would be more often targeted in governance.[65] There are layers of direct and indirect driving forces for pesticide problems and they differ depending on what risk factor is in focus. The strongest driving forces for using pesticides emerge at the global level, where the agrochemical industry, along with governments, promote incentives for modern high-input agriculture, and at the national level, where agricultural policy, research, and extension advice and marketing strategies of the agrochemical companies create similar incentives. However, these are the drivers least addressed in pesticide governance.

The strongest driving forces that determine the types of pesticides used are also located at the global and national levels. At the global level, multinational corporations develop and choose which of their products to market in developing countries. The Stockholm and Rotterdam Conventions target specific pesticides to regulate or ban; if implemented, these rules can reduce risks in developing countries. However, the number of substances included is very small, making this process a weak match of institutions and driving forces. At the national

Table 5a. Matching institutions

		Global		National		Local	
		Negative effects	Driving forces	Driving forces	Capacity to act	Negative effects	Driving forces
Example		Global transboundary pollution of pesticides	Institutions giving incentives for high-input agriculture, resulting in increased pesticide use	Regulations prescribing how toxic pesticides are allowed to be used in countries	Governmental authority to regulate pesticide storage, sales, and marketing	Local health and environmental effects	Influence that farmers and workers who handle pesticides have on how safely they are used
Degree example is matched in pesticide risk reduction		High: The Stockholm Convention addresses many persistent organic pollutant pesticides.	Low: Regulation is absent and there are few incentives for alternative agricultural systems.	Moderate: Many developing countries allow many highly toxic pesticides.	Moderate: These institutions are weak or very weakly enforced in many developing countries.	Low: Many places have poor healthcare facilities and no institutions that address potential environmental effects.	Low: Most countries have limited and/or ineffective safe-use education and training.

Table 5b. Changing institutions

Types of institution	Example	Proneness to change
Operational	Guidelines for safe use	• Relatively easy to change with few resources and within a short time span. • Difficult to implement/enforce, particularly at global scale.
Collective-choice	Regulations prescribing which pesticides are banned at national and global levels	• Moderately difficult to change with large variances in time and resources required.
Constitutional-choice	Favored type of agricultural system (including system of pest management)	• Very difficult to change, requiring significant political will and involving multiple sectors.

SOURCE: S. I. Karlsson, 2004.

level, the governments in Kenya and Costa Rica control which pesticides may be used in their respective countries—although this also depends on which products the pesticide industry chooses to market. Overall, developing countries have no influence on pesticide development: Most research in this regard centers around pests and crops in the temperate regions.[66] And while Kenya and Costa Rica have the authority to ban pesticides, the overwhelming majority are approved.[67] In terms of how pesticides are actually used, there are a number of driving forces, including images presented in marketing campaigns. Ultimately, however, individual farmers or workers are the ones who determine how they are used. This is where the major mismatch between institutions and driving forces lies. Even if many efforts are made at national and higher levels to make farmers use pesticides more safely, such measures have a long way to go before they actually reach individuals who spray—or to a sufficient degree effect a change in their behavior.

• *Matching capacity.* If there is no capacity to act at the level where effects or driving forces originate, then there is not much one can expect in terms of institution building.[68] Stakeholders at a governance level where there is capacity to act may not necessarily have contributed to the problem but could take on responsibility for governance because of a sense of concern and moral obligation.[69] The whole focus at the global level to assist developing countries with scientific and technical information on pesticide risk is largely due to the fact that IGOs have the capacity to assess those risks, while many developing countries have neither the expertise nor resources needed. The governments in Kenya and Costa Rica focus on deciding which products are allowed for use because they have the capacity to make those decisions, while local actors are considered not to have such capacity. Moreover, for the most part, farmers do not have the capacity to adopt risk-reduction strategies—either because they are not aware of risks or because they believe it is impossible to farm without pesticides and have no alternative pest-control strategies to adopt. Those stakeholders who do have the capacity to influence this—such as the pesticide industry and the international community—have not fully utilized their resources to this end.[70]

Changing Institutions

Another key set of criteria relates to the cost (in monetary or human-resource terms, for example) required to create a new institutional arrangement or to enforce or change an existing one. Three types of institutions are considered here: operational, collective-choice, and constitutional-choice institutions.[71]

- *Operational institutions.* These institutions provide structures for making day-to-day decisions in a wide diversity of operational situations, which means a large number of individual actors are involved.[72] In a local context, it is assumed that these are the institutions that can be most quickly changed. However, this may not always be the case on the global scale, where the sheer number and diversity of multiple localities make the picture more complex. When pesticides continue to be the primary pest-management tool, and risk reduction follows the mode strategy of ensuring safe use, an effective approach to effect change must include a strong focus on operational institutions at the local level. A global code of conduct is not enough nor are national laws that make unsafe use illegal. Any institu-

tions established at higher levels have to be implemented by a large number of stakeholders: farmers and workers—female and male alike; their families; and all others who handle pesticides throughout their life cycle. Even if changes can be initiated quickly, they are costly and time consuming to implement on a large scale.

Government officials in Costa Rica recognized herbicides had negative impacts, including erosion. Many local officials later discouraged their use.

- *Collective-choice institutions.* This category describes those institutions that indirectly affect the options for operational rules.[73] The number of stakeholders involved in designing these institutions are fewer but, depending on the governance level, can range from a single local NGO to all the member states of the United Nations. The cost of changing such institutions, and the time it takes to do so, will vary accordingly. When risk-reduction efforts target the types of pesticides used, the focus is on collective-choice institutions at the national level. Here, decisions on which pesticides farmers will have access to involve only very few individuals—and in Costa Rica, for example, the pesticide registration process takes 6-12 months.[74] Implementation of these governmental decisions involves a smaller number of stakeholders, such as customs officers and pesticide retailers, who are charged with ensuring that the unwanted pesticides do not reach the farmers. This strategy also involves measures at the global level, through, for example, sharing toxicity information about certain chemicals or banning specific substances. Such global processes can take a very long time: For instance, the process to establish the Stockholm Convention began in 1997 and will not have entered into force until

mid-May 2004. Change to collective-choice institutions can thus in some cases be made at relatively low cost within a short time frame, but in other cases the process is lengthy and cumbersome.

- *Constitutional-choice institutions.* These determine the specific institutions that create the collective-choice institutions.[75] These types of institutions may involve the smallest number of the most powerful and knowledgeable stakeholders when crafting formal institutions. Or, when institutions are part of deeply rooted structures in society, the number of stakeholders involved may be innumerable and not easy to pinpoint. These types of institutions are usually the slowest to change. The risk-reduction strategy to reduce or eliminate the use of pesticides requires changes in constitutional-choice institutions. These are institutions that favor one type of agricultural system: one dependent on pesticides, one less dependent on pesticides, or one completely independent of pesticides. Such institutions consist of consumer demands, national and international market and trade structures, government policies, and farmer attitudes (for example). The inter-linkages between sectors, the resistance from prevailing power structures, and the number of decisionmakers involved all present a considerable and time-consuming challenge for anyone attempting to change these institutions.

Correcting Mismatches and Facilitating Change

In very general terms, it appears that institutions must be assigned primarily to those levels where driving forces originate and where stakeholders have the capacity to establish and enforce institutions. Furthermore, if rapid institutional change is desired, the focus should be on enforcing changes in operational institutions at local levels—although this may not be the most cost-effective or long-lasting governance strategy. If slower—but likely more enduring—change is to be achieved it is the constitutional-choice institutions that need to be targeted—not only on global levels but across national and local levels as well. The more pragmatic and manageable approach in time and resources is to target collective-choice institutions.

Many agricultural workers, unaware of the risks or lacking adequate resources, do not always wear protective gear. Effective education programs could change this.

In more specific terms, to address the mismatch between institutions and effects, institutions need to be established that enable developing countries and the international community to incorporate the concerns for local health and environmental effects from pesticides in their specific climatic, ecological, economic, and social context. Currently, developing countries rely on global institutions and knowledge when establishing their own institutions.[76] There is a need to create institutions that facilitate the collection of data locally and nationally—for instance, those that prescribe the monitoring of health and environmental effects after registration is approved.

To address the mismatch between driving forces and institutions, there are specific needs for each risk-reduction strategy. Institutions at national and global levels are needed that discourage the use of pesticides and provide alternative, economically viable farming strategies. Some national and international NGOs are currently involved in small-scale initiatives promoting organic farming, but these are very limited in scope and hard to upscale. The situation is better for IPM initiatives. Such higher-level institutions could include changing or developing new agronomist education programs, reducing hidden pesticide subsidies, increasing the market channels for organic products, or even influencing consumer demand for these products through education campaigns that raise the general public's awareness. Institutions are also needed at the global level to support those countries that can neither create institutions nor enforce them—a situation that results in a substantial black market of smuggled and substandard pesticides—to prevent the most toxic products from entering their countries.[77] Only global-level phase-outs can, in such cases, keep the most toxic types of pesticides out of the hands of smallholder farmers. This implies that the criteria for the type of products to be banned globally would need to be expanded to include those that are known to produce locally occurring severe health and environmental effects. Finally, to strengthen safer modes of pesticide handling, existing higher-level institutions need to radically upscale or fundamentally change their implementation efforts. A local culture of safe pesticide use can only emerge with long-term, more effective education efforts that reach all groups who come in contact with pesticides. To address the mismatch between capacity and institutions, one step is to look at the pesticide industry—a stakeholder that is a strong driving force and that holds considerable capacity to effect change. They could be charged with greater responsibility to become a stronger player in risk reduction. An example of how this could be implemented is illustrated with the cases in Kenya and Costa Rica: Pesticide companies now pay for field tests of the efficacy of their products on the crops in both countries—the governments require this for registration.[78] National institutions could hold these companies accountable to contribute resources toward national data-gathering programs examining the impacts of their products in each respective country. The international community—the only entity that can regulate global production and trade—could also take on a larger share of responsibility by creating more encompassing institutions, for example by phasing out more substances globally (see above) and making such institutions more effective with stronger mechanisms of monitoring and enforcement. Combining the two sets of criteria puts in focus the collective-choice institutions targeted at eliminating the pesticides that pose the highest risks. Changing these institutions would take considerable time and effort, particularly in building up a better knowledge base and banning substances on a global scale. Nevertheless, it would be the most accessible "fast track" to reduce risks pending a global-scale mustering of resources for widespread implementation of safe use and changes in agricultural systems.

Conclusions

The final choice of strategies for risk reduction, institution building, and change will be heavily dependent on answers to such questions as

• What are the major contributing factors to risks?

• What are the inherent toxic properties of pesticides?

• What are the exposure patterns under conditions of use?

• What is the acceptable level of risk? and

• Who should be responsible to address the risks?

Some of these questions could be resolved with more monitoring and research, others are more value laden, and these kinds of issues divide stakeholder groups substantially. Stakeholder views range from NGO movements that consider all pesticides inherently toxic to many in the pesticide industry who assert that all pesticides—as long as used as prescribed—are safe. The former are convinced that the conditions in developing countries—especially considering the human and financial input necessary to effect change—make it impossible to change the operational institutions and ensure safe use. The latter fear that changing the constitutional-choice institutions—striving to establish IPM or organic farming on a global scale, for example—-would endanger food and economic security. There is no easy way to resolve the debate between the widely diverging knowledge bases and value judgments that underlie these different views. However, some of the recommendations above can help clarify and structure the available options for governance and the institution building and change that they would require.

In addition, the pesticide case and the policy-relevant conclusions drawn from it have much to teach us in other policy areas. Pesticides are one of the first groups of toxic chemicals introduced on a large scale in developing countries, and the complexities involved with their use illustrate many of the challenges and possibilities for the management of other groups of chemicals in these regions. As one of the environmental issues that exhibits a number of local-global linkages, the impacts of pesticide use illuminate directions that future research and policy discussions need to take. Governance needs to be analyzed and addressed with a much more holistic approach, viewing the efforts at all levels—local, national, regional, and global—as elements of one system of governance. Only then can we evaluate how individual policies operate in the context of a large, interconnected system. Only then can research start identifying the most important elements of establishing multi-layered governance, with a nested hierarchy of mutually supportive policies and institutions initiated at all governance levels.[79]

NOTES

1. The term "governance" has emerged as one of the most-used concepts when discussing measures taken in society to address a particular issue—specifically when stressing that there are many more actors than just governments involved. It is par-

ticularly useful for the global level where there is no world government but still a lot of governance. The Commission on Global Governance defined governance as "the sum of the many ways individuals and institutions, public and private, manage their common affairs." See Commission on Global Governance, *Our Global Neighbourhood* (Oxford, UK: Oxford University Press), 2.

2. This figure is for 2001, a year in which the market suffered a 7.4 percent decline. See Phillipps McDougall (2002) quoted in CropLife International, *Facts and Figures*, accessed via http://www.gcpf.org on 21 January 2004. In 1999, the market was more than US$30 billion. See "World Agrochemical Market Held Back by Currency Factors," *Agrow*, 11 June 1999, 19-20.

3. See, for example, World Health Organization (WHO), *Public Health Impact of Pesticides Used in Agriculture* (Geneva: WHO, 1990); P. N. Viswanathan and V. Misra, "Occupational and Environmental Toxicological Problems of Developing Countries," *Journal of Environmental Management* 28 (1989): 381-86; L. A. Thrupp, "Exporting Risk Analysis to Developing Countries," *Global Pesticide Campaigner* 4, no. 1 (1994): 3-5; Health Council of the Netherlands, *Risks of Dangerous Substances Exported to Developing Countries* (Den Haag: Health Council of the Netherlands, 1992); and C. Wesseling, R. McConnell, T. Partanen, and C. Hogstedt, "Agricultural Pesticide Use in Developing Countries: Health Effects and Research Needs," *International Journal of Health Services* 27, no. 2 (1997): 273-308. For a special analysis of the impact on women, see M. Jacobs and B. Dinham, eds., *Silent Invaders: Pesticides, Livelihoods and Women's Health* (New York: Zed Books, 2003).

4. Food and Agriculture Organization of the United Nations (FAO), *Analysis of Government Responses to the Second Questionnaire on the State of Implementation of the International Code of Conduct on the Distribution and Use of Pesticides* (Rome: FAO, 1996). For example, highly toxic insecticides is the

main pesticide category in use in many less developed countries. Wesseling, McConnell, Partanen, and Hogstedt, note 3 above, page 276.

5. WHO, note 3 above, pages 85-86. This estimate is calculated by using a 6:1 ratio between nonhospitalized (unreported) and hospitalized (reported) cases.

6. WHO, note 3 above, page 86. The background for this is that pesticides in rural areas are among the most accessible types of toxic substances. Because these data are based on hospital registers, they probably overestimate the proportion of suicides. Wesseling, McConnell, Partanen, and Hogstedt note 3 above, page 283.

7. The figures from the WHO 1990 report have been strongly challenged by the pesticide industry. Anonymous official, International Programme on Chemical Safety (IPCS), interview by author, Geneva, 25 June 1998. IPCS has a project to support developing countries to gather data on pesticide intoxications more systematically. IPCS, *Pesticide Project: Collection of Human Case Data on Exposure to Pesticides* (Geneva), accessed via http://www.intox.org/pagesource/intox%20area/other/pesticid.htm on 22 January 2004. However, the project has not yet resulted in new global estimates. In the first stage, studies were carried out in India, Indonesia, Myanmar, Nepal, and Thailand, based on hospital records (some results of these are available at http://www.nihs-go.jp/GINC/meeting/7th/profile.html), but the results were not satisfactory. In a second phase, they will use community-based studies in pilot countries. Dr. Nida Besbelli, IPCS, e-mail message to author, 10 February 2004. While there are some developing countries where intoxications have to be reported to the authorities, overall there is limited data on pesticide health impacts in developing countries. Wesseling, McConnell, Partanen, and Hogstedt, note 3 above, page 284.

8. The lack of data provides a significant obstacle for global estimates of the number of people suffering from chronic effects. WHO, note 3

above, page 87. The few studies that have been done have demonstrated neurotoxic, reproductive, and dermatologic effects. Wesseling, McConnell, Partanen, and Hogstedt, note 3 above, page 273.

9. See, for example, P. Bourdeau, J. A. Haines, W. Klein and C. R. K. Murti, eds., *Ecotoxicology and Climate With Special Reference to Hot and Cold Climates* (Chichester, UK: John Wiley and Sons Ltd., 1989); and T. E. Lacher and M. I. Goldstein, "Tropical Ecotoxicology: Status and Needs," *Environmental Toxicology and Chemistry* 16, no. 1 (1997): 100-11.

10. For example, 89 percent of the 60 developing countries who responded to an FAO questionnaire in 1993 reported that they are not studying the effects of pesticides on the environment. FAO, note 4 above, page 61. Research in disciplines that are essential for detecting and understanding environmental degradation—such as biology, ecology, and ecotoxicology—is very limited in sub-tropical and tropical regions compared to that in nontropical latitudes. Bourdeau, Haines, Klein and Murti, note 9 above; and Lacher and Goldstein, note 9 above. This situation reflects a general knowledge divide in the environmental field between developed and developing countries, which in turn reflects the generic divide in resources (human and financial) available for monitoring and research. S. Karlsson, "The North-South Knowledge Divide: Consequences for Global Environmental Governance," in D. C. Esty and M. H. Ivanova, eds., *Global Environmental Governance: Options & Opportunities* (New Haven, CT: Yale School of Forestry & Environmental Studies, 2002), 53-76. It is estimated that about 5 percent of the world's scientific production comes from developing countries. International Development Research Centre, "The Global Research Agenda: A South-North Perspective," *Interdisciplinary Science Reviews* 16, no. 4 (1991): 337-4. The number of scientists/engineers per million inhabitants in developed countries is 2,800 on average; in developing countries it is 200. T. H.

I. Serageldin, "The Social-Natural Science Gap in Educating for Sustainable Development," in T. H. I. Serageldin, J. Martin-Brown, G. López Ospina, and J. Dalmatian, eds., *Organizing Knowledge for Environmentally Sustainable Development* (Washington, DC: The World Bank, 1998).

11. See B. Dinham, *The Pesticide Trail: The Impact of Trade Controls on Reducing Pesticide Hazards in Developing Countries* (London: The Pesticide Trust, 1995), which summarizes case studies from a number of countries. One of the few larger studies on environmental impact is the Locustox project studying the impact from large-scale locust sprayings in Africa. See J. W. Everts, D. Mbaye, and O. Barry, *Environmental Side-Effects of Locust and Grasshopper Control, Volume I and II*, (Senegal: FAO and the Plant Protection Directorate, Ministry of Agriculture 1997, 1998). As part of that study ponds were treated with deltamethrin (a synthetic pyrethroid) and bendiocarb (a carbamate). Delthamethrin had considerable acute effects on most macroinvertebrates, and bendiocarb affected a number of zooplankton.

12. This does not mean that organochlorines have not caused environmental effects in the tropics. For some examples, see F. Bro-Rasmussen, "Contamination by Persistent Chemicals in Food Chain and Human Health," *The Science of the Total Environment* 188 Suppl. 1 (1996): S45-60.

13. The proposed process of long-range transport in the atmosphere of certain organic compounds is called "global distillation," or "global fractionation." F. Wania and D. Mackay, "Global Fractionation and Cold Condensation of Low Volatility Organochlorine Compounds in Polar Regions" *Ambio* 22, no.1 (1993): 10-18. The theory has received support by modeling and monitoring data. H. W. Vallack et al, "Controlling Persistent Organic Pollutants—What Next?" *Environmental Toxicology and Pharmacology* 6 (1998): 143-75; and S. N. Meijer, W. A. Ockenden, E. Steinnes, H. P. Corrigan, and K. C. Jones, "Spatial and Temporal Trends of POPs in

Norwegian and UK Background Air: Implications for Global Cycling," *Environmental Science and Technology* 37, no. 3 (2003): 454-61. Atmospheric deposition is considered to be the major source of POPs in the Arctic. A. Godduhn and L. K. Duffy, "Multigeneration Health Risks of Persistent Organic Pollution in the Far North: Use of the Precautionary Approach in the Stockholm Convention," *Environmental Science & Policy* 6 (2003): 341-53. There is thus a growing scientific consensus for the global distillation/ fractionation hypothesis. Vallack et al, this note. This has also been reflected in policy where the Stockholm Convention includes in its screening criteria for adding further substances to the convention the potential for long-range transport through air, for example (one of the criteria for such substances is that their half-life in air must be greater than two days). Stockholm Convention on Persistent Organic Pollutants (POPs): Texts and Annexes (Geneva: Interim Secretariat for the Stockholm Convention on Persistent Organic Pollutants, 2001), 46-47.

14. An alternative concept for what has been referred to here as governance levels is levels of social organization. O. Young, *Institutional Dimensions of Global Environmental Change (IDGEC) Science Plan* (Bonn: International Human Dimensions Programme on Global Environmental Change, 1999).

15. Several methods were combined to solicit the perspectives and approaches of these stakeholders, including semistructured interviews and policy document analyses. A total number of 204 interviews were carried out during 8.5 months of fieldwork in the years 1997-1999. For further details on methodology, theoretical framework, and results of the study that are associated with this article, see S. Karlsson, *Multilayered Governance. Pesticides in the South: Environmental Concerns in a Globalised World* (Linköping, Sweden: Linköping University, 2000).

16. In Costa Rica, the average quantity of imported formulated pes-

ticides rose from 8,100 tons to 15,300 tons between the second half of the 1980s and the first half of the 1990s. A. C. Rodríguez, R. van der Haar, D. Antich, and C. Jarquín, *Desarollo e Implementacion de un Sistema de Vigilancia de Intoxicaciones con Plaguicidas, Experenica en Costa Rica, Informe Tecnico Proyecto Plagsalud Costa Rica, Fase 1* (Development and Implementation of a Pesticide Intoxication Monitoring System, Costa Rica's Experience, Technical Report, Plagsalud Costa Rica Project, Phase I), (San José, Costa Rica: Ministerio de Salud Departamento de Sustancias Toxicas y Medicina del Trabajo (Ministry of Health, Department of Toxic Substances and Occupational Medicine), 1997). The nominal value of pesticide imports rose approximately 50 percent between 1990 and 1994. In 1994, the value of pesticide imports reached US$84.2 million and in the same year on average more than US$170 were spent on pesticides per ha agricultural land. S. Agne, *Economic Analysis of Crop Protection Policy in Costa Rica, Publication Series No. 4* (Hannover, Germany: Pesticide Policy Project, 1996): 6, 12. Because of the substantive formulating industry, the trade figures on the value of pesticide purchases are gross underestimates of the amount spent on pesticides in Costa Rican agriculture. Agne, this note, page 12. Statistics are limited on the use of pesticides in the African continent. J. J. Ondieki, "The Current State of Pesticide Management in Sub-Saharan Africa" *The Science of the Total Environment* 188, Suppl. no. I (1996): S30-34; and S. Williamson, *Pesticide Provision in Liberalised Africa: Out of Control? Network Paper No. 126* (London: Overseas Development Institute Agricultural Research & Extension Network, 2003). As many as 47 percent of African countries responding to an FAO questionnaire do not collect any statistics on pesticide import and use. FAO, note 4 above, page 71. One study reported that in Kenya the average annual import of pesticides for 1989-1993 was just over 5,000 tons at a value of US$28 million. H. Partow, *Pesticide Use and Management in Kenya* (Geneva: Institut Universitaire D'Etudes du Développement (Graduate Institute of Development Studies), 1995): 205-6. The data reported for Kenya and Costa Rica only goes to the mid-1990s. (See note 2 above on falling pesticide sales in 2001.) The increasing trend is not taking place in all developing countries. Furthermore, studies in Africa have shown significant variation in impacts of, for example, liberalization on pesticide prices, access, and use. See A. W. Shepherd and S. Farfoli, "Export Crop Liberalisation in Africa: A Review," *FAO Agricultural Services Bulletin No. 135* (Rome: FAO, 1999); and Williamson, this note.

17. Agne, ibid., page 8. One source listed the main groups of pesticides used in the Central American countries: the insecticides organophosphates, carbamates, and pyrethroids; fungicides, mainly dithiocarbamics; and the herbicides phenoxyacids, dipyridyls, and more recently, triazines. L. E. Castillo, E. de la Cruz, and C. Rupert, "Ecotoxicology and Pesticides in Tropical Aquatic Ecosystems of Central America," *Environmental Toxicology and Chemistry* 16, no. 1 (1997): 41-51.

18. Partow, note 16 above, page vii. Inorganic pesticides accounted for 21 percent of imported pesticides, organophosphates accounted for 15 percent, organochlorines accounted for 11 percent, thiocarbamates accounted for 7 percent, and phtalimides accounted for 7 percent. These figures do not distinguish between products imported as technical grade or formulated product. For example, around 25 percent of the organophosphates are imported as technical grade material. Partow, note 16 above, pages vii, 39.

19. Agne, note 16 above and C. Conejo, R. Díaz, E. Furst, E. Gitli, and L. Vargas, *Comercio y Medio Ambiente: El Caso de Costa Rica* (Trade and the Environment: The Case of Costa Pica) (San José, Costa Rica: Centro Internacional en Política Económica Para el Desarollo Sostenible (International Center of Political Economy for Sustainable Development), 1996). The banana sector uses 45 percent of all pesticides and at the end of the 1970s the disease commonly known in Spanish as *sigatoka negra* (the scientific name is *Mycosphaerella fijiensis*), arrived in the country, which affected production severely and led to significant pesticide use.

20. Castillo, de la Cruz, and Rupert note 17 above. Permanent crops like coffee, oilpalm, and cacao are sprayed less often (1-5 times/year) compared to annual crops such as tobacco, potatoes, or vegetables and products like banana, melon, watermelon, or flowers—extreme cases of which can be sprayed up to 39 times per cycle. J. E. García, *Introducción a los Plaguicidas* (Introduction to Pesticides) (San José, Costa Rica: Editorial Universidad Estatal a Distancia (Publishing Trust of the State University for Distance Education), 1997).

21. For a long time, coffee was the main export earner in agriculture, but tea has taken over the lead position. Since 1985, horticulture has grown substantially. In 1990 it came in third place as an export earner. General Agreement on Tariffs and Trade (GATT, *Trade Policy Review Kenya Volume 1* (Geneva: GATT, 1994). Between 1991 and 1994 the value of horticultural exports increased by 113 percent, the principle horticultural crops being cut flowers, French beans, mangoes, avocados, pineapples, and Asian vegetables. Republic of Kenya, *National Development Plan 1997-2001* (Nairobi: Government Printer, 1996).

22. While most pesticides are used on export crops, studies have shown that in some areas where the coffee economy has introduced a modernized agricultural system—for example, by using pesticides, their use on food crops has increased. A. Goldman, "Tradition and Change in Postharvest Pest Management in Kenya" *Agriculture and Human Values* 8, no. 1-2 (1991): 91-113. This is a phenomenon found in other countries as well. See, for example, Williamson, note 16 above.

23. Karlsson, note 15 above, pages 209-13 and 238—40.

24. In Naranjo, coffee farmers apply fungicides on average three times a year, nematicides a maximum of once per year and herbicides once or twice a year. Karlsson, note 15 above, page 262. About 7 percent of all pesticide purchases in Costa Rica are used on the 20 percent of agricultural land that is under coffee production. See Agne, note 16 above, page 13. In Meru, on the other hand, those who still spray their coffee apply fungicides either 2–4 times or 8–12 times a year. Herbicides are sprayed occasionally there. Karlsson, note 15 above, page 262. It is important to note that the comparisons here are only made from the self-reported numbers of sprayings per year. The dose in each application is not addressed.

25. This result is based on careful analysis of a large number of interviews with stakeholders and study of policy documents. Thus, the summary conclusions cannot be attributed to a single source. Karlsson, note 15 above. The problem categorization is not clear-cut since several of the areas are closely interrelated. The environment serves as a medium for transport of pesticides and their metabolites, which may expose humans to these substances via air and water (for example) and potentially affect human health. Pesticides, as a trade issue, emerge in the regulations established to address the concern of long-term, low-level exposure of pesticide residues in food. Pesticides that exert effects on non-target organisms—those on the farm and surrounding environment—can disrupt populations of natural enemies of the original pest, leading to increased and different pest attacks and thus production problems.

26. In addition to evidence from the interviews (see Karlsson, note 15 above, pages 144-45), references to environmental considerations were absent in pesticide management plans. Partow, note 16 above, page xii. A study from the early 1990s on the legislation and institutional framework for environmental protection and natural resource management in Kenya concluded that there was no stress on environmental problems in the agricultural areas. S. H. Bragdon, *Kenya's Legal and Institutional Structure for Environmental Protection and Natural Resource Management—An Analysis and Agenda for the Future* (Washington, DC: Economic Development Institute of the World Bank, 1992). However, the Kenya National Environment Action Plan urges the adoption of as many nonchemical measures as possible and it urges the use of the least toxic chemicals as a last resort. Ministry of Environment and Natural Resources, *The Kenya National Environment Action Plan* (NEAP) (Nairobi, 1994).

27. For example, in Kenya, stricter European Union pesticide residue limits on agricultural products had caused significant concern both in government and in the pesticide industry. Standing Committee on the Use of Pesticides, *Interim Report of the Standing Committee on the Use of Pesticides* (Nairobi, 1996); and "Editorial," *Newsletter for the Pesticide Chemicals Association of Kenya*, February 1995. This was also one reason the existence of 100 tons of pesticide wastes (including many from banned substances) raised concern that these may find their way on to horticultural produce. Standing Committee on the Use of Pesticides, this note. It should be emphasized that the trade issue is very sensitive for national governments because of the high economic stakes involved, and they are likely to be very reluctant to discuss possible problems openly.

28. Republic of Kenya, "Chemical Safety Aspects in Kenya" (A country paper presented by the Kenyan delegation during IPCS Intensive Briefing Session on Toxic Chemicals, Environment and Health for Developing Countries, held in Arusha, United Republic of Tanzania, 1997).

29. These figures are quoted in M. A. Mwanthi and V. N. Kimani, "Patterns of Agrochemical Handling and Community Response in Central Kenya," *Journal of Environmental Health* 55, no. 7 (1993): 11-16. This figure—7 percent of the agricultural population suffering poisonings annually—lies within the range between 2 and 9 percent reported in various studies in developing countries. Wesseling, McConnell, Partanen, and Hogstedt, note 3 above, page 283.

30. V. W. Kimani, "Studies of Exposure to Pesticides in Kibirigwi Irrigation Scheme, Kirinyaga District" (Submitted Ph.D. thesis, Department of Crop Science, University of Nairobi, 1996).

31. Ondieki, note 16 above, page S32.

32. Kimani, note 30 above, page 23.

33. A study in the 1980s of organochlorine residues in domestic fowl eggs in Central Kenya showed levels of dichlorodiphenyltrichloroethane (DDT) and dieldrin especially high, exceeding the acceptable daily intake (ADI) for children. (This ADI standard was developed by a panel of experts linked to WHO as part of the FAO/WHO Joint Meeting on Pesticide Residues (JMPR). Residues of dieldrin exceeded ADI for adults. J. M. Mugambi, L. Kanja, T. E. Maitho, J. U. Skaare, and P. Lökken, "Organochlorine Pesticide Residues in Domestic Fowl (*Gallus domesticus*) Eggs from Central Kenya," *Journal of the Science of Food and Agriculture* 48, no. 2 (1989): 165-76. Another study found organochlorines in mothers' milk exceeding the ADI for infants; except for lindane the exposure occurred long ago. Kimani note 30 above, pages 224-25. Because these substances have been banned, the levels will decline.

34. Regulation No. 20345-S in E. Wo-Ching Sancho and R. Castro Córdoba, *Compendio de Legislacion Sobre Plaguicidas* (Compendium of Pesticide Legislation) (San José, Costa Rica: Organizacion Panamericana de la Salud Proyecto Plag-Salud, Centro de Derecho Ambiental y de los Recursos Naturales (CEDARENA) (Pan American Health Organization Plag-Salud Project, Center for Environmental Rights and Natural Resources), 1996). A project was started in 1993 to improve reporting and help to

prevent intoxications. R. Castro Córdoba, N. Morera González, and C. Jarquín Nuñez, *Sistema de Vigilancia Epidemiologica de Intoxicaciones con Plaguicidas, la Experiencia de Costa Rica, 1994-1996* (Epidemiological Monitoring System of Pesticide Intoxications, Costa Rica's Experience, 1994-1996) (San José, Costa Rica: Departamento de Registro y Control de Sustancias y Medicina del Trabajo del Ministerio de Salud (Ministry of Health Department of Registration and Control of Occupational Substances and Medicine), 1998).

35. Rodriguez, van der Haar, Antich, and Jarquín, note 16 above.

36. Ministerio de Salud, División de Saneamiento Ambiental, Departamento de Registro y Control de Sustancias Tóxicas y Medicina del Trabajo (Ministry of Health, Division of Environmental Health, Department of Registration and Control of Toxic Substances and Occupational Medicine), *Reporte Oficial intoxicaciones con Plaguicidas 1996* (Official Report of Pesticide Intoxications 1996) (Costa Rica, 1997).

37. Castro Córdoba, Morera González, and Jarquín Núñez, note 34 above. A study of the percentage of underreporting of intoxications in the system of monitoring gave an underreporting of 43 percent of symptoms that should have been linked to pesticides. Most of these were either dermal lesions or intoxications occurring outside the working environment, Rodríguez, van der Haar, Antich, and Jarquín, note 16 above, page 19.

38. Ministerio de Salud, note 36 above, page 13.

39. In the 1980s, even workers younger than 15 had very high occupational incidence rates. Today such instances are less frequent due to better enforcement of legislation that prohibits children under 18 to work with pesticides. Catharina Wesseling, Instituto Regional de Estudios en Sustancias Tóxicas (Central American Institute for Studies on Toxic Substances), e-mail message to author, 22 March 2004.

40. C. Wesseling, *Health Effects From Pesticide Use in Costa Rica—An Epidemiological Approach* (Stockholm: Institute of Environmental Medicine, Karolinska Institute, 1997).

41. C. Wesseling, L. Castillo, and C.F. Elinder, "Pesticide Poisoning in Costa Rica," *Scandinavian Journal of Work and Environmental Health* 19 (1993): 227-35. The data on hospitalizations and occupational accidents included a significant number of cases where the pesticide substance had not been identified.

42. Wesseling, note 40 above, page 42

43. Wesseling, note 40 above page 50.

44. Wesseling, note 40 above, page 50

45. For example, a study of organochlorine pesticides along the Kenyan coast only revealed very low levels compared to other areas, including tropical areas, J. M. Everaarts, E. M. van Weerlee, C. V. Fischerm, and T J. Hillebrand, "Polychlorinated Byphenyls and Cyclic Pesticides in Sediments and Macro-invertebrates from the Coastal Zone and Continental Slope of Kenya," *Marine Pollution Bulletin* 36, no. 6 (1998): 492-500. In a study of the concentration in water of organochlorine pesticides in coffee- and tea-growing areas that was made in 1994-1995, the mean pesticide levels did not exceed WHO or U.S. Environmental Protection Agency limits for drinking water. M. A. Mwanthi, "Occurrence of Three Pesticides in Community Water Supplies, Kenya" *Bulletin of Environmental Contamination and Toxicology* 60, no. 4 (1998): 601-8. However, the Development Plan of 1989-1993 reported that agrochemicals had led to severe pollution effects in the Tana and Athi Rivers regimes. B. D. Ogolla, "Environmental Management Policy and Law" *Environmental Policy and Law* 22, no. 3 (1992): 164-75. There had also been a case of paraquat contamination in the water supply for one town (Anonymous official, Ministry of Land Reclamation, Regional Development and Water, interview by author, Nairobi, Kenya, 15 October 1997).

46. L. Corrales and A. Salas, *Diagnóstico Ambiental de la Actividad Bananera en Sarapiquí Tortuguero y Talamanca, Costa Rica 1990-1992 (Con Actualizaciones Parciales a 1996)* (Environmental Assessment of Banana Cultivation in Sarapiquí, Tortuguero, and Talamanca, Costa Rica 1990-1992 (with Partial Updates until 1996) (San José, Costa Rica: Oficina Regional para Mesoamérica, Unión Mundial para la Naturaleza (Regional Office for Mesoamerica, International Union for the Conservation of Nature (IUCN)), 1997). The environmental pollution reported included: high concentrations of heavy metals in coral reef along the coast that could partly be due to pesticides; detection of DDT residues in some fish species in the late 1980s; detection of hexachlorobenzene (HCB), dieldrin, DDT, 1, 1-dichloro-2,2-bis(p-chlorophenyl)ethylene (DDE), paraquat and lindane in soil; detection of organochlorines and organophosphates in rivers and along the coast of the Atlantic; residues of various pesticides in sediments primarily chlorothalonil; the detection of chlorothalonil in ground water around banana plantations.

47. Karlsson, note 15 above, page 264. The study was not a quantitative survey with the rigidity that is implied in random farmer selection. The more than 30 farmers who were interviewed in each district were selected with purposeful sampling, with the goal of seeking the broadest range of views rather than averages. It is thus not possible to give figures on the number of experienced intoxications and pesticide products used (for example), but qualitative judgments on differences between districts can be made.

48. When looking at how individuals and organizations address pesticide-associated problems, the study focuses on those measures that aim to reduce primarily the health and environmental risks. Evidently, such measures can be of relevance for reducing trade, production, and economic problems, and conversely measures taken to address these may have positive effects on health and environment as well. Karlsson, note 15 above.

49. In addition to the Rotterdam and Stockholm Conventions described in Table 4, there is, for example, the Basel Convention on the Control of Transboundary Movements of Hazardous Waste and Their Disposal (www.basel.int) and the ILO Convention on Safety in the Use of Chemicals (www.ilo.org/public/english/protection/safework/cis/products/safetytm/c170.htm). Furthermore, the Codex Alimentarius (linked to WHO and FAO) establishes recommended maximum residue limits (MRLs) of pesticides in traded food products, but with the Agreement of Sanitary and Phytosanitary Measures these standards have indirectly become legally binding on the member countries of the WTO. Karlsson, note 15 above, page 96.

50. The FAO Code was first adopted by the FAO General Assembly in 1985 and has since been amended twice, in 1989 and 2002. FAO, International Code of Conduct on the Distribution and Use of Pesticides, Revised Version (Rome: FAO, 2003), accessed via http://www.fao.org/ag/agp/agpp/pesticid/ on 19 January 2004.

51. Examples of guidelines include FAO, *Guidelines for Legislation on the Control of Pesticides* (Rome: FAO, 1989); and FAO, *Revised Guidelines on Environmental Criteria for the Registration of Pesticides* (Rome: FAO, 1989).

52. Support for agricultural pesticide use has been a commonplace element in development projects, supported by various multi- and bilateral donors, but data on these are often missing from government import data. Williamson, note 16 above, page 3. While this practice is now less common, it is still taking place. Williamson, note 16 above, page 11. The IPM concept had been present in IGO discussions for many decades. Originally it was focused on controlling pest populations through a combination of all suitable techniques below thresholds that would cause economic damage. FAO, *International Code of Conduct on the Distribution and Use of Pesticides, Amended Version*, (Rome:

FAO, 1990). In the 1990s, however, the IPM concept came to be understood as a means to minimize the use of pesticides and increase reliance on alternative pest management technologies. See "Agenda 21," Rio de Janeiro, 2002, in Report of the United Nations Conference on Environment and Development, Rio de Janeiro, 3-14 June 1992, Volume I Resolutions Adopted by the Conference, A/CONF.151/26.Rev.1.

53. For example, it took until 1998 for FAO to have a first meeting with the International Federation of Organic Agriculture Movements (IFOAM), whose member organizations around the world are involved in research on alternatives to pesticides and in the design and manufacture of technology for controlling weeds, Karlsson, note 15 above, pages 107-8.

54. Organizations and IGOs left out of Table 4, but which were part of the study, include the International Labour Organization (ILO), United Nations Institute for Training and Research (UNITAR), World Health Organization (WHO), and the World Trade Organization (WTO).

55. Kenya's Pest Control Products Act came into force in 1984 and is often referred to as one of the most comprehensive laws in Africa. Republic of Kenya, The Pest Control Products Act Chapter 346 (Nairobi: Government Printer, 1985). In many respects it conforms to the guidelines for pesticide legislation from FAO. In 1997, 241 pesticide products had been registered and the rest were being screened. Republic of Kenya, note 28 above. The first effort to regulate pesticides in Costa Rica was made in 1954 and a modern pesticide registration process was established in 1976. R. Castro Córdoba, *Estudio Diagnostico Sobre la Legislación de Plaguicidas en Costa Rica* (Diagnostic Study of Pesticide Legislation in Costa Rica) (San José, Costa Rica: CEDARENA, 1995). The most recent revision was made in 1995 when law No. 24337MAG-S was published. Wo-Ching Sancho and Castro Córdoba, note 34 above. In 1993, there were 1,213 pesticides registered in Costa Rica, 347 generics,

and 38 mixtures. Garcia, note 20 above, page 235.

56. As a basis for the government agency's decision, the companies need to submit data that supports the efficacy of the product to control pests on the crops it was intended to be used as well as a long list of physical, chemical, toxicological and ecotoxicological data. Pest Control Products Board (PCPB), *Data Requirements for Registration of Pest Control Products, Legal Notice N46* (Nairobi, 1994); and Ministerio de Salud (Ministry of Health), *Pesticide Registration in Costa Rica* (Costa Rica, n.d.).

57. Kenya has banned the following pesticides: in 1986, dibromochloropropane, ethylene dibromide, 2,4,5 Trichlorophenoxyacetic acid (2,4,5-T), chlordimeform, hexachlorocychlohexane (HCH), chlordane, heptachlor, endrin, toxaphene; in 1988, parathion; and in 1989, captafol. In addition, in 1986, lindane was restricted use for seed dressing only; aldrin and diledrin were restricted for termite control in the building industry and DDT was restricted for use in public health. PCPB, Banned/Restricted Pesticides in Kenya (Nairobi, n.d.). Costa Rica has banned the following pesticides: in 1987, 2,4,5-T; in 1988, alchin, captafol, chlordecone, chlordimeform, DDT, dibromochloropropane (DBCP), dinoseb, ethylendibromide (EDB), nitrofen, toxafen; in 1990, lead arsenate, cyhexatin, endrin, pentachlorophenol; in 1991, chlordane, heptachlor; and in 1995, lindane. Ministerio de Salud, *Lista de Plaguicidas Prohibidos y Restringidos en Costa Rica* (List of Banned and Restricted Pesticides in Costa Rica) (Costa Rica, n.d).

58. In Kenya, the pesticide industry, through the Global Crop Protection Federation (GCPF), cooperated with the government's extension system in a safe-use project, and between 1991 and 1993, about 280,000 people were trained, including 2,800 retailers. Anonymous official, GCPF-Kenya, interview by author, Nairobi, Kenya, 23 September 1997. In Costa Rica, cooperation between the government and the pesticide industry

association has taken place throughout the 1990s, reaching 110,000 people. Anonymous official, Cámara Insumos Agropecuarios (Chamber of Agricultural and Livestock Inputs), interview by author, San Jose, Costa Rica, 4 February 1998. In 1991, the industry association and two ministries initiated the project "Teach," which was geared at training teachers in the rural areas so that they can teach children about pesticide issues. Conejo, Díaz, Furst, Gitli, and Vargas, note 19 above. Despite these efforts, there were a number of stakeholders in both countries who were concerned about the low effectiveness of the training measures. Karlsson, note 15 above. In Kenya, for example, results in the form of increased understanding of the toxic effects of pesticides were noted, but less than 30 percent of the farmers trained adopted the safety measures prescribed. Kimani, note 30 above, page 49. The Kenya Safe Use Project was also criticized for neglecting to train pesticide workers in the plantation sector. Partow, note 16 above, page xiv.

59. The Pest Control Products (Labeling, Advertising and Packaging) Regulation 3 (2) (n) requires each pesticide label to state that it is against the law to "use or store pest control products under unsafe conditions." Republic of Kenya, note 55 above. In Costa Rica, the banana sector is regulated by a special law (No. 7147), which obliges employers to train workers in appropriate use of pesticides and their associated risks. Castro Córdoba, note 55 above, page 21.

60. The Kenyan government has been involved in research on IPM and there were some pilot projects (often supported by donors). P. C. Matteson and M. I. Meltzer, *Environmental and Economic Implications of Agricultural Trade and Promotion Policies in Kenya: Pest and Pesticide Management* (Arlington, VA: Winrock International Environmental Alliance, 1995); O. Zethner, "Practice of Integrated Pest Management in Sub-Tropical Africa: An Overview of Two Decades (1970-1990)," in A. N. Mengech, K. N. Saxena and H. N. B. Gopalan, eds., *Inte-*

grated Pest Management in the Tropics, Current Status and Future Prospects (New York: John Wiley & Sons, 1995); and Organisation for Economic Co-operation and Development (OECD), *Report of the OECD/FAO Workshop on Integrated Pest Management and Pesticide Risk Reduction, Neuchâtel, Switzerland, 28 June—2 July 1998. ENV/JM/MONO (99) 7* (Paris: OECD, 1999). In Costa Rica, the official extension service promotes IPM. Ague, note 16 above, page 27. Several programs have been introduced in the country to reduce the volume of pesticide use. A. Faber, *Study on Investigations on Pesticides and the Search for Alternatives in Costa Rica* (Guápiles, Costa Rica: Wageningen Agricultural University, 1997). However, these initiatives remained isolated in a dominating agricultural system where pesticide use has become considered essential to increase productivity in most crops Organización Panamericana de la Salud, Programa Medío Ambiente y Salud en el Istmo CentroAmericano (Pan American Health Organization, Program of Environment and Health in the Central American Isthmus), *Aspectos Ocupacionales y Ambientales de la Exposición a Plaguicidas en el Istmo Centroamericano PLAGSALUD-Fase II* (Environmental and Occupational Aspects of Pesticide Exposure in the Central American Isthmus PLAGSALUD-Phase II), PLG97ESP.POR (San José, Costa Rica: 1997).

61. In Kenya, the Kenya Institute of Organic Farming (KIOF) began in 1986, and in the first years they met resistance from the government. Anonymous official, KIOF, interview by author, Nairobi, Kenya, 26 September 1997. KIOF trains farmers' groups in the field and arrange exchange visits among groups. KIOF, *Organic Farming, A Sustainable Method of Agriculture* (Nairobi, n.d.). Organic farming has been through a period of significant growth in Costa Rica with estimates of about 3,000 ha being under organic agriculture. Faber, ibid., page 14. There was high demand for organic products in the export market but virtually no demand in the domestic market. Karlsson, note 15

above, page 180. The Costa Rican government supported organic farming, for instance, it created a law for organic agriculture included in the environmental law in 1995, and it established a special office in the Ministry of Agriculture. The Asociación Nacional de Agricultura Organica (ANAO) (National Association for Organic Agriculture) had received funds to establish a nationally based organic certification system in the second half of the 1990s. Karlsson, note 15 above, page 181.

62. This definition of institutions is often used in social science and economics and differs somewhat from the common-language use that defines institutions as organizations. For a discussion on the role of institutions in environmental governance, see, for example, Young, note 14 above.

63. The issue of fit is extensively explored in relation to common property resource management. See, for example, R. J. Oakerson, "Analyzing the Commons: A Framework," in D. W. Bromley, ed., *Making the Commons Work: Theory, Practice and Policy* (San Francisco: ICS Press, 1992); E. Ostrom, "Designing Complexity to Govern Complexity," in S. Hanna and M. Munasinghe, eds., *Property Rights and the Environment, Social and Ecological Issues* (Washington, DC: Beijer International Institute of Ecological Economics and the World Bank, 1995); M. McGinnis and E. Ostrom "Design Principles for Local and Global Commons," in O. R. Young, ed., *The International Political Economy and International Institutions Volume II* (Cheltenham, UK: Edward Elgar Publishing Ltd., 1996). For a general discussion of the concept of fit, see C. Folke, L. Jr. Pritchard, F. Berkes, J. Coiling and U. Svedin, *The Problem of Fit between Ecosystems and Institutions* (Bonn, Germany: IHDP, 1998) accessed via www.uni-bonn.de/ihdp/wp02main.htm, on 27 June 2000.

64. Anonymous official, CEDARENA, interview by author, San José, Costa Rica, 10 March 1998. There have been efforts to harmonize pesticide registration in Central America

under the umbrella of Organismo Internacional Regional de Sanidad Agropecuaria (OIRSA) (International Regional Organization for Plant and Animal Health) with some support from FAO, but at the time of the study (1998/1999) efforts seemed to have halted. See Karlsson, note 15 above, page 199.

65. This strategy is not straightforward, however, as there are usually layers of driving forces and responsible stakeholders, often at different levels. It can be difficult to identify the original causes (or ultimate drivers), and this confuses the allocation of responsibility between levels, J. Saurin, "Global Environmental Degradations, Modernity and Environmental Knowledge," in C. Thomas, ed., *Rio Unravelling the Consequences* (Essex, UK: Frank Cass, 1994).

66. Karlsson, note 15 above, page 87.

67. Karlsson, note 15 above, pages 156-60 and 184-89.

68. It has been argued that it is difficult to be held responsible for a problem if one does not have the means to respond to it. T. Princen, "From Property Regime to International Regime: An Ecosystems Perspective," *Global Governance* 4, no. 4 (1998): 395-413.

69. For a discussion on altruistic motivations for behavior, see, for example, J. J. Mansbridge, ed. *Beyond Self-Interest* (London: University of Chicago Press, 1990).

70. In an FAO questionnaire, 46 percent of the responding developing countries felt that the pesticide industry acted only partly responsibly or not responsibly in adhering to the provisions of the FAO Code of Conduct as a standard for the manufacture, distribution, and advertising of pesticides. FAO, note 4 above, page 9.

71. L. L. Kiser and E. Ostrom, "The Three Worlds of Action: A Metatheoretical Synthesis of Institutional Approaches," in E. Ostrom, ed., *Strategies of Political Inquiry* (London: Sage Publications, 1982); and E. Ostrom, *Governing the Commons, The Evolution of Institutions for Collective Action* (New York: Cambridge University Press, 1990). While these authors refer to them as rules, they fall within the definition of institutions and that term is used here for clarity.

72. C. C. Gibson, E. Ostrom, and T K. Ahn "The Concept of Scale and the Human Dimensions of Global Change: A Survey," *Ecological Economics* 32, no. 2 (2000): 217-39.

73. Ostrom, note 71 above, page 52.

74. Karlsson, note 15 above, page 185.

75. Ostrom, note 71 above, page 52.

76. Karlsson, note 15 above; and S. Karlsson, "Institutionalized Knowledge Challenges in Pesticide Governance—The End of Knowledge and Beginning of Values in Governing Globalized Environmental Issues," *International Environmental Agreements* (forthcoming in 2004).

77. Williamsson, note 16 above.

78. In Kenya, the efficacy tests were made under some kind of cost-sharing arrangement with the government, while in Costa Rica the companies had to cover the full costs. Karlsson, note 15 above, pages 157, 186.

79. The term "multilayered governance" was constructed and defined as a system of coordinated and collective governance across levels in Karlsson, note 15 above, page 40. See also P. Hirst and G. Thompson, *Globalization in Question* (Cambridge, UK: Polity Press, 1996), 184, who argue for more 'sutured' governance across levels when discussing economic aspects of globalization, and Young, note 14 above, page 34, who argues for the need to influence behavior at all levels of governance.

Sylvia I. Karlsson is a postdoctoral fellow at Yale University's Center for Environmental Law and Policy and is a research fellow with the Institutional Dimensions of Global Environmental Change (IDGEC) project. Her current research focuses on cross-level aspects of global sustainable development governance. In 2001-2003 she worked as International Science Project Coordinator at the International Human Dimensions Programme on Global Environmental Change (IHDP) in Bonn, Germany. Sylvia I. Karlsson has worked for a short time at UNEP Chemicals in Geneva and the Economic Development Institute of the World Bank and as program officer for an action research project in Eastern Africa. Parallel to her studies and research, she has been actively engaged in the NGO processes of the Rio Conference in 1992, the World Summit for Social Development in 1995, and most recently, the World Summit on Sustainable Development, where she headed the delegation of the International Environment Forum, a scientific NGO accredited to the summit. The author wishes to thank Dr. Arthur L. Dahl and Ms. Agneta SundénByléhn and several anonymous reviewers for their helpful comments on earlier drafts. She further wishes to express her gratitude to all the people around the world who patiently answered questions during interviews—from the UN halls of Geneva to the soft grass of a Kenyan *shamba* (farm). The research was made possible through a grant from the Swedish International Development Agency (Sida). Karlsson's work has appeared in the peer-reviewed journals *International Environmental Agreements* and *The Common Property Resource Digest* and in a book edited by D.C. Esty and M. Ivanova, *Strengthening Global Environmental Governance: Options and Opportunities* (New Haven, CT: Yale School of Forestry & Environmental Studies, 2002). Karlsson can be contacted via e-mail at sylvia.karlsson@yale.edu.

From *Environment*, May 2004, pp. 24-41. Reprinted by permission of the Helen Dwight Reid Educational Foundation. Published by Heldref Publications, 1319, Eighteenth St., NW, Washington, DC 20036-1802. Copyright © 2004.

The Quest for Clean Water

As water pollution threatens our health and environment, we need to implement an expanding array of techniques for its assessment, prevention, and remediation.

Joseph Orlins and Anne Wehrly

In the 1890s, entrepreneur William Love sought to establish a model industrial community in the La Salle district of Niagara Falls, New York. The plan included building a canal that tapped water from the Niagara River for a navigable waterway and a hydroelectric power plant. Although work on the canal was begun, a nationwide economic depression and other factors forced abandonment of the project.

By 1920, the land adjacent to the canal was sold and used as a landfill for municipal and industrial wastes. Later purchased by Hooker Chemicals and Plastics Corp., the landfill became a dumping ground for nearly 21,000 tons of mixed chemical wastes before being closed and covered over in the early 1950s. Shortly thereafter, the property was acquired by the Niagara Falls Board of Education, and schools and residences were built on and around the site.

In the ensuing decades, groundwater levels in the area rose, parts of the landfill subsided, large metal drums of waste were uncovered, and toxic chemicals oozed out. All this led to the contamination of surface waters, oily residues in residential basements, corrosion of sump pumps, and noxious odors. Residents began to question if these problems were at the root of an apparent prevalence of birth defects and miscarriages in the neighborhood.

Eventually, in 1978, the area was declared unsafe by the New York State Department of Health, and President Jimmy Carter approved emergency federal assistance. The school located on the landfill site was closed and nearby houses were condemned. State and federal agencies worked together to relocate hundreds of residents and contain or destroy the chemical wastes.

That was the bitter story of Love Canal. Although not the worst environmental disaster in U.S. history, it illustrates the tragic consequences of water pollution.

Water quality standards

In addition to toxic chemical wastes, water pollutants occur in many other forms, including pathogenic microbes (harmful bacteria and viruses), excess fertilizers (containing compounds of phosphorus and nitrogen), and trash floating on streams, lakes, and beaches. Water pollution can also take the form of sediment eroded from stream banks, large blooms of algae, low levels of dissolved oxygen, or abnormally high temperatures (from the discharge of coolant water at power plants).

The United States has seen a growing concern about water pollution since the middle of the twentieth century, as the public recognized that pollutants were adversely affecting human health and rendering lakes unswimmable, streams unfishable, and rivers flammable. In response, in 1972, Congress passed the Federal Water Pollution Control Act Amendments, later modified and referred to as the Clean Water Act. Its purpose was to "restore and maintain the chemical, physical, and biological integrity of the nation's waters."

The Clean Water Act set the ambitious national goal of completely eliminating the discharge of pollutants into navigable waters by 1985, as well as the interim goal of making water clean enough to sustain fish and wildlife, while being safe for swimming and boating. To achieve these goals, certain standards for water quality were established.

The "designated uses" of every body of water subject to the act must first be identified. Is it a source for drinking water? Is it used for recreation, such as swimming? Does it supply agriculture or industry? Is it a significant habitat for fish and other aquatic life? Thereafter, the water must be tested for pollutants. If it fails to meet the minimum standards for its designated uses, then steps must be taken to limit pollutants entering it, so that it becomes suitable for those uses.

On the global level, the fundamental importance of clean water has come into the spotlight. In November 2002, the UN Committee on Economic, Cultural and Social Rights declared access to clean

Tragedy at Minamata Bay

The Chisso chemical factory, located on the Japanese island of Kyushu, is believed to have discharged between 70 and 150 tons of methylmercury (an organic form of mercury) into Minamata Bay between 1932 and 1968. The factory, a dominant presence in the region, used the chemical to manufacture acetic acid and vinyl chloride.

Methylmercury is easily absorbed upon ingestion, causing widespread damage to the central nervous system. Symptoms include numbing and unsteadiness of extremities, failure of muscular coordination, and impairment of speech, hearing and vision. Exposure to high levels of the substance can be fatal. In addition, the effects are magnified for infants exposed to methylmercury through their mothers, both before birth and while nursing.

In the 1960s and '70s, it was revealed that thousands of Minamata Bay residents had been exposed to methylmercury. The chemical had been taken up from the bay's waters by its fish and then made its way into the birds, cats, and people who ate the fish. Consequently, methylmercury poisoning came to be called Minamata disease.

Remediation, which took as long as 14 years, involved removing the mercury-filled sediments and containing them on reclaimed land in Minamata Bay. Fish in the bay had such high levels of methylmercury that they had to be prevented from leaving the bay by a huge net, which was in place from 1947 to 1997.

Mercury poisoning has recently appeared in the Amazon basin, where deforestation has led to uncontrolled runoff of natural accumulations of mercury from the soil into rivers and streams. In the United States, testing has revealed that predator fish such as bass and walleye in certain lakes and rivers contain enough mercury to justify warnings against consuming them in large amounts.

—*J.O. and A.W.*

Minamata Bay residents who were exposed to methylmercury have been suffering from such problems as loss of muscular control, numbing of extremities, and impairment of speech, hearing, and vision.

water a human right. Moreover, the United Nations has designated 2003 to be the International Year of Freshwater, with the aim of encouraging sustainable use of freshwater and integrated water resources management.

Here, there, and everywhere

Implementing the Clean Water Act requires clarifying the sources of pollutants. They are divided into two groups: "point sources" and "nonpoint sources." Point sources correspond to discrete, identifiable locations from which pollutants are emitted. They include factories, wastewater treatment plants, landfills, and underground storage tanks. Water pollution that originates at point sources is usually what is associated with headline-grabbing stories such as those about Love Canal.

Nonpoint sources of pollution are diffuse and therefore harder to control. For instance, rain washes oil, grease, and solid pollutants from streets and parking lots into storm drains that carry them into bays and rivers. Likewise, irrigation and rainwater leach fertilizers, herbicides, and insecticides from farms and lawns and into streams and lakes.

In the United States, the Clean Water Act requires that industrial wastes be neutralized or broken down before being released into rivers and lakes.

The direct discharge of wastes from point sources into lakes, rivers, and streams is regulated by a permit program known as the National Pollutant Discharge Elimination System (NPDES). This program, established through the Clean Water Act, is administered by the Environmental Protection Agency (EPA) and authorized states. By regulating the wastes discharged, NPDES has helped reduce point-source pollution dramatically. On the other hand, water pollution in the United States is now mainly from nonpoint sources, as reported by the EPA.

In 1991, the U.S. Geological Survey (USGS, part of the Department of the Interior) began a systematic, long-term program to monitor watersheds. The National Water-Quality Assessment Program (NAWQA), established to help manage surface and groundwater supplies, has involved the collection and analysis of water quality data in over 50 major river basins and aquifer systems in nearly all 50 states.

The program has encompassed three principal categories of investigation: (1) the current conditions of surface water and groundwater; (2) changes in those conditions over time; and (3) major factors—such as climate, geography, and land use—that affect water quality. For each of these categories, the water and sediment have been tested for such pollutants as pesticides, plant nutrients, volatile organic compounds, and heavy metals.

The NAWQA findings were disturbing. Water quality is most affected in watersheds with highest population density and urban development. In agricultural areas, 95 percent of tested streams and 60 percent of shallow wells contained herbicides, insecticides, or both. In urban areas, 99 percent of tested streams and 50 percent of shallow wells had herbicides, especially those used on lawns and golf courses. Insecticides were found more frequently in urban streams than in agricultural ones.

The study also found large amounts of plant nutrients in water supplies. For instance, 80 percent of agricultural streams and 70 percent of urban streams were found to contain phosphorus at concentrations that exceeded EPA guidelines.

Moreover, in agricultural areas, one out of five well-water samples had nitrate concentrations higher than EPA standards for drinking water. Nitrate contamination can result from nitrogen fertilizers or material from defective septic systems leaching into the groundwater, or it may reflect defects in the wells.

Effects of pollution

According to the UN World Water Assessment Programme, about 2.3 billion people suffer from diseases associated with polluted water, and more than 5 million people die from these illnesses each year. Dysentery, typhoid, cholera, and hepatitis A are some of the ailments that result from ingesting water contaminated with harmful microbes. Other illnesses—such as malaria, filariasis, yellow fever, and sleeping sickness—are transmitted by vector organisms (such as mosquitoes and tsetse flies) that breed in or live near stagnant, unclean water.

A number of chemical contaminants—including DDT, dioxins, polychlorinated biphenyls (PCBs), and heavy metals—are associated with conditions ranging from skin rashes to various cancers and birth defects. Excess nitrate in an infant's drinking water can lead to the "blue baby syndrome" (methemoglobinemia)—a condition in which the child's digestive system cannot process the nitrate, diminishing the blood's ability to carry adequate concentrations of oxygen.

Besides affecting human health, water pollution has adverse effects on ecosystems. For instance, while moderate amounts of nutrients in surface water are generally not problematic, large quantities of phosphorus and nitrogen compounds can lead to excessive growth of algae and other nuisance species. Known as *eutrophication*, this phenomenon reduces the penetration of sunlight through the water; when the plants die and decompose, the body of water is left with odors, bad taste, and reduced levels of dissolved oxygen.

Low levels of dissolved oxygen can kill fish and shellfish. In addition, aquatic weeds can interfere with recreational activities (such as boating and swimming) and can clog intake by industry and municipal systems.

Some pollutants settle to the bottom of streams, lakes, and harbors, where they may remain for many years. For instance, although DDT and PCBs were banned years ago, they are still found in sediments in many urban and rural streams. They occur at levels harmful to wildlife at more than two-thirds of the urban sites tested.

Prevention and remediation

As the old saying goes, an ounce of prevention is worth a pound of cure. This is especially true when it comes to controlling water pollution. Several important steps taken since the passage of the Clean Water Act have made surface waters today cleaner in many ways than they were 30 years ago.

For example, industrial wastes are mandated to be neutralized or broken down before being discharged to streams, lakes, and harbors. Moreover, the U.S. government has banned the production and use of certain dangerous pollutants such as DDT and PCBs.

In addition, two major changes have been introduced in the handling of sewage. First, smaller, less efficient sewage treatment plants are being replaced with modern, regional plants that include biological treatment, in which microorganisms are used to break down organic matter in the sewage. The newer plants are releasing much cleaner discharges into the receiving bodies of water (rivers, lakes, and ocean).

Second, many jurisdictions throughout the United States are building separate sewer lines for storm water and sanitary wastes. These upgrades are needed because excess water in the older, "combined" sewer systems would simply bypass the treatment process, and untreated sewage would be discharged directly into receiving bodies of water.

To minimize pollutants from nonpoint sources, the EPA is requiring all municipalities to address the problem of runoff from roads and parking lots. At the same time, the use of fertilizers and pesticides needs to be reduced. Toward this end, county extension agents are educating farmers and homeowners about their proper application and the availability of nutrient testing.

To curtail the use of expensive and potentially harmful pesticides, the approach known as *integrated pest management* can be implemented [see "Safer Modes of Pest Control," THE WORLD & I, May 2000, p. 164]. It involves the identification of specific pest problems and the use of nontoxic chemicals and chemical-free alternatives whenever possible. For instance, aphids can be held in check by ladybug beetles and caterpillars can be controlled by applying neem oil to the leaves on which they feed.

Moreover, new urban development projects in many areas are required to implement storm-water management practices. They include such features as: oil and grease traps in storm drains; swales to slow down runoff, allowing it to infiltrate back into groundwater; "wet" detention basins (essentially artificial ponds) that allow solids to settle out of runoff; and artificial wetlands that help break down contaminants in runoff. While such additions may be costly, they significantly improve water quality. They are of course much more expensive to install after those areas have been developed.

Once a waterway is polluted, cleanup is often expensive and time consuming. For instance, to increase the concentration of dissolved oxygen in a lake that has undergone eutrophication, fountains and aerators may be necessary. Specially designed boats may be needed to harvest nuisance weeds.

At times, it is costly just to identify the source of a problem. For example, if a body of water contains high levels of coliform bacteria, expensive DNA testing may be needed to determine whether the bacteria came from leakage of human sewage, pet waste, or the feces of waterfowl or other wildlife.

Testing the effect of simulated rainfall on fresh manure, scientists with the Agricultural Research Service have found that grass strips are highly effective at preventing manure-borne microbes from being washed down a slope and contaminating surface waters.

Contaminated sediments are sometimes difficult to treat. Available techniques range from dredging the sediments to "capping" them in place, to limit their potential exposure. Given that they act as reservoirs of pollutants, it is often best to remove the sediments and burn off the contaminants. Alternatively, the extracted sediments may be placed in confined disposal areas that prevent the pollutants from leaching back into groundwater. Dredging, however, may create additional problems by releasing pollutants back into the water column when the sediment is stirred up.

The future of clean water

The EPA reports that as a result of the Clean Water Act, millions of tons of sewage and industrial waste are being treated before they are discharged into U.S. coastal waters. In addition, the majority of lakes and rivers now meet mandated water quality goals.

Yet the future of federal regulation under the Clean Water Act is unclear. In 2001, a Supreme Court decision (*Solid Waste Agency of Northern Cook County v. United States Army Corps of Engineers, et al.*) brought into question the power of federal agencies to regulate activities affecting water quality in smaller, nonnavigable bodies of water. This and related court decisions have set the stage for the EPA and other federal agencies to redefine which bodies of water can be protected from unregulated dumping and discharges under the Clean Water Act. As a result, individual states may soon be faced with much greater responsibility for the protection of water resources.

Worldwide, more than one billion people presently lack access to clean water sources, and over two billion live without basic sanitation facilities. A large proportion of those who die from water-related diseases are infants. We would hope that by raising awareness of these issues on an international level, the newly recognized right to clean water will become a reality for a much larger percentage of the world's population.

Joseph Orlins, professor of civil engineering at Rowan University in Glassboro, New Jersey, specializes in water resources and environmental engineering. Anne Wehrly is an attorney and freelance writer.

On the Internet

2003: INTERNATIONAL YEAR OF FRESHWATER (UN)
www.wateryear2003.org

CLEAN WATER ACT (EPA)
www.epa.gov/watertrain/cwa

DO'S AND DON'T'S AROUND THE HOME (EPA)
www.epa.gov/owow/nps/dosdont.html

LOVE CANAL COLLECTION (SUNY BUFFALO)
ublib.buffalo.edu/libraries/projects/lovecanal

NATIONAL WATER-QUALITY ASSESSMENT PROGRAM (USGS)
water.usgs.gov/nawqa

OFFICE OF WATER (EPA)
www.epa.gov/ow

THE WORLD'S WATER
www.worldwater.org

Human Biomonitoring of Environmental Chemicals

*Measuring chemicals in human tissues is the "gold standard"
for assessing people's exposure to pollution.*

Ken Sexton, Larry L. Needham and James L. Pirkle

What chemicals in your daily routine should you be most concerned about? The volatile organic compounds from your carpet? The exhaust fumes on the road to work? The pesticide residues in the apple in your lunch? Most of us are exposed to low levels of thousands of toxic chemicals every day. How can a person—or a nation—decide which substances should be controlled most rigorously?

One strategy is to go after the largest sources of pollution. This approach certainly makes sense when those pollutants have obvious and widespread consequences, such as warming the globe, causing algal blooms, eroding the ozone layer or killing off wildlife. But for protecting human health, this strategy does not serve so well, because the link between a given compound and its biological effects can be difficult to gauge. For epidemiologists to correlate environmental pollutants with health problems, they need to know who has been exposed and at what level.

This knowledge is exceptionally difficult to gain when there is a lag between exposure and the manifestation of illness. In such cases, the data are seldom—if ever—sufficient to determine the precise agent, the details of contact and the full extent of the affected population. Complicating matters, the scientific understanding of the mechanisms of exposure, such as how various compounds are carried through the air and changed along the way, is often incomplete. As a result, epidemiologists often find it difficult to establish cause-and-effect relationships for environmentally induced sicknesses. Without reliable information some pollutants may be unfairly blamed, whereas others exert their dire effects without challenge. Fortunately, there is hope: a method of accurately measuring not only contact with, but also absorption of toxic chemicals from, the environment—human biomonitoring.

Is It in Me?
Each person's risk of developing an environmentally related disease,

such as cancer, results from a unique combination of exposure, genes, age, sex, nutrition and lifestyle. Science doesn't fully understand how these variables interact, but exposure is clearly a key factor. Thus, a fundamental goal of environmental health policy is to prevent (or at least reduce) people taking in chemicals that lead to any of the five *D*s—discomfort, dysfunction, disability, disease or death.

Exposure to an environmental chemical is minimally defined as contact with the skin, mouth or nostrils—a meaning that includes breathing, eating and drinking. For the purposes of assessing risk, the most important attributes of exposure are magnitude (what is the concentration?), duration (how long does contact last?), frequency (how often do exposures occur?) and timing (at what age do exposures occur?). The calculation of actual exposure also requires complex detective work to discover all kinds of details, including the chemical identity (for example, the pesticide chlorpyrifos), source (nearby agricultural use), medium of transport (groundwater) and route (drinking contaminated well water). Scientists must consider this information on exposure against the background of people's activity patterns, eating and drinking habits, and lifestyle, and they must also evaluate the influence of other chemicals in the air, water, beverages, food, dust and soil. Overall, this is a daunting challenge.

Historically, those scientists who undertook such a complex task have relied on indirect methods: questionnaires, diaries, interviews, centralized monitoring of community air or water, and a record of broad activity patterns among the population. But the results were often disappointing. Although these circumstantial approaches have the advantages of practicality and frugality, they can also introduce substantial uncertainty into resulting exposure estimates. This shortcoming multiplies the potential for a fundamental error—classifying a person as "not exposed" when he or she has been or vice versa.

A second approach, the direct measurement of an individual's environment, is sometimes a possibility—for example, a person might carry a portable monitor to record contact with airborne chemicals. Although this technique offers an unequivocal record of chemical contact, it is technologically infeasible or prohibitively expensive to measure most pollutants this way. Also, although such monitors document exposure, they tell nothing about the person's uptake of these airborne chemicals—how much truly gets into his or her body, which is, of course, the most relevant information for assessing health risk. Fortunately, technological advances in biomedicine and analytical chemistry now make it possible to get exactly this information. Biomonitoring measures the actual levels of suspected environmental chemicals in human tissues and fluids. This third approach has come to be the "gold standard" for assessing exposure to chemicals.

Blood (and Urine) Will Tell

Biomonitoring is not new. It has its roots in the analysis of biological samples for markers for various pharmaceutical compounds and occupational chemicals, efforts that sought to prevent the harmful accumulation of dangerous substances. Although it had a different name at the time, the general idea was first applied about 130 years ago when doctors monitored the amount of salicyluric acid in the urine of rheumatics who were being treated with large doses of salicylic acid (the precursor of aspirin). And as early as the 1890s, factory workers who were exposed to lead had their blood and urine screened to forestall the elevated levels that produced acute lead poisoning.

These investigators soon learned that the degree of contact with a substance doesn't necessarily determine the biologically relevant exposure to that chemical. As a result, this measure didn't help much in predicting the risks of lead poisoning. However, they did find that the amount of a compound that crosses the body's boundaries (called the internal or absorbed dose, or sometimes the body burden) has considerable value for estimating the risk to health. Today, it is relatively affordable to measure the absorbed doses for hundreds of chemicals by looking for biomarkers of exposure in accessible human tissues and fluids, including saliva, semen, urine, sputum, hair, feces, breast milk and fingernails (all of which can be collected readily), and blood, lung tissue, bone marrow, follicular fluid, adipose tissue and blood vessels (which require incursion into the body). Although procedures to collect any of the first set would, technically, be considered "noninvasive," in fact, that categorization rests on cultural, psychological and social factors. So obtaining the right material can sometimes be awkward. Fortunately for those of us in the biomonitoring field, it's never necessary to collect all of those samples—blood and urine are typically suffi-

cient. These are analyzed for the presence of biological markers of exposure—generally the targeted chemical, its primary metabolites or the products of its reaction with certain natural compounds in the body, such as proteins.

Choosing the appropriate tissue or fluid for biological monitoring is based primarily on the chemical and physical properties of the chemical of interest and, in some cases, the time interval since the last exposure. For example, some chemicals including dioxins, polychlorinated biphenyls and organochlorine pesticides have long biological residence times in the body (months or years) because they are sequestered in fatty tissues. They are thus said to be fat-loving or, to use the proper term, lipophilic. By contrast, other chemicals such as organophosphate pesticides and volatile organic compounds, which don't accumulate in fats (being lipophobic), have relatively short biological residence times (hours or days) and tend to be metabolized rapidly and excreted in the urine.

The time since the last exposure can also play a key role in determining the best biological specimen for analysis. For example, a persistent chemical, such as a dioxin, remains present in blood for a much longer period (years) than does a nonpersistent compound such as benzene (hours), but dioxin does not form significant urinary metabolites, whereas benzene does. For these reasons, persistent chemicals are typically measured in blood, and nonpersistent chemicals are measured in urine (as soon after exposure as possible), although they can also be detected in blood soon after exposure if the analytical methods are sufficiently sensitive—and they usually are. Specialists can now detect extremely low levels—parts-per-billion, parts-per-trillion, even parts-per-quadrillion—of multiple markers using a relatively small sample, say, 10 milliliters or less.

Clearly, the sensitivity of the analysis is important in choosing what to measure—but it's not everything.

Other issues must be considered before the results can be considered meaningful. Well before attempting to discern trace amounts of target chemicals, an investigator should be able to answer three broad questions: How is the measurement related to the magnitude, duration, frequency and timing of exposure? How do subsequent processes within the body—such as absorption, distribution, metabolism and excretion—influence the targeted biomarker? And is this particular marker specific for a certain chemical or does it indicate an entire class of substances?

Because the science underpinning human biomonitoring has improved significantly in recent years, these questions are now easier to answer. The rapid advancement in knowledge of what the body does to chemicals that are inhaled, ingested or absorbed through the skin has led to better interpretation of the range of concentrations for various biomarkers. And the number of testable compounds has increased dramatically: Sensitive and specific biomarkers are available for many environmental chemicals, including metals, dioxins, furans, polychlorinated biphenyls, pesticides, volatile organic compounds, phthalates, phytoestrogens and environmental tobacco smoke. As research continues, the list will surely continue to grow.

Exposure *and* Uptake

Biomonitoring has many advantages over traditional methods. For example, biological samples reveal the integrated effects of repeated contact. Also, this approach documents all routes of exposure—inhalation, absorption through the skin and ingestion, including hand-to-mouth transfer by children. Such specimens also reflect the modifying influences of physiology, bioavailability and bioaccumulation, which can magnify the concentrations of some environmental chemicals enough to raise them above the detection threshold. Perhaps most importantly, these tests can help establish correlations between exposure and subsequent illness in individuals—which is often the key observation in proving whether or not a link exists.

A great strength of biomonitoring is that it provides unequivocal evidence that both exposure and uptake have taken place. In some cases these data can confirm the findings of traditional exposure estimates. For example, in 1979, residents of Triana, Alabama, were notified that fish from a nearby creek had forty times more DDT than the allowable limit, even though the local DDT manufacturing plant had been inactive since 1971. The announcement was especially concerning because many people in that area caught and ate the fish regularly. In response to this discovery, the Centers for Disease Control and Prevention (CDC) constructed an evaluation based on DDT concentrations in fish and the amount of fish eaten per week. This estimate indeed correlated with levels of DDT and its metabolites, DDE and DDD, in the blood of Triana residents. In a similar story that unfolded in the late 1980s, chemical-plant workers in New Jersey and Missouri discovered that they had been exposed to dioxin-contaminated compounds up to the early 1970s. They had come into contact with the dioxin in various ways—breathing it, swallowing it and taking it in through the skin. Despite the complexities of their interaction with this dangerous substance—and the time interval since exposure—a scheme that used occupational records to calculate the duration of potential exposure was able to accurately estimate internal doses. This finding was confirmed by the correlation of these results with the concentration of dioxins in their blood.

Having information about exposure *and* uptake is more than a *pro forma* detail: There are many cases in which traditional estimates of exposure (questionnaires, proximity to sources, environmental concentrations, constructed scenarios) are not correlated with measured biomarkers. For example, from 1962 to 1971, the U.S. Air Force sprayed the defoliant known as "Agent Orange" in Vietnam. Many service members who participated in that operation touched or breathed the herbicide, potentially exposing themselves to high levels of dioxin. The Air Force first estimated the risk to soldiers using a scenario approach, which included the average dioxin concentration in Agent Orange, the number of gallons used during a soldier's tour of duty, and the frequency and duration of potential contact based on job description. Despite a considerable scientific effort that went into these predictions, CDC studies in the late 1980s proved that none of the exposure estimates were correlated with the measured blood levels of dioxin in at-risk troops. A subsequent investigation of personnel with the highest dioxin levels did identify some patterns that explained their increased contact—for example, small-statured enlisted men often climbed into the chemical tanks to clean out residual Agent Orange.

A more striking example of the value of biomonitoring came in the mid-1970s when the United States elected to start phasing out leaded gasoline. Prior to this decision, traditional models had suggested that eliminating lead in gasoline would have only a slight effect on people's uptake of that metal. However, biomonitoring data from the CDC's Second National Health and Nutrition Examination Survey revealed that from 1976 to 1980 (as unleaded fuel was first introduced and gasoline lead decreased by approximately 55 percent) there was a parallel decline in the amount of lead coursing through the veins of the U.S. population. Overall, average blood concentrations decreased from about 16 to less than 10 micrograms of lead per deciliter of blood. These data demonstrated the effectiveness of removing lead from gasoline, and they were a dominant factor in the decision by the Environmental Protection Agency (EPA) to remove lead from gasoline more rap-

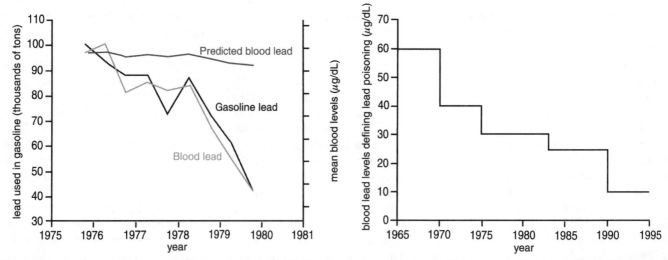

Figure 1. Leaded gasoline began to be phased out in the 1970s. Although the predicted effect on blood lead was minimal, actual exposure in the U.S. population (measured in micrograms of lead per deciliter of blood) sharply declined between 1976 and 1980, paralleling the changes in gasoline (*left*). Blood lead and gas lead continued to follow nearly identical decreases up to 1990. At the same time, a series of studies on lead toxicity showed that lower doses could still cause adverse effects, prompting a steady decline in the level defining lead poisoning (*right*).

idly—a task that was effectively complete by 1991. Today, the average blood-lead level in the U.S. population is less than 2 micrograms per deciliter

Exposure Disclosure

The study that revealed the tight connection between the lead in people's gas tanks and the lead in their blood was mounted by the CDC, which conducts the National Health and Nutrition Examination Surveys (NHANES for short). Although no environmental chemicals were measured as part of NHANES I (1971-1975), starting with NHANES II (1976-1980), the CDC began measuring blood lead levels in the U.S. population, ironically enough, after the Food and Drug Administration voiced concerns about possible exposures from eating food stored in lead-soldered cans, which turned out to be a very minor risk compared with leaded gasoline. As part of NHANES II, the EPA tested for certain persistent pesticides in people's blood and nonpersistent pesticides or their metabolites in urine. After an eight-year hiatus, NHANES III was conducted in two three-year phases from 1988 to 1994. In that iteration, the CDC measured lead and cadmium and began

testing for cotinine, the major metabolite of nicotine, in blood. Additionally, the CDC began a separate pilot program to measure new compounds, testing for trace amounts of 32 volatile organic chemicals in blood and 12 pesticides or their metabolites in urine from approximately 1,000 of the NHANES III participants.

Then came another long gap in coverage. But thankfully, in 1999, NHANES became a continuous survey of the noninstitutionalized U.S. population. (It is thought that excluding members of isolated organizations, such as military personnel, college students and prisoners, provides a better cross-section of America.) In the current design, a new national sample is collected every two years. Although some other studies have focused on specific populations or on more restricted data, NHANES is the only national survey that includes both a medical examination and collection of biological samples from participants. Individuals selected for NHANES are representative of the U.S. population, meaning that they do not necessarily have high or unusual exposures. About 5,000 participants are examined annually from 15 locations throughout the country.

Reporting For Duty

In March 2001, the CDC released the National Report on Human Exposure to Environmental Chemicals, which included data from 1999 on 27 chemicals. A second report was published in January 2003 that examined 116 chemicals in samples from 1999-2000. Both studies used biomonitoring to provide an ongoing assessment of exposure to a variety of substances. Although various studies of workplace exposure, for example, had raised concerns about the health effects of such chemicals, most of them had never before been measured in a representative slice of the U.S. population.

The inventory of tested substances in the second CDC report includes lead, mercury, cadmium and other metals; persistent (organochlorine-based) and nonpersistent (organophosphate- and carbamate-based) insecticides, herbicides and other pesticides; pest repellents and disinfectants; cotinine; phthalates; polycyclic aromatic hydrocarbons; dioxins, furans and polychlorinated biphenyls; and phytoestrogens. Results from the general population are subdivided by age, gender and ethnicity.

An important feature of the CDC report is that it provides reference

Figure 2. One important function of biomonitoring is that it can identify specific subpopulations that may be more vulnerable to exposure from a particular chemical. For example, p,p'-DDE, a long-lasting metabolite of DDT, is more than twice as high in Mexican-Americans compared with the general population, By contrast, cotinine levels are the lowest among this group, indicating that they have the least exposure to environmental tobacco smoke. For both cotinine and lead, non-hispanic blacks showed the highest levels. DDE (in nanograms per gram of lipid) and lead (in micrograms per deciliter of blood serum) data are from the CDC's Second National Report on Human Exposure to Environmental Chemicals, published in 2003. Information on cotinine (in nanograms per milliliter of blood) is from the third National Health and Nutrition Examination Survey (NHANES III), 1988-1991.

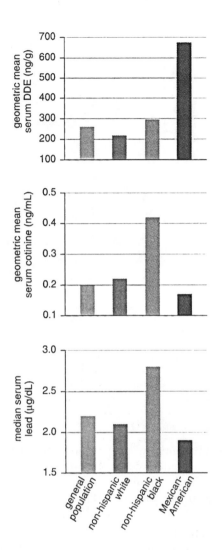

ranges for exposure among the general U.S. population. If people are concerned that they may have been excessively exposed to an environmental chemical, they can compare their biomarker levels to those standards. These reference ranges are immensely beneficial to public-health scientists who must decide if certain high-exposure groups need follow-up action. If average levels among the cohort are similar to those of the general public, then the group's exposure is unlikely to cause unique problems. On the other hand, if levels are substantially higher than national norms, epidemiologists can confirm the unusual exposure, identify the sources and provide continuing health care as appropriate. The reference ranges provide indirect fi-

nancial advantages too, because distinguishing common from unusual chemical contact helps direct resources to the most pertinent exposure situations.

The overarching purpose of these reports is to help scientists, physicians and health officials to prevent, reduce and treat environmentally induced illnesses. However, some caution must be exercised in interpreting the findings: It is important to remember that detecting a chemical in a person's blood or urine does not by itself mean that the exposure causes disease. Separate scientific studies in animals and humans are required to determine which levels are likely to do harm. For most chemicals, toxicologists simply don't have this information.

But even if scientists are not sure of the overall level of risk, they can make concrete statements about whether situations are getting better or worse. The latest CDC report, in addition to listing current biomarker levels in the population, also highlights some interesting exposure trends gleaned from earlier NHANES findings. For example, from 1991 to 1994, 4.4 percent of children between the ages of one and five had levels of blood lead greater than or equal to 10 micrograms per deciliter, the Federal action level. By the second collection period (1999 and 2000), only 2.2 percent of this age group exceeded this threshold. This decrease suggests that efforts to reduce lead exposure for children have been successful. It also serves

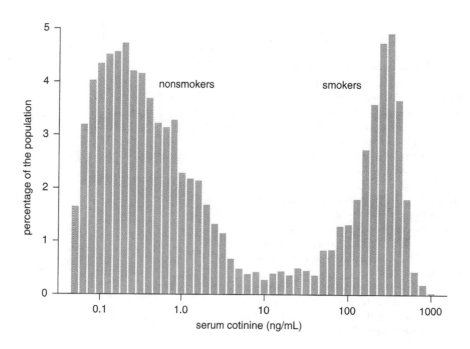

Figure 3. U.S. population clearly segregates into smokers and nonsmokers based on the level of cotinine in blood. The working threshold for distinguishing the two groups is 10 nanograms per milliliter of blood serum. Among nonsmokers, the highest values of cotinine were found in children under 12, and they were strongly reflective of the number of smokers in the home. The data are from NDANES III, 1988-1991.

as a reminder that some children, including those living in homes with lead-based paint or lead-contaminated dust, remain at unacceptably high risk.

The last report also indicates a hopeful trend in the exposure to environmental tobacco smoke, as shown by tests for the biomarker cotinine in the blood of nonsmokers. Median levels of cotinine fell more than 70 percent in roughly a decade—that is, between the second (1988 to 1991) and third (1999 and 2000) periods of data collection. This drop provides objective evidence of reduced exposure to environmental tobacco smoke for the general U.S. population. Nevertheless, the fact that more than half of American youth continue to be exposed to environmental tobacco smoke remains a public-health concern.

The CDC plans to release future reports that document their biomonitoring efforts every two years. In the next edition, they will also add the findings from separate studies of special populations, such as the laborers who apply pesticides to

crops, people living near hazardous-waste sites and workers in lead smelters, all of which are likely to have higher-than-average exposures to certain environmental chemicals.

Annual Check-Up With Biomarkers?

As the 21st century unfolds, the CDC surveys and other well-designed biomonitoring studies will continue to build an understanding of people's exposure to toxic environmental chemicals. Nonetheless, these data will not obviate the need to collect other kinds of relevant information—to monitor sources of pollution, to conduct surveys of toxic substances in the environment and to study human activities and behaviors that contribute to exposure. Moreover, further research in toxicology and epidemiology is necessary before specialists can interpret the health significance of exposure biomarkers for most environmental chemicals. Particularly as detection methods improve—enabling investigators to measure lower concentrations of more chemi-

cals from smaller samples at less cost—scientific understanding of what the body does to the chemical (and vice versa) must keep pace. If this effort is successful, a full screen of exposure biomarkers may be a part of every routine physical exam in the not-too-distant future.

Bibliography

DeCaprio, A. P. 1997. Biomarkers: coming of age for environmental health and risk assessment. *Environmental Science & Technology* 31:1837-1848.

Mendelsohn, M. L., J. P. Peeters and M. J. Normandy, eds. 1995. *Biomarkers and Occupational Health: Progress and Perspectives.* Washington, DC: Joseph Henry Press.

Mendelsohn, M. L., L. C. Mohr and J. P. Peeters, eds. 1998. *Biomarkers: Medical and Workplace Applications.* Washington, DC: Joseph Henry Press.

Needham, L. L., and K. Sexton. 2000. Assessing children's exposure to hazardous environmental chemicals: An overview of selected research challenges and complexities. *Journal of Exposure Analysis and Environmental Epidemiology* 10 (Part 2):611-629.

Needham, L. L., D. G. Patterson, Jr., V. W. Burse, D. C. Paschal, W. E. Turner and

R. H. Hill, Jr. 1996. Reference range data for assessing exposure to selected environmental toxicants. *Toxicology and Industrial Health* 12:507-513.

Pirkle, J. L., E. J. Sampson, L. L. Needham, D. G. Patterson, Jr., and D. L. Ashley. 1995. Using biological monitoring to assess human exposure to priority toxicants. *Environmental Health Perspectives* 103 (supplement 3):45-48.

Ken Sexton is a professor of environmental sciences at the University of Texas School of Public Health, Brownsville Regional Campus, and past president of the International Society of Exposure Analysis (ISEA). Larry L. Needham is Chief of the Organic Analytical Toxicology Branch in the National Center for Environmental Health of the Centers for Disease Control and Prevention (CDC) and the current ISEA president. James L. Pirkle is the Deputy Director for Science at the CDC's Environmental Health Laboratory. Sexton's address is University of Texas School of Public Health, Brownsville Regional Campus, RAHC Building, 80 Fort Brown, Brownsville, TX 78520. Internet: ksexton@utb.edu

A LITTLE ROCKET FUEL WITH YOUR SALAD?

By Gene Ayres

I SPENT TWO DECADES living in Southern California, during which time I married, fathered a child, got divorced, and became a single parent. I finally left the Golden State in 1989, taking my young son with me because I was concerned about his health. It wasn't just the smog, although that was bad enough around our house in the San Fernando Valley, where chronic inversions turned the sky over the San Gabriel Mountains the color of dried blood. My son had been born with myriad allergies, and had experienced a terrifying reaction to the DPT vaccine (screaming fits suggestive of being tortured).

Then came the medfly. It was 1988, and I can still recall the sight of the State agricultural commissioner going on TV and dramatically drinking a glass of malathion to reassure a panicky public that, despite the conspicuous presence of a skull and crossbones on all malathion containers (the pesticide was readily available at most garden centers), it was really quite harmless. Of course, we only had the man's word on that, as well as on what was really in his glass, but it was a classic case of the "as-seen-on-TV" school of credibility.

What had prompted this extraordinary demonstration was the discovery of a single medfly in the Port of Long Beach, some 20 miles south of Los Angeles. The big citrus producers had few, if any, commercial groves south of Ventura County, which was two mountain ranges north of Los Angeles. No matter. The medfly was capable of destroying entire groves of citrus, which would cost state growers untold millions. Governor George Deukmejian declared a state of emergency. The Air National Guard was called out to begin spraying with malathion, despite howls of protest from the 11 million inhabitants of Los Angeles. Within a matter of weeks, as the panic in Sacramento spread south along the Central Valley, scores of choppers swarmed the skies over southern California, coming out at night like bats, to saturation-spray the entire urban area. It soon became apparent that malathion could take the paint off your car, but we were assured it was harmless to humans.

At the time I was recently divorced, and my son's mother and I had separate households a few miles apart.

We would trade phone calls, warning, "They're coming!" "Move your car, they're headed your way!" And then a nervous, "They're here!" Our son was acting up by then, having increasingly bad attacks of asthma, and developing behavioral problems. By age four, to make matters worse, he'd been diagnosed with attention deficit disorder (ADD).

We decided we'd had enough. It was agreed I would pack up, take the boy, and head for Florida, where my parents lived.

AN INTRODUCTION TO ROCKET FUEL

Florida, however, proved to be anything but a safe haven. Within a few years of our move, a medfly turned up in the Port of Miami. Florida's response, while it didn't include drinking pesticide on TV, was if anything even more draconian. Florida, like California, is the turf of large agribusiness concerns, especially citrus growers. Like their West Coast counterparts before them, they flew into a panic. This, of course, led to the now no-longer-unprecedented decision to mount a preemptive strike. The state legislature quickly passed a new law, written (as since has become common practice) by the affected industry. The Florida law required that all citrus trees within 1,800 feet of an infected tree (the evident roving range of the average medfly) be destroyed. In the years since, the flies have continued to creep northward, and commercial groves and private yard trees alike have fallen before them. Lawsuits, protests, and petitions have done nothing to slow this preemptive juggernaut.

Not long after my move to Florida, my thyroid gland ceased functioning. I learned later that my ex-wife, as well as my brother Ed, both of whom had lived in California in those years, had also suffered hypothyroid disease. Doctors still don't know the cause, although—as we'd eventually learn—they have some suspects. In my case they blamed an unknown virus.

It wasn't until years later that we found that malathion wasn't the only toxic chemical to which we'd been exposed during our time in California. On December 16,

2003, *The Wall Street Journal* published an article by investigative reporter Peter Waldman on the history of California's experience with a chemical called perchlorate, a component of rocket fuel dating back to the first solid-fuel rockets of World War II. Perchlorates are actually a group of salts—ammonium perchlorate, potassium perchlorate, sodium perchlorate, cobalt perchlorate, and a score of others. They were developed mainly as oxidizer components for propellants and other explosive materials (including flares and fireworks) in the 1940s, emerging into a full-bore industry during the Cold War buildup of the 1950s. They have more recently turned up in such diverse products as automobile airbags and certain fertilizers, particularly those produced in Chile.

Despite repeated efforts by California water managers and regulators to stop the dumping of rocket fuel and related toxic chemicals into the state's groundwater and wastewater systems, defense contractors such as Lockheed Martin and Aerojet General pumped and dumped millions of gallons of these chemicals into unlined pits or holding ponds, or injected them deep into the ground. They did this with impunity, considering themselves answerable only to the U.S. Department of Defense (DOD), whose view on the subject, according to Waldman, was that "its job is national security, not environmental safety." That view seems to persist today, as reflected in the recent push by DOD to attain wide-ranging exemption from environmental regulation and restriction.

The U.S. Environmental Protection Agency (EPA) has a very different view than DOD, although no one in either organization denies that perchlorates are highly toxic. No one from DOD has volunteered to drink a glassful of the stuff, to my knowledge. Nineteen recent studies tracked by the Washington-based Environmental Working Group (EWG) between 1997 and 2002 have associated perchlorates with thyroid damage ranging from metabolic and hormone disruption to cancer in adults, and with impaired neurological and bone development in fetuses. They have linked the chemical to reduced IQs, mental retardation, loss of hearing or speech, deficits in motor skills, and (surprise, surprise) learning disorders and ADD in children. Studies in rats have found tumors developing at extremely early stages. And these studies only focus on the apparent effects of perchlorates in water. EPA scientists now believe that the levels of this chemical to which Californians and others are being exposed are far too high. And they wonder what happens when huge quantities of this water pass through the roots of irrigated vegetable crops and end up concentrating in someone's salad.

WHO WORRIES ABOUT MISSILE FUEL IN OUR FOOD, WHEN THERE'S A MISSILE CRISIS IN OUR BACK YARD?

Perchlorates were first developed by a group of aeronautics engineers at the California Institute of Technology in Pasadena, led by a Hungarian immigrant professor named Theodore von Karman. He and a group of colleagues from the university founded Aerojet, which pioneered "jet-assisted" takeoff rockets that enabled the new generation of military jets to take off from the decks of aircraft carriers. They also developed the Minuteman missile. The developers dubbed perchlorates "powdered oxygen" for their rapid and intense combustibility.

As it happens, these chemicals break down over time, and require replacement—hence the large-scale dumping over the decades. Aerojet was first warned to stop dumping as early as 1949, at its Azusa manufacturing plant east of Pasadena, by Los Angeles County engineers who even then were aware of the likely dangers to groundwater. Aerojet ignored those warnings and many others that followed, and received no sanctions. At one point, to further facilitate its perchlorates disposal, the company hooked itself up to a public sewer line.

In 1951, Aerojet moved north to Sacramento and the suburb of Rancho Cordova. According to Peter Waldman, the company's unlined holding ponds and pits leached up to 1,000 gallons of liquid waste and 300 pounds of ammonium perchlorate into the local aquifer every day. Today, a number of families there are suffering from cancer and other ailments alleged to be attributable to perchlorates. With Southern California left eating Aerojet's dust, other authorities took up the pursuit. In 1952, the California Central Valley Regional Water Pollution Control Board issued a resolution specifically intended to block further dumping of perchlorates into local groundwater or the nearby American River. Nothing changed. Aerojet's defense was that, according to guidelines issued by the DOD, its unlined holding ponds and pits were quite adequate methods of disposal.

By 1957, an underground toxic plume had spread across several square miles east of Sacramento. According to a national task force and *The Wall Street Journal*, the plume ranged in perchlorates concentration from 3.5 to 5 parts per billion (ppb). Surely, scientists began to hypothesize, this had to be bad for you. That year, a study at Harvard University found that perchlorates passed through the placenta of guinea pigs and affected the development of the thyroid and its hormones that regulate growth and development.

Still, Aerojet continued its stonewalling, even refusing to disclose exactly what chemicals it was using. In 1962, the board tried again, passing a resolution prohibiting Aerojet from disposing of anything "deleterious to human, animal, plant, or aquatic life."

At this point, national security once again assumed priority—in the form of the Cuban Missile Crisis. Nothing I have seen in the accounts of that time suggests that anyone saw any irony in the possibility that American missiles might be poisoning the American people in order to protect them from a theoretical attack by missiles from another country.

And so, again, nothing changed. Over the ensuing decades, thousands of tons of these toxic chemicals were deposited into open ditches, canals, holding ponds, and pits. Only in 1985 did perchlorates finally become a "drinking water problem," when they were detected by the EPA in wells serving 42,000 households in the vicinity of Aerojet's original plant in Azusa, back in Southern California.

In 1992, the EPA turned to the Centers for Disease Control in Atlanta for help. CDC declared that "the effects of low level perchlorate ingestion need to be described as soon as possible." So the EPA went back to the 1952 Harvard study linking the chemicals to thyroid damage, and issued its first health assessment, recommending an "initial reference dose" of no more than 4 ppb in drinking water. In response, DOD insisted that the reference dose should be 42,000 ppb. The dispute remains unresolved, although DOD has since shifted its estimate of acceptable levels sharply downward. While EPA holds firmly to its recommended level of 4 ppb, the people who brought us Agent Orange and Gulf War Syndrome now tell us that 200 ppb is safe.

THE UNLUCKY EMPLOYEES OF LUCKY FARMS

The concept of bioaccumulation has become chillingly familiar to those who have followed the stories of PCBs, mercury, and other contaminants that may be found in low levels in insects or algae, but that concentrate as they rise through the food chain. The same kind of concentration can occur as polluted ground water is consumed by irrigated plants.

The first clues that perchlorates were accumulating in vegetables emerged around October 1996, when a commercial grocers' farm, Lucky Farms of San Bernardino, California, began handing a release form to its employees—"todos los empleados"—which it required them to sign. The disclaimer stated, in English and Spanish:

> I have been informed of the dangers if I drink irrigation water from the sprinklers, valves, or faucets that are marked red. This water may cause cancer or birth defects. I know that I am only to drink water from the orange coolers. And drink only from the water specified as "good" water, which is the faucet located by the shop.

Each of the predominately Mexican farm workers was requested to sign that form, presumably to protect Lucky Farms from bad luck in the form of lost labor due to illness, or in the form of lawsuits.

Someone at the farm had been savvy enough to become suspicious of the water supply there and had had some water tested. Apparently, there was a growing awareness in the Redlands area of San Bernardino County that there was something wrong with the water. Maybe it was not mere coincidence that Lockheed Martin had operated missile testing facilities in that area for many years, using a large amount of rocket fuel. The facilities had been long since closed, but they'd left a stain of perchlorates in the ground.

On December 19, 1997, a test report was delivered to farm manager Robert Liso of Lucky Farms by Weck Laboratories of Industry, California. The preliminary report indicated that Week had tested a "vegetable" (later reported as lettuce) and found 110 micrograms (mcg) of perchlorate in the sample. While perchlorates were (and are still) an unregulated commodity, and this was not an astronomical quantity, it was well in excess of EPA's recommended level for a lettuce leaf. It was the first indication that perchlorates had migrated from the water supply to the food chain.

Then, on February 9, 1998, Weck delivered a bombshell. A second study, analyzing four lettuce samples, revealed the presence of massive amounts of perchlorates:

Leafy vegetable no. 401635: 3,260 mcg

Leafy vegetable no. 401636: 6,590 mcg

Leafy vegetable no. 401637: 6,900 mcg

Leafy vegetable no. 401638: 3,210 mcg

But none of this information reached the public. Both Weck Laboratories and Lucky Farms refused to comment on these findings when I called for clarification. Unless decimal points migrated along with the perchlorates, these were highly alarming numbers. According to Kevin Mayer of the EPA's San Francisco office, if water were the only source of perchlorate intake (which evidently it is not), the EPA's 4 ppb standard would translate to about 1 microgram per liter, or 2.1 micrograms per day for a 150-pound (70 kilogram) person drinking 2 liters of water a day. One of the Weck samples, it seemed, would deliver several thousand times that amount.

A flurry of behind-the-scenes legal activity quickly followed. By May 7, 1998, Lockheed Martin had engaged the services of the Los Angeles law firm of Gibson, Dunn & Crutcher to represent it in a secret settlement agreement with Lucky Farms. Evidently intimidated by the defense contractor's heavy guns being brought to bear, Lucky Farms backed off making any threat of a lawsuit. Lockheed Martin, in turn, generously offered to pay for the cost of Weck Laboratories' testing—an offer that was apparently accepted, because no further action was taken. Instead, Lucky Farms continued to require signed releases from its employees and continued with business as usual, at least until 1999. Lettuce continued to be shipped to markets throughout the country.

In 1999, the EPA swung ponderously into action, ordering a study by its National Environmental Research Laboratory in Athens, Georgia. The new study found that perchlorate accumulated in leaves by factors of 100 or more times the agency's recommended levels. The researchers also found that lettuce leaves were capable of absorbing and storing up to 95 percent of the perchlorates in the water supply. This meant that even small levels of

HAVE YOU HEARD THIS ONE BEFORE?

The perchlorate story brings together three all-too-familiar themes of the modern military-industrial saga:

First, there's the technological hubris that so often surrounds a major new invention. Immediate beneficiaries become so focused on the invention's primary intended use that they overlook the secondary, often slower-acting and less visible, effects on health or the environment. The dramatic Cold War takeoff of the rocket industry, with little thought to its eventual, insidious effects on the health of millions of people, parallels the histories of many other lucrative technologies—from thalidomide to DDT to breast implants—that have been rushed into use. Not coincidentally, the fiery power of rockets, and the apparent lack of thought about what longer-term effects might follow, was replicated on a grand scale in the "shock and awe" with which the U.S. launched its rocket-raining war on Iraq.

Second, there's the secrecy and denial—and attempted coverup—that repeatedly accompanies the rise or perpetuation of highly risky but lucrative industries. Think of the decades-long coverup of the health effects of tobacco smoke, or the chemical industry effort to destroy the credibility of Rachel Carson after the publication of her book Silent Spring, or the Cheney-Bush administration's collusion with fossil-fuel industries to distract public attention from the dangers of climate change, which far exceed those of terrorism but pose an unwanted challenge to those industries. In the case of perchlorates, it's likely that a large part of the coverup has yet to be exposed. But already, there's a telling pattern—in the Air Force's dog-ate-my-homework story that someone stole its study results; in the refusals of Lucky Farms or Weck Labs to speak; in the Department of Agriculture's unexplained cancellation of a perchlorate study for other irrigated crops (what about our tomatoes, chard, and grapes?); and most of all, in the White House gag order that has made both the EPA and the Pentagon clam up. We talked with a top official of EPA's San Francisco office, who said his agency had worked long and hard to determine its recommended level of 4 ppb, and was prepared to make those recommendations final, when "someone" stepped in and requested that the data be turned over to the National Academy of Science for review. And there, so far, it has remained.

Third, and perhaps most dangerous of all, there is the familiar presumption—once unthinkable but now pervasive—that scientific findings are not matters of public knowledge but proprietary industrial or government information that the holders need disclose only if it is to their financial or political advantage to do so. While the rocket-fuel contamination is mainly an American problem (it's mainly Americans who make and shoot off rockets), the water that flows through our lettuce—and through the Earth's hydrological cycle—is a global commons. The knowledge of what happens to it is an essential part of the global public domain. So, why are all the unanswered questions about perchlorates being treated like a national security secret? The need to keep our water clean is common sense, not rocket science.

perchlorate in the water could concentrate into extremely high levels in the lettuce leaves.

Curiously, the EPA discounted this study due to the fact that the water used in the study had excessive amounts of perchlorates in it and was presumably not representative of typical California irrigation water, though these doses weren't anywhere near the concentrations of those found in the Rancho Cordova and Azusa plumes. Even more curiously, EPA also concluded that "foods do not contribute to" perchlorate accumulation in the human body—an assertion that appears to be directly contradicted by the 1952 Harvard study and the 19 studies that followed.

SHADES OF SILKWOOD

By 2002, sources of perchlorates in irrigation water were being traced to the Colorado River, which was found to be seriously contaminated from Las Vegas to the Mexican border. This was especially troublesome news, since Colorado River water was the sole source of irrigation water for the entire Coachella and Imperial Valley regions, which produce 90 percent of America's lettuce. On July 14, 2001, The Sacramento Bee reported that 20 million people in California, Arizona, and Nevada had some level of perchlorates in their drinking water, averaging between 5 and 10 ppb. According to The Wall Street Journal, scientists have traced perchlorates found in the Los Angeles water supply 400 miles up the Colorado River to Lake Mead, above Hoover Dam. From there, according to the Journal's Waldman, they tracked the plume 10 miles west up a desert riverbed called the Las Vegas Wash, to a giant ammonium perchlorate plant in Henderson, Nevada, owned and operated by Oklahoma City-based Kerr-McGee Corporation. The Las Vegas Wash is the main drain leading into Lake Mead, the primary source of water for Las Vegas and the lower Colorado River.

Kerr-McGee is the company that was featured in the movie *Silkwood*, based on the story of Karen Silkwood, a chemical technician at the company who claimed that her employer was exposing its unwitting employees to plutonium radiation. Silkwood was killed in an unwitnessed one-car crash while gathering evidence implicating the company. This time, rather than trying to hide its involvement and providing the plot for another movie, Kerr-McGee has sued the Pentagon for reimbursement of cleanup costs. Its plant is now closed but continues to leak 900 pounds of perchlorate a day into the Las Vegas Wash.

Other lawsuits have been filed as well, including a class action suit by the residents of Rancho Cordova, where Aerojet operated with impunity all those years. Many residents there have developed thyroid and other cancers. These suits, against Lockheed Martin and others, have gone nowhere, except to the extent that the State of California agreed to pay the cost of a suit by a Rancho Cordova water company that, according to *The Sacramento Bee*, "accuses pollution enforcers of having willfully allowed Aerojet Corp. to contaminate the ground water with rocket fuel." So, Aerojet continues to escape unscathed, while the people who tried for decades to stop it from perchlorate dumping end up taking the hit.

In April 1999 the EPA convened an "eco-summit" of representatives from the Air Force (the prime perchlorates consumer, other than NASA), a coalition of perchlorate manufacturers and users called the Perchlorate Study Group, and members of five Indian tribes whose livelihoods are based on produce farming along the lower Colorado River. The DOD promised a grant of $650,000, or less than one-fourth the cost of a single cruise missile, according to *The Los Angeles Times*, for a so-called "real world" study that would test a variety of crops through the auspices of the U.S. Department of Agriculture. Then, after consultations with the Food and Drug Administration, the project was indefinitely postponed.

Instead, the Air Force, according to records obtained through the Freedom of Information Act by *The Riverside (CA) Press-Enterprise*, obtained a $500,000 grant from DOD earmarked for two studies: perchlorates in crops, and perchlorates in wild plants and animals. Yet the first study, it seems, was never done—or if it was, the findings were never disclosed. Other documents, obtained by the Environmental Working Group, indicate that the Air Force did in fact conduct a second study of greenhouse-grown lettuce. In October 2002, at an industry-sponsored perchlorate conference in Ontario, California (not far from Aerojet's original plant in Azusa), EWG questioned Air Force spokesman David Mattie about the second lettuce study and was told that the study had in fact been completed, but "someone walked away with the data."

When prodded for the results under the Freedom of Information Act, the Air Force then claimed that any findings are "fully exempt from disclosure until the formally sponsored EPA peer review is complete."

In the spring of 2003, the lettuce finally hit the fan. On April 2, California's U.S. Senator Diane Feinstein accused the Defense Department of "dragging its feet in the cleanup of rocket fuel from old military facilities, which state officials say has contaminated hundreds of wells in California." On April 27, reporters David Danelski and Douglas Beeman of *The Riverside Press-Enterprise* released the results of a study commissioned by the newspaper of 18 winter lettuce samples and one mustard greens sample harvested in the Imperial and Coachella Valleys, both irrigated by the Colorado River. All 19 samples tested were found to be contaminated with perchlorates. A spokesman for the U.S. Department of Agriculture insisted that the levels found by *The Press-Enterprise* were too low to pose a health risk, but nevertheless expressed concern that other crops might also be contaminated.

The day after the *Press-Enterprise* story, the EWG released its own study of 22 samples of lettuce purchased at Northern California supermarkets. EWG had found four to be contaminated with perchlorates. This study received more national attention than the *Press-Enterprise* one, possibly because some of the EWG findings, based on tests conducted at Texas Tech University, showed levels of contamination "as much as 20 times as high as the amount California considers safe for drinking water."

In the meantime, renewed attention was focused on the EPA, which while talking tough about recommended levels, has yet to make any firm decision pending a peer review by the National Academy of Sciences. EPA said it needed to see the review before it could issue any regulations, reports, or guidelines on the subject of perchlorates.

So what has become of that review? EPA has been less than forthcoming on this subject. Sometime in the weeks since the most recent revelations of the presence of perchlorates in the U.S. food supply, the White House issued a gag order to the EPA prohibiting its researchers or scientists from discussing perchlorates with the press. Interestingly, during the unfolding of these events, EPA chief Christine Todd Whitman resigned. Calls requesting information on this matter have gone unanswered, but Whitman's home state of New Jersey is one of many states threatened by the discovery of perchlorates in its water. Furthermore, New Jersey—"The Garden State"—is one of the country's biggest sources both of vegetable produce and of perchlorate manufacturing.

California Congresswoman Lois Capps, in quick response to the two lettuce studies, wrote to the White House—together with 57 other members of the House of Representatives—demanding explanation and rescinding of the reported EPA gag order, noting that "perchlorate is known or suspected to be a contaminant in hundreds of locations in 43 states. It has been confirmed in more than 100 drinking water sources in 19 states including Texas, California, Arizona, Nebraska, Iowa, New York, Maryland, and Massachusetts.... It is highly disturbing to think that the agency charged with protecting our environmental health and safety could be barred from discussing an increasingly prevalent and potentially dangerous chemical contaminant like perchlorate. Americans deserve to get information from the scientists who work on their behalf. President Bush should see to it that this 'gag order' is lifted immediately." The White House did not respond.

Meanwhile, while the EPA has since shown a willingness to talk with *World Watch* about the issue and openly acknowledges the gravity of the situation, it has still not established any firm safety standards for perchlorates, which remain completely unregulated at both the state and Federal levels. At a time when the secretary of defense often seems to hold as much sway as the Congress, which makes the laws, the Pentagon doesn't want standards. So far, Kerr-McGee is the only defense contractor to have voluntarily embarked on cleaning up its mess, which in its case was inherited from an old Navy lab at the same site.

The DOD's stonewalling harks back half a century to its 1950s positions about matters of munitions and pollution: the Army, Navy, and Air Force are in the business of national security, not environmental protection. The military hasn't yet come around to the radical idea that when it comes to national security—or even global security, for that matter—protecting our food and water may be the first line of defense.

Gene Ayres is a writer based in St. Petersburg, Florida.

Global Warming as a Weapon of Mass Destruction

We are carbonizing the oceans with dire consequences

By Bruce E. Johansen

Lord Peter Levene, board chair of Lloyd's of London, says that terrorism is not the insurance industry's biggest worry, despite the fact that his company was the largest single insurer of the World Trade Center. Levene says that Lloyd's, like other large international insurance companies, is bracing for an increase in weather disasters related to global warming. Likewise, following his assignment as chief weapons inspector in Iraq, Hans Blix said: "To me the question of the environment is more ominous than that of peace and war. We will have regional conflicts and use of force, but world conflicts I do not believe will happen any longer. But the environment, that is a creeping danger. I'm more worried about global warming than I am of any major military conflict." Sir John Houghton, co-chair of the Intergovernmental Panel on Climate Change, agrees. "Global warming is already upon us," he said. "The impacts of global warming are such that I have no hesitation in describing it as a weapon of mass destruction." So what do they know that George W. Bush doesn't?

Weather is the story; climate is the plot. We are carbonizing the oceans, with dire implications for life in them. As the 21st century dawned, carbon-dioxide levels were rising in the oceans more rapidly than any time since the age of dinosaurs. In a report published September 25, 2003 in *Nature*, oceanographers Ken Caldeira and Michael E. Wickett wrote: "We find that oceanic absorption of CO^2 from fossil fuels may result in larger pH changes over the next several centuries than any inferred in the geological record of the past 300 million years, with the possible exception of those resulting from rare, extreme events such as bolide impacts or catastrophic methane hydrate degassing." (A "bolide" is a large extraterrestrial body, usually at least a half mile in diameter, perhaps much larger, that impacts the earth at a speed roughly equal to that of a bullet in flight.)

Rising carbon dioxide levels in the oceans could threaten the health of many marine organisms, beginning with the plankton at the base of the food chain. "If we continue down the path we are going, we will produce changes greater than any experienced in the past 300 million years—with the possible exception of rare, extreme events such as comet impacts," Caldeira, of the Lawrence Livermore National Laboratory, warned. Since carbon dioxide levels began to be measured on a systemic basis worldwide in 1958, its concentration in the atmosphere has risen 17 percent.

Until now, some climate experts have asserted that the oceans would help to control the rise in carbon dioxide by acting as a filter. However, Caldeira and Michael Wickett said that carbon dioxide that is removed from the atmosphere enters the oceans as carbonic acid, gradually altering the acidity of ocean water. According to their studies, the change over the last century already matches the magnitude of the change that occurred in the entire 10,000 years preceding the industrial age. Caldeira pointed to acid rain from industrial emissions as a possible precursor of changes in the oceans. "Most ocean life resides near the surface, where the greatest change would be expected to come, but deep ocean life may prove to be even more sensitive to changes," Caldeira said.

Marine plankton and other organisms whose skeletons or shells contain calcium carbonate, which is dissolved by acid solutions, may be particularly vulnerable. Coral reefs—already suffering from pollution, rising ocean temperatures, and other stresses—are comprised almost entirely of calcium carbonate. "It's difficult to predict what will happen because we haven't really studied the range of impacts," Caldeira said. "But we can say that if we continue business as usual, we are going to see some significant changes in the acidity of the world's oceans."

Along the same line, warming seas also are devastating plankton, eroding the ocean's food chain. Global warming is contributing to an "ecological meltdown," with devastating implications for fisheries and wildlife. The "meltdown" begins at the base of the food chain, as increasing sea temperatures kill plankton. Fish stocks and sea-bird populations are declining as well.

Scientists at the Sir Alistair Hardy Foundation for Ocean Science in Plymouth, England, which has been monitoring plankton growth in the North Sea for more than 70 years, have said that an unprecedented warming of the North Sea has driven plankton hundreds of miles to the north. They have been replaced by smaller, warm-water species that are less nutritious. Over-fishing of cod and other species has played a role, but fish stocks have not recovered after cuts in fishing quotas.

The number of salmon returning to British waters are now half of what they were 20 years ago, and a decline in plankton populations is a major factor. "A regime shift has taken place and the whole ecology of the North Sea has changed quite dramatically," said Dr. Chris Reid, the foundation's director. "We are seeing a collapse in the system as we knew it. Catches of salmon and cod are already down and we are getting smaller fish. We are seeing visual evidence of climate change on a large-scale ecosystem. We are likely to see even greater warming, with temperatures becoming more like those off the Atlantic coast of Spain or further south, bringing a complete change of ecology."

Research by the British Royal Society for the Protection of Birds has established that seabird colonies off the Yorkshire coast and the Shetlands this year suffered their worst breeding season since records began, with many abandoning nesting sites. Sea-bird populations are falling in large part because sand eels are declining. The sand eels feed on plankton. This survey concentrated on kittiwakes, one breed of sea birds, but other species that feed on the eels, including puffins and razorbills, also have been seriously affected.

Sand eels also comprise a third to half of the North Sea catch, by weight. They have heretofore been caught in huge quantities by Danish factory ships, which turn them into food pellets for pigs and fish. During the summer of 2003, the Danish fleet caught only 300,000 English tons of its 950,000-ton quota, a record low.

Beware the Methane Burp

Yesterday's SUV exhaust does not become today's rising temperature, not immediately. Through an intricate feedback loop, fossil fuel burned today is expressed in warming 30 to 50 years later. Today we are seeing temperatures related to fossil-fuel emissions from roughly 1960, when fossil fuel consumption was much lower. Today's fossil-fuel emissions will be expressed in the atmosphere about 2040.

Increasing levels of greenhouse gases near the surface hold heat there, impeding radiation into the upper layers of the atmosphere. As the surface warms, the stratosphere cools. The chemical reactions that consume the ozone that protects us from ultraviolet radiation accelerate as the air chills. Thus, the area of depleted ozone over Antarctica remains at near-record size despite the fact that chlorofluorocarbons (CFCs), the culprits on ozone depletion, have now been banned for more than 15 years.

In his book, *When Life Nearly Died: The Greatest Mass Extinction of All Time* (London: Thames and Hudson, 2003), Michael J. Benton describes a mass extinction at the end of the Permian period, about 250 million years ago, when at least 90 percent of life on Earth died. The extinction probably was initiated by massive volcanic eruptions in Siberia. According to present theories, the eruptions injected massive amounts of carbon dioxide into the atmosphere, causing a number of biotic feedbacks that accelerated global warming of about 6 degrees Celsius. In a chapter titled "What Caused the Biggest Catastrophe of all Time?" Benton sketches how the warming (which was accompanied by anoxia) may have fed upon itself: "The end-Permian runaway greenhouse may have been simple. Release of carbon dioxide from the eruption of the Siberian Traps [volcanoes] led to a rise in global temperatures of 6 degrees Celsius or so. Cool polar regions became warm and frozen tundra became unfrozen. The melting might have penetrated to the frozen gas hydrate reservoirs located around the polar oceans, and massive volumes of methane may have burst to the surface of the oceans in huge bubbles.

This further input of carbon into the atmosphere caused more warming, which could have melted further gas hydrate reservoirs. So the process went on, running faster and faster. The natural systems that normally reduce carbon dioxide levels could not operate, and eventually the system spiraled out of control, with the biggest crash in the history of life."

The oxygen-starved aftermath of this immense global belch of methane left land animals gasping for breath and caused the Earth's largest mass extinction, suggests new research. Greg Retallack, an expert in ancient soils at the University of Oregon, has speculated that the same methane "belch" was of such a magnitude that it caused mass extinction via oxygen starvation of land animals. Bob Berner of Yale University has calculated that a cascade of effects on wetlands and coral reefs may have reduced oxygen levels in the atmosphere from 35 percent to just 12 percent over 20,000 years. Marine life also may have suffocated in the oxygen-poor water.

Events 250 million years ago are of more than academic interest today because the 6 degrees Celsius that Benton estimates triggered these events is roughly the same temperature rise forecast for the Earth by the IPCC by the end of this century.

In *Abrupt Climate Change* (2002), Richard B. Alley wrote that climate may change rapidly (as much as 16 degrees Celsius within a decade or two) "when gradual causes push the Earth system across a threshold. Just as the slowly increasing pressure of a finger eventually flips a switch and turns on a light...." Half the North Atlantic warming since the last ice age was achieved, writes Alley, within one decade. The temperature record for Greenland, according to Alley's research, more resembles a jagged row of very sharp teeth than a gradual passage from one epoch to another. According to Alley: "Model projections of global warming find increased global precipitation, increased variability in precipitation, and summertime drying in many continental interiors, including "grain belt" regions. Such changes might produce more floods and more droughts." Human emissions of greenhouse gases may provide enough of a change to trigger such a rapid change.

By 2000, the hydrological cycle seemed to be changing more quickly than temperatures. Warmer air holds more moisture, making rain (and sometimes snow) more intense. Warmer air also increases evaporation, paradoxically intensifying drought at the same time. With sustained warming, usually wet places generally seem to be receiving more rain than before; dry places often receive less rain and become subject to more persistent drought. In many places, drought or deluge is becoming the weather regime du jour. Atmospheric moisture increases more rapidly than temperature; over the United States and Europe, atmospheric moisture increased 10 to 20 percent from 1980 to 2000. "That's why you see the impact of global warming mostly in intense storms and flooding like we have seen in Europe," Kevin Trenberth, a scientist with National Center for Atmospheric Research (NCAR) told London's *Financial Times*.

As if on cue to support climate models, the summer of 2002 featured a number of climatic extremes, especially regarding precipitation. Excessive rain deluged Europe and Asia, swamping cities and villages and killing at least 2,000 people, while drought and heat scorched the United States' west and eastern cities. Climate skeptics argued that weather is always variable, but other observers noted that extremes seemed to be more frequent than before. A year later, following episodic floods during the summer of 2002, Europe experienced some of it highest (and longest-sustained) temperatures in recorded history, causing (by various estimates) between 19,000 and 35,000 excess deaths. As much as 80 percent of the grain crop died in eastern Germany, site of some of 2002's worst floods.

"In a hotter climate, your chances of being caught with either too much or too little are higher," said Dr. John M. Wallace, a professor of atmospheric sciences at the University of Washington. Government scientists have measured a rise in downpour-style storms in the United States during the last century. "Over the past 50 years, said Wallace, winter precipitation in the Sierra Nevada has been falling more and more in the form of rain, increasing flood risks, instead of as snow, which supplies farmers and taps alike as it melts in the spring."

The World Water Council report compiled statistics indicating that between 1971 and 1995 floods affected more than 1.5 billion people worldwide, or 100 million people a year. An estimated 318,000 were killed and more than 18 million left homeless. The economic costs of these disasters rose to an estimated $300 billion in the 1990s from about $35 billion in the 1960s. Global warming is causing changes in weather patterns as growing populations migrate to vulnerable areas, increasing costs of individual weather events, said William Cosgrove, vice president of the World Water Council. Scientists cited by the World Water Council expect that climate changes during the 21st century will lead to shorter and more intense rainy seasons in some areas, as well as longer, more intense droughts in others, endangering some crops and species and causing a drop in global food production.

Examples abound of increasing extremes in precipitation. November 2002, December 2002, and January 2003 were Minneapolis-St. Paul's driest in recorded history. These followed the wettest June through October there in more than 100 years. In December 2002, Omaha recorded its first month with no measurable precipitation. In March 2003, having endured its driest year in recorded history during 2002, Denver, Colorado recorded 30 inches of snow in one storm. Some areas of the drought-parched Front Range received as much as ten feet of snow in the same storm. After that one storm, drought conditions returned.

Roughly half the United States was under serious drought conditions during the summer of 2002. The drought was occasionally punctuated by torrential rains. On September 13, 2002, for example, drought-stricken Denver was inundated by floods from a fast-moving thunderstorm that caused widespread flooding. Similar events took place south of Salt Lake City. Ten days later, a flooding cloudburst inundated similarly drought-stricken Atlanta. On September 10, 2002, six months' worth of rain fell in a few hours in the Gard, Herault, and Vaucluse departments in the south of France, drowning at least 20 people. In the village of Sommieres, near Nimes, a usually-tiny stream exploded to a width of 300 meters, cutting off road traffic.

The suburbs of Chicago received 8 to 13 inches of rain the night of August 12, 2002, in a summer that included devastating floods in Prague and Dresden, as well as parts of southern China. India had a variable monsoon—some areas flooded, while others went dry. Severe summer floods in Europe during 2002 may be an indicator of an emerging pattern, according to Jens H. and Ole B. Christensen, who modeled precipitation patterns in Europe under warming conditions of a type that may be prominent in the area by 2070 to 2100. "Our results," they wrote in *Nature*, "indicate that episodes of severe flooding may become more frequent, despite a general trend toward drier summer conditions." The trend toward drought or deluge will intensify as warming distorts the hydrological cycle. A warming atmosphere will contain more water vapor, which will provide "further potential for latent-heat release during the buildup of low-pressure systems, thereby possibly both intensifying the systems and making more water available for precipitation," Christensen and Christensen wrote.

Annual mean precipitation amounts over the United States have been increasing at two to five percent per decade, according to atmospheric scientists Ken Trenberth and colleagues (writing in the *Bulletin of the American*

Meteorological Society), with "most of the increase related to temperature and hence in atmospheric water-holding capacity.... There is clear evidence that rainfall rates have changed in the United States.... The prospect may be for fewer but more intense rainfall—or snowfall—events." Individual storms may be further enhanced by latent heat release, which supplies even more moisture during individual storms.

Generally, higher temperatures enhance evaporation, with some compensatory cooling when water is available. Increased evaporation also intensifies drought, which, to some degree, compounds itself as moisture is depleted, leading "to increased risk of heat waves and wildfires in association with such droughts; because once the soil moisture is depleted then all the heating goes into raising temperatures and wilting plants."

In mountain areas, wrote Trenberth, "The winter snowpack forms a vital resource, not only for skiers, but also as a freshwater resource in the spring and summer as the snow melts. Yet warming makes for a shorter snow season with more precipitation falling as rain rather than snow, earlier snowmelt of the snow that does exist, and greater evaporation and ablation. These factors all contribute to diminished snowpack. In the summer of 2002, in the western parts of the United States, exceptionally low snowpack and subsequent low soil moisture likely contributed substantially to the widespread intense drought because of the importance of recycling [in the hydrological cycle]. Could this be a sign of the future?"

The insurance companies, whose business is making book on the future, are watching the weather—and they are worried.

Bruce E. Johansen, Frederick W. Kayser professor of Journalism at the University of Nebraska at Omaha, is author of the Global Warming Desk Reference (Greenwood Press, 2002).

Glossary

This glossary of environmental terms is included to provide you with a convenient and ready reference as you encounter general terms in your study of environment that are unfamiliar or require a review. It is not intended to be comprehensive, but taken together with the many definitions included in the articles themselves, it should prove to be quite useful.

A

Abiotic   Without life; any system characterized by a lack of living organisms.

Absorption   Incorporation of a substance into a solid or liquid body.

Acid   Any compound capable of reacting with a base to form a salt; a substance containing a high hydrogen ion concentration (low pH).

Acid Rain   Precipitation containing a high concentration of acid.

Adaptation   Adjustment of an organism to the conditions of its environment, enabling reproduction and survival.

Additive   A substance added to another in order to impart or improve desirable properties or suppress undesirable ones.

Adsorption   Surface retention of solid, liquid, or gas molecules, atoms, or ions by a solid or liquid.

Aerobic   Environmental conditions where oxygen is present; aerobic organisms require oxygen in order to survive.

Aerosols   Tiny mineral particles in the atmosphere onto which water droplets, crystals, and other chemical compounds may adhere.

Air Quality Standard   A prescribed level of a pollutant in the air that should not be exceeded.

Alcohol Fuels   The processing of sugary or starchy products (such as sugar cane, corn, or potatoes) into fuel.

Allergens   Substances that activate the immune system and cause an allergic response.

Alpha Particle   A positively charged particle given off from the nucleus of some radioactive substances; it is identical to a helium atom that has lost its electrons.

Ammonia   A colorless gas comprised of one atom of nitrogen and three atoms of hydrogen; liquefied ammonia is used as a fertilizer.

Anthropocentric   Considering humans to be the central or most important part of the universe.

Aquaculture   Propagation and/or rearing of any aquatic organism in artificial "wetlands" and/or ponds.

Aquifers   Porous, water-saturated layers of sand, gravel, or bedrock that can yield significant amounts of water economically.

Atom   The smallest particle of an element, composed of electrons moving around an inner core (nucleus) of protons and neutrons. Atoms of elements combine to form molecules and chemical compounds.

Atomic Reactor   A structure fueled by radioactive materials that generates energy usually in the form of electricity; reactors are also utilized for medical and biological research.

Autotrophs   Organisms capable of using chemical elements in the synthesis of larger compounds; green plants are autotrophs.

B

Background Radiation   The normal radioactivity present; coming principally from outer space and naturally occurring radioactive substances on Earth.

Bacteria   One-celled microscopic organisms found in the air, water, and soil. Bacteria cause many diseases of plants and animals; they also are beneficial in agriculture, decay of dead matter, and food and chemical industries.

Benthos   Organisms living on the bottom of bodies of water.

Biocentrism   Belief that all creatures have rights and values and that humans are not superior to other species.

Biochemical Oxygen Demand (BOD)   The oxygen utilized in meeting the metabolic needs of aquatic organisms.

Biodegradable   Capable of being reduced to simple compounds through the action of biological processes.

Biodiversity   Biological diversity in an environment as indicated by numbers of different species of plants and animals.

Biogeochemical Cycles   The cyclical series of transformations of an element through the organisms in a community and their physical environment.

Biological Control   The suppression of reproduction of a pest organism utilizing other organisms rather than chemical means.

Biomass   The weight of all living tissue in a sample.

Biome   A major climax community type covering a specific area on Earth.

Biosphere   The overall ecosystem of Earth. It consists of parts of the atmosphere (troposphere), hydrosphere (surface and ground water), and lithosphere (soil, surface rocks, ocean sediments, and other bodies of water).

Biota   The flora and fauna in a given region.

Biotic   Biological; relating to living elements of an ecosystem.

Biotic Potential   Maximum possible growth rate of living systems under ideal conditions.

Birthrate   Number of live births in one year per 1,000 midyear population.

Breeder Reactor   A nuclear reactor in which the production of fissionable material occurs.

C

Cancer   Invasive, out-of-control cell growth that results in malignant tumors.

Carbon Cycle   Process by which carbon is incorporated into living systems, released to the atmosphere, and returned to living organisms.

Carbon Monoxide (CO)   A gas, poisonous to most living systems, formed when incomplete combustion of fuel occurs.

Carcinogens   Substances capable of producing cancer.

Carrying Capacity   The population that an area will support without deteriorating.

Chlorinated Hydrocarbon Insecticide   Synthetic organic poisons containing hydrogen, carbon, and chlorine. Because they are fat-soluble, they tend to be recycled through food chains, eventually affecting nontarget systems. Damage is normally done to the organism's nervous system. Examples include DDT, Aldrin, Deildrin, and Chlordane.

Chlorofluorocarbons (CFCs)   Any of several simple gaseous compounds that contain carbon, chlorine, fluorine, and sometimes hydrogen; they are suspected of being a major cause of stratospheric ozone depletion.

Circle of Poisons   Importation of food contaminated with pesticides banned for use in this country but made here and sold abroad.

Clear-Cutting   The practice of removing all trees in a specific area.

Climate   Description of the long-term pattern of weather in any particular area.

Climax Community   Terminal state of ecological succession in an area; the redwoods are a climax community.

Coal Gasification   Process of converting coal to gas; the resultant gas, if used for fuel, sharply reduces sulfur oxide emissions and particulates that result from coal burning.

Commensalism   Symbiotic relationship between two different species in which one benefits while the other is neither harmed nor benefited.

Community Ecology   Study of interactions of all organisms existing in a specific region.

Competitive Exclusion   Resulting from competition; one species forced out of part of an available habitat by a more efficient species.

Conservation   The planned management of a natural resource to prevent overexploitation, destruction, or neglect.

Conventional Pollutants   Seven substances (sulfur dioxide, carbon monoxide, particulates, hydrocarbons, nitrogen oxides, photochemical oxidants, and lead) that make up the largest volume of air quality degradation, as identified by the Clean Air Act.

Core   Dense, intensely hot molten metal mass, thousands of kilometers in diameter, at Earth's center.

Cornucopian Theory   The belief that nature is limitless in its abundance and that perpetual growth is both possible and essential.

Corridor   Connecting strip of natural habitat that allows migration of organisms from one place to another.

Crankcase Smog Devices (PCV System)   A system, used principally in automobiles, designed to prevent discharge of combustion emissions into the external environment.

Critical Factor   The environmental factor closest to a tolerance limit for a species at a specific time.

Cultural Eutrophication   Increase in biological productivity and ecosystem succession resulting from human activities.

D

Death Rate   Number of deaths in one year per 1,000 midyear population.

Decarbonization   To remove carbon dioxide or carbonic acid from a substance.

Decomposer   Any organism that causes the decay of organic matter; bacteria and fungi are two examples.

Deforestation   The action or process of clearing forests without adequate replanting.

Degradation (of water resource)   Deterioration in water quality caused by contamination or pollution that makes water unsuitable for many purposes.

Demography   The statistical study of principally human populations.

Desert   An arid biome characterized by little rainfall, high daily temperatures, and low diversity of animal and plant life.

Desertification   Converting arid or semiarid lands into deserts by inappropriate farming practices or overgrazing.

Detergent   A synthetic soap-like material that emulsifies fats and oils and holds dirt in suspension; some detergents have caused pollution problems because of certain chemicals used in their formulation.

Detrivores   Organisms that consume organic litter, debris, and dung.

Dioxin   Any of a family of compounds known chemically as dibenzo-p-dioxins. Concern about them arises from their potential toxicity as contaminants in commercial products. Tests on laboratory animals indicate that it is one of the more toxic anthropogenic (man-made) compounds.

Diversity   Number of species present in a community (species richness), as well as the relative abundance of each species.

DNA (Deoxyribonucleic Acid)   One of two principal nucleic acids, the other being RNA (Ribonucleic Acid). DNA contains information used for the control of a living cell. Specific segments of DNA are now recognized as genes, those agents controlling evolutionary and hereditary processes.

Dominant Species   Any species of plant or animal that is particularly abundant or controls a major portion of the energy flow in a community.

Drip Irrigation   Pipe or perforated tubing used to deliver water a drop at a time directly to soil around each plant. Conserves water and reduces soil waterlogging and salinization.

E

Ecological Density   The number of a singular species in a geographical area, including the highest concentration points within the defined boundaries.

Ecological Succession   Process in which organisms occupy a site and gradually change environmental conditions so that other species can replace the original inhabitants.

Ecology   Study of the interrelationships between organisms and their environments.

Ecosystem   The organisms of a specific area, together with their functionally related environments; considered as a definitive unit.

Ecotourism   Wildlife tourism that could damage ecosystems and disrupt species if strict guidelines governing tours to sensitive areas are not enforced.

Edge Effects   Change in ecological factors at the boundary between two ecosystems. Some organisms flourish here; others are harmed.

Effluent   A liquid discharged as waste.

El Niño   Climatic change marked by shifting of a large warm water pool from the western Pacific Ocean toward the East.

Electron   Small, negatively charged particle; normally found in orbit around the nucleus of an atom.

Eminent Domain   Superior dominion exerted by a governmental state over all property within its boundaries that authorizes it to appropriate all or any part thereof to a necessary public use, with reasonable compensation being made.

Glossary

Endangered Species   Species considered to be in imminent danger of extinction.

Endemic Species   Plants or animals that belong or are native to a particular ecosystem.

Environment   Physical and biological aspects of a specific area.

Environmental Impact Statement (EIS)   A study of the probable environmental impact of a development project before federal funding is provided (required by the National Environmental Policy Act of 1968).

Environmental Protection Agency (EPA)   Federal agency responsible for control of air and water pollution, radiation and pesticide problems, ecological research, and solid waste disposal.

Erosion   Progressive destruction or impairment of a geographical area; wind and water are the principal agents involved.

Estuary   Water passage where an ocean tide meets a river current.

Eutrophic   Well nourished; refers to aquatic areas rich in dissolved nutrients.

Evolution   A change in the gene frequency within a population, sometimes involving a visible change in the population's characteristics.

Exhaustible Resources   Earth's geologic endowment of minerals, nonmineral resources, fossil fuels, and other materials present in fixed amounts.

Extinction   Irrevocable elimination of species due to either normal processes of the natural world or through changing environmental conditions.

F

Fallow   Cropland that is plowed but not replanted and is left idle in order to restore productivity mainly through water accumulation, weed control, and buildup of soil nutrients.

Fauna   The animal life of a specified area.

Feral   Refers to animals or plants that have reverted to a non-cultivated or wild state.

Fission   The splitting of an atom into smaller parts.

Floodplain   Level land that may be submerged by floodwaters; a plain built up by stream deposition.

Flora   The plant life of an area.

Flyway   Geographic migration route for birds that includes the breeding and wintering areas that it connects.

Food Additive   Substance added to food usually to improve color, flavor, or shelf life.

Food Chain   The sequence of organisms in a community, each of which uses the lower source as its energy supply. Green plants are the ultimate basis for the entire sequence.

Fossil Fuels   Coal, oil, natural gas, and/or lignite; those fuels derived from former living systems; usually called nonrenewable fuels.

Fuel Cell   Manufactured chemical systems capable of producing electrical energy; they usually derive their capabilities via complex reactions involving the sun as the driving energy source.

Fusion   The formation of a heavier atomic complex brought about by the addition of atomic nuclei; during the process there is an attendant release of energy.

G

Gaia Hypothesis   Theory that Earth's biosphere is a living system whose complex interactions between its living organisms and nonliving processes regulate environmental conditions over millions of years so that life continues.

Gamma Ray   A ray given off by the nucleus of some radioactive elements. A form of energy similar to X rays.

Gene   Unit of heredity; segment of DNA nucleus of the cell containing information for the synthesis of a specific protein.

Gene Banks   Storage of seed varieties for future breeding experiments.

Genetic Diversity   Infinite variation of possible genetic combinations among individuals; what enables a species to adapt to ecological change.

Geothermal Energy   Heat derived from the Earth's interior. It is the thermal energy contained in the rock and fluid (that fills the fractures and pores within the rock) in the Earth's crust.

Germ Plasm   Genetic material that may be preserved for future use (plant seeds, animal eggs, sperm, and embryos).

Global Warming   An increase in the near surface temperature of the Earth. Global warming has occurred in the distant past as the result of natural influences, but the term is most often used to refer to the warming predicted to occur as a result of increased emissions of greenhouse gases. Scientists generally agree that the Earth's surface has warmed by about 1 degree Fahrenheit in the past 140 years.

Green Revolution   The great increase in production of food grains (as in rice and wheat) due to the introduction of high-yielding varieties, to the use of pesticides, and to better management techniques.

Greenhouse Effect   The effect noticed in greenhouses when shortwave solar radiation penetrates glass, is converted to longer wavelengths, and is blocked from escaping by the windows. It results in a temperature increase. Earth's atmosphere acts in a similar manner.

Gross National Product (GNP)   The total value of the goods and services produced by the residents of a nation during a specified period (such as a year).

Groundwater   Water found in porous rock and soil below the soil moisture zone and, generally, below the root zone of plants. Groundwater that saturates rock is separated from an unsaturated zone by the water table.

H

Habitat   The natural environment of a plant or animal.

Habitat Fragmentation   Process by which a natural habitat/landscape is broken up into small sections of natural ecosystems, isolated from each other by sections of land dominated by human activities.

Hazardous Waste   Waste that poses a risk to human or ecological health and thus requires special disposal techniques.

Herbicide   Any substance used to kill plants.

Heterotroph   Organism that cannot synthesize its own food and must feed on organic compounds produced by other organisms.

Hydrocarbons   Organic compounds containing hydrogen, oxygen, and carbon. Commonly found in petroleum, natural gas, and coal.

Hydrogen   Lightest-known gas; major element found in all living systems.

Hydrogen Sulfide   Compound of hydrogen and sulfur; a toxic air contaminant that smells like rotten eggs.

Hydropower   Electrical energy produced by flowing or falling water.

I

Infiltration   Process of water percolation into soil and pores and hollows of permeable rocks.

Intangible Resources   Open space, beauty, serenity, genius, information, diversity, and satisfaction are a few of these abstract commodities.

Integrated Pest Management (IPM)   Designed to avoid economic loss from pests, this program's methods of pest control strive to minimize the use of environmentally hazardous, synthetic chemicals.

Invasive   Refers to those species that have moved into an area and reproduced so aggressively that they have replaced some of the native species.

Ion   An atom or group of atoms, possessing a charge; brought about by the loss or gain of electrons.

Ionizing Radiation   Energy in the form of rays or particles that have the capacity to dislodge electrons and/or other atomic particles from matter that is irradiated.

Irradiation   Exposure to any form of radiation.

Isotopes   Two or more forms of an element having the same number of protons in the nucleus of each atom but different numbers of neutrons.

K

Keystone Species   Species that are essential to the functioning of many other organisms in an ecosystem.

Kilowatt   Unit of power equal to 1,000 watts.

L

Leaching   Dissolving out of soluble materials by water percolating through soil.

Limnologist   Individual who studies the physical, chemical, and biological conditions of aquatic systems.

M

Malnutrition   Faulty or inadequate nutrition.

Malthusian Theory   The theory that populations tend to increase by geometric progression (1, 2, 4, 8, 16, etc.) while food supplies increase by arithmetic means (1, 2, 3, 4, 5, etc.).

Metabolism   The chemical processes in living tissue through which energy is provided for continuation of the system.

Methane   Often called marsh gas (CH^4); an odorless, flammable gas that is the major constituent of natural gas. In nature it develops from decomposing organic matter.

Migration   Periodic departure and return of organisms to and from a population area.

Monoculture   Cultivation of a single crop, such as wheat or corn, to the exclusion of other land uses.

Mutation   Change in genetic material (gene) that determines species characteristics; can be caused by a number of agents, including radiation and chemicals, called mutagens.

N

Natural Selection   The agent of evolutionary change by which organisms possessing advantageous adaptations leave more offspring than those lacking such adaptations.

Niche   The unique occupation or way of life of a plant or animal species; where it lives and what it does in the community.

Nitrate   A salt of nitric acid. Nitrates are the major source of nitrogen for higher plants. Sodium nitrate and potassium nitrate are used as fertilizers.

Nitrite   Highly toxic compound; salt of nitrous acid.

Nitrogen Oxides   Common air pollutants. Formed by the combination of nitrogen and oxygen; often the products of petroleum combustion in automobiles.

Nonrenewable Resource   Any natural resource that cannot be replaced, regenerated, or brought back to its original state once it has been extracted, for example, coal or crude oil.

Nutrient   Any nutritive substance that an organism must take in from its environment because it cannot produce it as fast as it needs it or, more likely, at all.

O

Oil Shale   Rock impregnated with oil. Regarded as a potential source of future petroleum products.

Oligotrophic   Most often refers to those lakes with a low concentration of organic matter. Usually contain considerable oxygen; Lakes Tahoe and Baikal are examples.

Organic   Living or once living material; compounds containing carbon formed by living organisms.

Organophosphates   A large group of nonpersistent synthetic poisons used in the pesticide industry; include parathion and malathion.

Ozone   Molecule of oxygen containing three oxygen atoms; shields much of Earth from ultraviolet radiation.

P

Particulate   Existing in the form of small separate particles; various atmospheric pollutants are industrially produced particulates.

Peroxyacyl Nitrate (PAN)   Compound making up part of photochemical smog and the major plant toxicant of smog-type injury; levels as low as 0.01 ppm can injure sensitive plants. Also causes eye irritation in people.

Pesticide   Any material used to kill rats, mice, bacteria, fungi, or other pests of humans.

Pesticide Treadmill   A situation in which the cost of using pesticides increases while the effectiveness decreases (because pest species develop genetic resistance to the pesticides).

Petrochemicals   Chemicals derived from petroleum bases.

pH   Scale used to designate the degree of acidity or alkalinity; ranges from 1 to 14; a neutral solution has a pH of 7; low pHs are acid in nature, while pHs above 7 are alkaline.

Phosphate   A phosphorous compound; used in medicine and as fertilizers.

Glossary

Photochemical Smog Type of air pollution; results from sunlight acting with hydrocarbons and oxides of nitrogen in the atmosphere.

Photosynthesis Formation of carbohydrates from carbon dioxide and hydrogen in plants exposed to sunlight; involves a release of oxygen through the decomposition of water.

Photovoltaic Cells An energy-conversion device that captures solar energy and directly converts it to electrical current.

Physical Half-Life Time required for half of the atoms of a radioactive substance present at some beginning to become disintegrated and transformed.

Phytoplankton That portion of the plankton community comprised of tiny plants, e.g., algae, diatoms.

Pioneer Species Hardy species that are the first to colonize a site in the beginning stage of ecological succession.

Plankton Microscopic organisms that occupy the upper water layers in both freshwater and marine ecosystems.

Plutonium Highly toxic, heavy, radioactive, manmade, metallic element. Possesses a very long physical half-life.

Pollution The process of contaminating air, water, or soil with materials that reduce the quality of the medium.

Polychlorinated Biphenyls (PCBs) Poisonous compounds similar in chemical structure to DDT. PCBs are found in a wide variety of products ranging from lubricants, waxes, asphalt, and transformers to inks and insecticides. Known to cause liver, spleen, kidney, and heart damage.

Population All members of a particular species occupying a specific area.

Predator Any organism that consumes all or part of another system; usually responsible for death of the prey.

Primary Production The energy accumulated and stored by plants through photosynthesis.

R

Rad (Radiation Absorbed Dose) Measurement unit relative to the amount of radiation absorbed by a particular target, biotic or abiotic.

Radioactive Waste Any radioactive by-product of nuclear reactors or nuclear processes.

Radioactivity The emission of electrons, protons (atomic nuclei), and/or rays from elements capable of emitting radiation.

Rain Forest Forest with high humidity, small temperature range, and abundant precipitation; can be tropical or temperate.

Recycle To reuse; usually involves manufactured items, such as aluminum cans, being restructured after use and utilized again.

Red Tide Population explosion or bloom of minute single-celled marine organisms (dinoflagellates), which can accumulate in protected bays and poison other marine life.

Renewable Resources Resources normally replaced or replenished by natural processes; not depleted by moderate use.

Riparian Water Right Legal right of an owner of land bordering a natural lake or stream to remove water from that aquatic system.

S

Salinization An accumulation of salts in the soil that could eventually make the soil too salty for the growth of plants.

Sanitary Landfill Land waste disposal site in which solid waste is spread, compacted, and covered.

Scrubber Antipollution system that uses liquid sprays in removing particulate pollutants from an airstream.

Sediment Soil particles moved from land into aquatic systems as a result of human activities or natural events, such as material deposited by water or wind.

Seepage Movement of water through soil.

Selection The process, either natural or artificial, of selecting or removing the best or less desirable members of a population.

Selective Breeding Process of selecting and breeding organisms containing traits considered most desirable.

Selective Harvesting Process of taking specific individuals from a population; the removal of trees in a specific age class would be an example.

Sewage Any waste material coming from domestic and industrial origins.

Smog A mixture of smoke and air; now applies to any type of air pollution.

Soil Erosion Detachment and movement of soil by the action of wind and moving water.

Solid Waste Unwanted solid materials usually resulting from industrial processes.

Species A population of morphologically similar organisms, capable of interbreeding and producing viable offspring.

Species Diversity The number and relative abundance of species present in a community. An ecosystem is said to be more diverse if species present have equal population sizes and less diverse if many species are rare and some are very common.

Strip Mining Mining in which Earth's surface is removed in order to obtain subsurface materials.

Strontium-90 Radioactive isotope of strontium; it results from nuclear explosions and is dangerous, especially for vertebrates, because it is taken up in the construction of bone.

Succession Change in the structure and function of an ecosystem; replacement of one system with another through time.

Sulfur Dioxide (SO^2) Gas produced by burning coal and as a by-product of smelting and other industrial processes. Very toxic to plants.

Sulfur Oxides (SO^x) Oxides of sulfur produced by the burning of oils and coal that contain small amounts of sulfur. Common air pollutants.

Sulfuric Acid ($H2 SO^4$) Very corrosive acid produced from sulfur dioxide and found as a component of acid rain.

Sustainability Ability of an ecosystem to maintain ecological processes, functions, biodiversity, and productivity over time.

Sustainable Agriculture Agriculture that maintains the integrity of soil and water resources so that it can continue indefinitely.

T

Technology Applied science; the application of knowledge for practical use.

Tetraethyl Lead Major source of lead found in living tissue; it is produced to reduce engine knock in automobiles.

Thermal Inversion A layer of dense, cool air that is trapped under a layer of less dense warm air (prevents upward flowing air currents from developing).

Thermal Pollution   Unwanted heat, the result of ejection of heat from various sources into the environment.

Thermocline   The layer of water in a body of water that separates an upper warm layer from a deeper, colder zone.

Threshold Effect   The situation in which no effect is noticed, physiologically or psychologically, until a certain level or concentration is reached.

Tolerance Limit   The point at which resistance to a poison or drug breaks down.

Total Fertility Rate (TFR)   An estimate of the average number of children that would be born alive to a woman during her reproductive years.

Toxic   Poisonous; capable of producing harm to a living system.

Tragedy of the Commons   Degradation or depletion of a resource to which people have free and unmanaged access.

Trophic   Relating to nutrition; often expressed in trophic pyramids in which organisms feeding on other systems are said to be at a higher trophic level; an example would be carnivores feeding on herbivores, which, in turn, feed on vegetation.

Turbidity   Usually refers to the amount of sediment suspended in an aquatic system.

U

Uranium 235   An isotope of uranium that when bombarded with neutrons undergoes fission, resulting in radiation and energy. Used in atomic reactors for electrical generation.

Z

Zero Population Growth   The condition of a population in which birthrates equal death rates; it results in no growth of the population.

Index

Test Your Knowledge Form

We encourage you to photocopy and use this page as a tool to assess how the articles in *Annual Editions* expand on the information in your textbook. By reflecting on the articles you will gain enhanced text information. You can also access this useful form on a product's book support Web site at *http://www.dushkin.com/online/*.

NAME: _____ DATE: _____

TITLE AND NUMBER OF ARTICLE: _____

BRIEFLY STATE THE MAIN IDEA OF THIS ARTICLE:

_____

LIST THREE IMPORTANT FACTS THAT THE AUTHOR USES TO SUPPORT THE MAIN IDEA:

_____

WHAT INFORMATION OR IDEAS DISCUSSED IN THIS ARTICLE ARE ALSO DISCUSSED IN YOUR TEXTBOOK OR OTHER READINGS THAT YOU HAVE DONE? LIST THE TEXTBOOK CHAPTERS AND PAGE NUMBERS:

_____

LIST ANY EXAMPLES OF BIAS OR FAULTY REASONING THAT YOU FOUND IN THE ARTICLE:

_____

LIST ANY NEW TERMS/CONCEPTS THAT WERE DISCUSSED IN THE ARTICLE, AND WRITE A SHORT DEFINITION:

We Want Your Advice

ANNUAL EDITIONS revisions depend on two major opinion sources: one is our Advisory Board, listed in the front of this volume, which works with us in scanning the thousands of articles published in the public press each year; the other is you—the person actually using the book. Please help us and the users of the next edition by completing the prepaid article rating form on this page and returning it to us. Thank you for your help!

ANNUAL EDITIONS: Environment 05/06

ARTICLE RATING FORM

Here is an opportunity for you to have direct input into the next revision of this volume.
We would like you to rate each of the articles listed below, using the following scale:

1. **Excellent: should definitely be retained**
2. **Above average: should probably be retained**
3. **Below average: should probably be deleted**
4. **Poor: should definitely be deleted**

Your ratings will play a vital part in the next revision.
Please mail this prepaid form to us as soon as possible.
Thanks for your help!

RATING	ARTICLE	RATING	ARTICLE
_____	1. How Many Planets? A Survey of the Global Environment		
_____	2. Five Meta-Trends Changing the World		
_____	3. Crimes of (a) Global Nature		
_____	4. Advocating For the Environment: Local Dimensions of Transnational Networks		
_____	5. Rescuing a Planet Under Stress		
_____	6. Globalizing Greenwash		
_____	7. Population and Consumption: What We Know, What We Need to Know		
_____	8. An Economy for the Earth		
_____	9. Factory Farming in the Developing World		
_____	10. Why Race Matters in the Fight for a Healthy Planet		
_____	11. Will Frankenfood Save the Planet?		
_____	12. Powder Keg		
_____	13. Personalized Energy: The Next Paradigm		
_____	14. Renewable Energy: A Viable Choice		
_____	15. Hydrogen: Waiting for the Revolution		
_____	16. What Is Nature Worth?		
_____	17. Strangers in Our Midst: The Problem of Invasive Alien Species		
_____	18. Markets for Biodiversity Services: Potential Roles and Challenges		
_____	19. On the Termination of Species		
_____	20. Where Have All the Farmers Gone?		
_____	21. What's a River For?		
_____	22. A Human Thirst		
_____	23. Will the World Run Dry? Global Water and Food Security		
_____	24. Fire Fight		
_____	25. Agricultural Pesticides in Developing Countries		
_____	26. The Quest for Clean Water		
_____	27. Human Biomonitoring of Environmental Chemicals		
_____	28. A Little Rocket Fuel with Your Salad?		
_____	29. Global Warming as a Weapon of Mass Destruction		

(Continued on next page)

BUSINESS REPLY MAIL
FIRST CLASS MAIL PERMIT NO. 551 DUBUQUE IA

POSTAGE WILL BE PAID BY ADDRESEE

McGraw-Hill/Dushkin
2460 KERPER BLVD
DUBUQUE, IA 52001-9902

ABOUT YOU

Name

Date

Are you a teacher? ❏   A student? ❏
Your school's name

Department

Address   City   State   Zip

School telephone #

YOUR COMMENTS ARE IMPORTANT TO US!

Please fill in the following information:
For which course did you use this book?

Did you use a text with this ANNUAL EDITION?   ❏   yes ❏   no
What was the title of the text?

What are your general reactions to the *Annual Editions* concept?

Have you read any pertinent articles recently that you think should be included in the next edition? Explain.

Are there any articles that you feel should be replaced in the next edition? Why?

Are there any World Wide Web sites that you feel should be included in the next edition? Please annotate.

May we contact you for editorial input?   ❏   yes ❏   no
May we quote your comments?   ❏   yes ❏   no